PALEOINDIAN OR PALEOARCHAIC?

Paleoindian or Paleoarchaic?

Great Basin Human Ecology at the Pleistocene/Holocene Transition

Edited by

Kelly E. Graf and Dave N. Schmitt

The University of Utah Press
Salt Lake City

First paperback printing 2010

The Defiance House Man colophon is a registered trademark of the University of Utah Press. It is based upon a four-foot-tall, Ancient Puebloan pictograph (late PIII) near Glen Canyon, Utah.

14 13 12 11 10 1 2 3 4 5

LIBRARY OF CONGRESS CATALOGING-IN-PUBLICATION DATA

Paleoindian or paleoarchaic? : Great Basin human ecology at the pleistocene/holocene transition / edited by Kelly E. Graf and Dave N. Schmitt.
p. cm.
Papers previously presented at a symposium during the 70th Annual Meeting of the Society for American Archaeology held in 2005 in Salt Lake City, Utah.
Includes bibliographical references and index.
ISBN 978-1-60781-027-8 (paper : alk. paper) 1. Paleo-Indians—Great Basin—Congresses. 2. Indigenous peoples—Ecology—Great Basin—Congresses. 3. Human ecology—Great Basin—History—Congresses. 4. Paleoecology—Great Basin—Congresses. 5. Paleoecology—Holocene—Congresses. 6. Nature—Effect of human beings on—Great Basin—Congresses. 7. Plant remains (Archaeology)—Great Basin—Congresses. 8. Hunting and gathering societies—Great Basin—Congresses. 9. Great Basin—Antiquities—Congresses. I. Graf, Kelly E. II. Schmitt, Dave N. III. Society for American Archaeology. Meeting (70th : 2005 : Salt Lake City, Utah)
E78.G67P34 2007
304.20979'09012—dc22 2007020393

Contents

Figures

Tables

Introduction

Kelly E. Graf and Dave N. Schmitt

Current studies of the Great Basin's earliest inhabitants are centered on several key issues. To begin with, few buried sites exist that provide reliable radiometric age estimates, leading to significant gaps in the chronological record of this period. We have little hard evidence to support a pre-11,000 ^{14}C BP (13,000 cal BP) Clovis occupation of the region, and it remains unclear how fluted point technologies temporally relate to stemmed points and associated toolkits. This virtual absence of a solid chronology makes it extremely difficult to reconcile the archaeological record with the evolution of climate and environment across the Pleistocene/Holocene boundary. Further, reconstructions of human organization have been limited since the bulk of the archaeological record is surface in nature, and since macrobotanical and faunal remains are generally absent except for those preserved in dry rockshelters and caves (but see Pinson, this volume).

One of the major complications resulting from these chronological and contextual problems is discerning occupations with fluted points from those with stemmed points. At a few sites, fluted and stemmed points are found together with no clear stratigraphic separation, leading some to suggest that the two industries are contemporaneous (Beck and Jones 1997, this volume; Bedwell 1973; Bryan 1988; Jones and Beck 1999; Taylor 2003). Conversely, other researchers point out that when these two types of projectile points are found at the same sites, they tend to be on the surface and spatially separated, thus representing successive occupations dating to different time periods (Fagan 1988; Goebel et al. 2004; Willig 1989, 1991).

The other obvious complication is properly identifying and reconstructing human behavioral adaptations. Traditionally in North America, the term *Paleoindian* has been used to designate the earliest time period of humans, in which these populations are represented archaeologically by both lanceolate and stemmed projectile point forms and associated toolkits (Anderson and Sassaman 1996; Goodyear 1989; Hoffecker et al. 1993; Bever 2001; Meltzer 1988;

Schott 1993). The classic view of Paleoindians is one of highly mobile foragers whose subsistence focused on the exploitation of a few species of extinct megafauna (Basgall 1988; Butler 1973; Haynes 1982; Heizer and Baumhoff 1970; Kelly and Todd 1988; Mosimann and Martin 1975; Watters 1979; Wormington 1957). It is becoming increasingly clear, however, that Paleoindians across the continent exploited a much wider set of fauna and flora, and that adaptive strategies of the earliest settlers were much more complex than implied by the big-game hunting characterization (Connolly 1999; Dincauze 1993a; Grayson 1993; Hockett, this volume; Meltzer 1993; Pinson, this volume). As Haynes (2002) cautions, though, there is no such thing as a single-species specialist, and we should expect Paleoindians to have taken other taxa when they were encountered, whether they were small mammals, reptiles, nuts, or berries. What we need to know is when it was that the early inhabitants expanded their diet to include low-ranked, hard-to-process resources such as seeds (see Rhode and Louderback, this volume). Nowhere in North America has this problem been a focal point of research longer than in the Great Basin.

Early on, Great Basin researchers argued that the frequent occurrence of early-period sites at or near pluvial lake and marshland margins implied that the early populations of the Great Basin grew and became tied to lake resources (Bedwell 1973; Campbell et al. 1937). Such reconstructions were based primarily on association of surface finds and pluvial landforms and typically not with the associations of marshland resource remains (e.g., cattail seeds, waterfowl, fish found with fluted and stemmed points). As a result, few researchers questioned the appropriateness of the term *Paleoindian* to refer to the early archaeological time period of the Great Basin. They argued that the terms *Proto-Archaic*, *Pre-Archaic*, or *Western Pluvial Lakes Tradition* may be more appropriate (Bedwell 1973; Elston 1982; Krieger 1964).

Since the early 1980s, others have followed suit, arguing that early inhabitants of the Great Basin were different from those in other areas of North America. Since this time it has become increasingly clear that pluvial lakes came and went, early folks utilized ecotones other than marshlands, and they transported lithic raw materials great distances (Beck and Jones 1997; Madsen 1999; Willig et al. 1988). Some researchers, therefore, argued that localized distributions of wetland-marsh habitat during the Pleistocene/Holocene transition made food resources unpredictable in time and space and thus played a significant role in human organization. In response, early inhabitants of the Great Basin probably adopted a broad-spectrum adaptive strategy, not unlike that of Archaic era peoples. Therefore, subsistence focus was on exploitation of diverse resources found in mesic patches (Beck and Jones 1997; Jones and Beck 1999; Willig 1988, 1991). In fact, Willig (1989) coined the term *paleo-Archaic* to refer to these hunter-gatherers, arguing that the region's earliest inhabitants, although far-ranging and mobile, were tethered to marshland resources. Today, this generalized Paleoarchaic adaptive strategy finds continued support among archaeologists, although typically in a modified form (Madsen, this volume;

Beck and Jones, this volume; Schmitt et al., this volume; Duke and Young, this volume; Pinson, this volume). For example, Beck and Jones (1997) argue that hunter-gatherers at the Pleistocene/Holocene boundary in the Great Basin were moving great distances, but doing so to reach widely dispersed mesic wetland resource patches.

This volume has two major goals that generally align with the two major problems discussed above. The first goal is to provide a selection of recent works that attempt to reconstruct the lifeways of these early hunter-gatherers. To this end, contributions provide varied looks at the record, including reports from current excavation projects and associated chronostratigraphy, cladistic analysis of lithic variability, lithic technology and sourcing, and faunal and floral analyses from buried site locations.

The second goal of this volume is to provide a forum of the ongoing discussion of the human organization and behavioral patterns that are coming to light with continued research. Related to this goal, of course, is a concern with the status of the use of terms such as *Paleoindian*, *Pre-Archaic*, and *Paleoarchaic*. Innovative lithic studies, coupled with new subsistence data recovered from stratified contexts, have contributed to a clearer understanding of human organization during this time. These make up a major component of the present volume.

As editors of this volume, we became interested in these problems from two very different avenues. First, Dave Schmitt became involved through his analyses of several Great Basin small mammal bone assemblages and his experiences studying human ecology among living hunter-gatherers in Africa. Kelly Graf became involved in these issues through both her interests in the peopling of the Americas and her studies of lithic remains from several early sites in both the western and east-central Great Basin. Through our combined interests, we recognized the need to bring together the diverse studies within this volume. We hope that the data presented will contribute to our knowledge of the colonization of the region and the evolution of human adaptive strategies across the Pleistocene/Holocene boundary.

This volume stemmed chiefly from a symposium held at the 70th Annual Meeting of the Society for American Archaeology in Salt Lake City in 2005. The symposium had its beginnings when Charlotte Beck, Tom Jones, Dave Rhode, Dave Madsen, and Ted Goebel took turns twisting our arms into organizing a session on the Paleoarchaic of the Great Basin. It had been nearly 20 years since the last volume of this type was published (Willig et al. 1988). Though a major contribution for the time, that volume is now admittedly out of date and does not adequately reflect current understanding of early adaptive strategies in the Great Basin.

The following is a brief overview of the organization of the volume. Part 1 provides important background information about the early archaeology of the Great Basin as well as key definitions related to the region's earliest peoples. Madsen presents a thorough overview of foraging strategies used by Great

Basin inhabitants, paleoclimatic and paleoecological reconstructions of the Pleistocene/Holocene transition, and his ideas concerning Paleoarchaic and Archaic human organization.

Part 2 presents new information on chronological and spatial context of early peoples in the Great Basin. Beck and Jones present a cladistic study of known fluted points from the region to develop a hypothetical phylogeny or evolutionary sequence of development of these artifact types and associated movement of these point forms across the Great Basin. Henrikson and Long present a review of the early record from Kelvin's Cave in southern Idaho. They review the nature of late Pleistocene sediments, including the potential associations between several extinct faunal species and lithic materials found together in these sediments. Jenkins reports on recent work at Paisley Caves in southern Oregon. Similar to Henrikson and Long, he discusses geological context, age estimates from both radiocarbon and obsidian hydration analyses, and association of late Pleistocene extinct fauna and human cultural remains. Graf reports on recent geoarchaeological studies conducted at Bonneville Estates Rockshelter in eastern Nevada, including a description of the rockshelter's stratigraphy, radiocarbon sequence, and associated cultural materials. Finally, Schmitt, Madsen, Oviatt, and Quist review recent work along the Old River Bed in western Utah. They present both recent paleoecological and archaeological work, discussing site distributions and geomorphic contexts.

Part 3 provides several studies related to technological organization and settlement systems. Duke and Young present recent work in western Utah, focusing specifically on settlement patterns, raw material provisioning, and toolstone selection, paying special attention to the frequent use of basalt over obsidian in this region. Smith presents his new work on the stemmed point sites from Lake Parman in northwestern Nevada. He reports on the technological organization and provisioning and settlement strategies employed by the foragers represented by these sites. Likewise, Goebel concentrates on the technological organization and provisioning strategies employed by the early inhabitants at Bonneville Estates Rockshelter, comparing those with strategies employed by subsequent middle Holocene inhabitants of the shelter.

Part 4 presents recent results of studies focusing on the reconstruction of subsistence behavior. Pinson reports on her work in the Dietz Basin in southern Oregon, presenting faunal data from several late Pleistocene to late Holocene sites to reconstruct how subsistence organization evolved over time. Hockett presents his study of the faunal record from the early deposits at Bonneville Estates Rockshelter. He tracks the late Pleistocene through middle Holocene changes in subsistence organization in the framework of hunter-gatherer nutrition. Rhode and Louderback present results from the study of seed use over the Pleistocene/Holocene boundary at two important eastern Great Basin sites, Danger Cave and Bonneville Estate Rockshelter.

Finally, Part 5 provides a commentary. Haynes presents a thoughtful review of the chapters in the volume, paying close attention to use of the terms *Paleo-*

indian, *Pre-Archaic*, and *Paleoarchaic*, and the associated implications of these labels relative to settlement and subsistence organization. His discussion also relates Great Basin studies to the problem of the peopling of North America.

We hope that Great Basin specialists will find this volume especially valuable and timely in that it provides up-to-date research into early Great Basin archaeology. Further, several chapters provide new paleoecological data important for reconstructing Pleistocene/Holocene transition environments in the Great Basin. We also anticipate that Paleoindian archaeologists working elsewhere in the Americas will find the information provided important and useful. Our hope is that this will lead to more communication between regions to gain a holistic perspective on the colonization of the Americas.

PART I

Background

1

The Paleoarchaic to Archaic Transition in the Great Basin

David B. Madsen

GREAT BASIN ENVIRONMENTS

THE GREAT BASIN IS A VAST COMPLEX OF INTERNAL DRAINAGE SYSTEMS lying in interior western North America (Figure 1.1).[1] These systems are composed primarily of high, north-to-south trending mountains and long, narrow valleys into which mountain streams and rivers drain. Depending on how it is defined (and it can be defined in a number of ways), the Great Basin extends as far north as 44° north latitude, as far south as 33° north latitude, as far east as 110° west longitude, and as far west as 121° west longitude. Although many definitions of the Great Basin are appropriate for specific purposes, none exceeds John C. Frémont's (1845:175) original description for simple utility:

> The Great Basin—a term I apply to the intermediate region between the Rocky mountains and the next range, containing many lakes, with their own system of rivers and creeks, (of which the Great Salt is the principal,) and which have no connexion [sic] with the ocean, or the great rivers which flow into it.

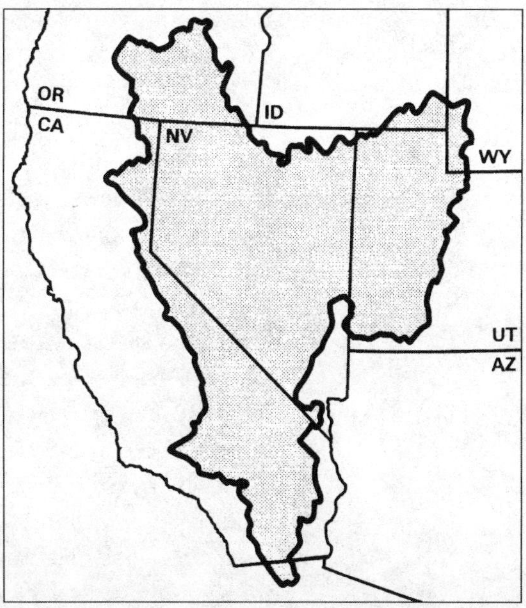

FIGURE 1.1. Location of the Great Basin in western North America (from Madsen et al. 2002).

The Great Basin is essentially an inverted equilateral triangle with a geomorphology that inhibits easy generalizations about climatic and cultural changes spanning the Pleistocene/Holocene transition and the change from Paleoarchaic to Archaic foraging strategies. The complexity of this triangular relationship is compounded by elevational differences between the central Great Basin and the basin margins to east, south, and west. The floor of the basin is basically a dome, and valley

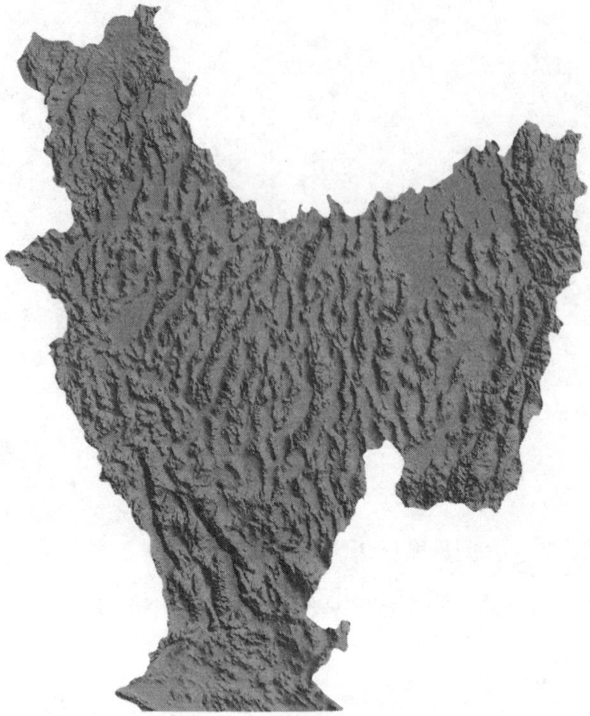

FIGURE 1.2. Shaded digital image of the physiographic Great Basin showing the north-south trending fault-block mountain and valley systems.

floors in the central Great Basin are 600 m or more above those on the periphery. These elevational differences, together with the differences in precipitation, temperature, and seasonality they engender, make it difficult to generalize about prehistoric Great Basin environments and the adaptive strategies required to live in them. For example, the expansion of numerous plant species, including pinyon, during the early Holocene was markedly different on the eastern and western sides of the basin, with a correspondingly differing effect on foragers in these areas.

Physiographically, the Great Basin is part of the Basin and Range Province (Hunt 1967) that, although coterminous with the hydrographic Great Basin in the north, extends into southern Arizona, New Mexico, and parts of northern Mexico. The north-south fault-block mountains that characterize the province are part of an extensional process related to plate tectonics (Zoback et al. 1981) and are separated by long, narrow graben or half graben valleys filled with Quaternary alluvium (Figure 1.2). The larger valleys exceed 160 km in length and 50 km in width. The basin is bounded by higher mountains—on the west by the Sierra Nevada and the Cascade ranges, and on the east by the western Rocky Mountains—but many of the small interior ranges reach almost as high (3,600–4,000 m) and extend 1,500–1,850 m above valley floors.

The biotic Great Basin is defined primarily by the distribution of plant communities, with the most widely employed floristic definition being that of Cronquist et al. (1972) (Figure 1.3). It is this definition of the Great Basin that is used primarily in the following discussion. Plant communities in the floristic Great

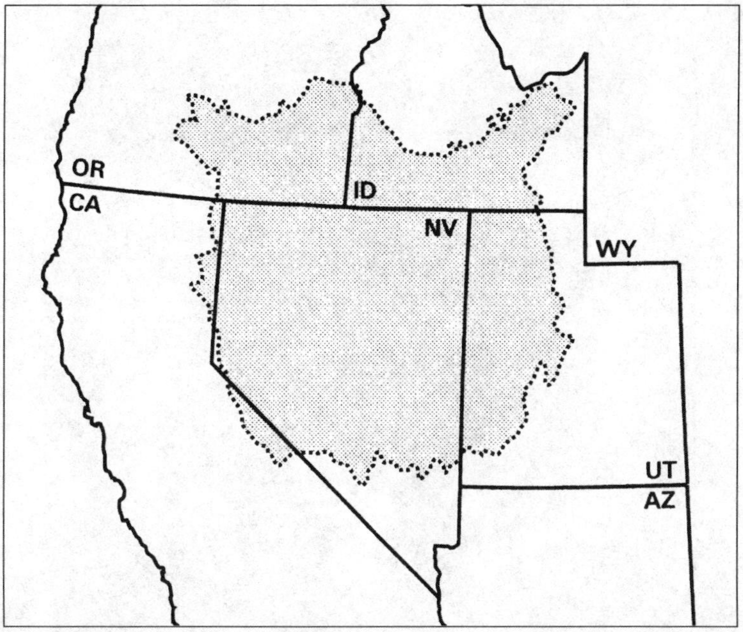

FIGURE 1.3. The floristic Great Basin as defined by Cronquist et al. (1972; from Madsen et al. 2002).

Basin are essentially zonal in character and are dominated by big sagebrush (*Artemisia tridentata*), shadscale (*Atriplex confertifolia*), and saltbush (*Atriplex canescens*) on valley floors; juniper (*Juniperus osteosperma*) and single-leaf pinyon (*Pinus monophylla*) on lower mountain slopes; and white fir (*Abies concolor*), Engelmann spruce (*Picea englemannii*), blue spruce (*Picea pungens*), Douglas fir (*Pseudotsuga menziesii*), and quaking aspen (*Populus tremuloides*) at higher elevations. Marsh communities of various sizes, dominated by cattails (*Typha* spp.) and bulrush (*Scirpus* spp.), abound at the terminal ends of the many streams and rivers in Great Basin valleys.

Hydrographically, the Great Basin consists of 187 separate internal drainage systems, and more than 80 percent of these contain modern lakes or playas (King 1986; Williams and Bedinger 1984). During the last glacial period, most of these basins contained lakes, many of which coalesced into larger lakes (Figure 1.4). The largest of these basins are the Lahontan Basin on the west and the Bonneville Basin on the east, where the relatively more massive Sierra Nevada and Wasatch Mountains, respectively, capture substantial amounts of precipitation. At their maximum, the waters of the Lahontan Basin covered nearly 22,500 km^2, while Lake Bonneville extended more than 800 km north to south, covered more than 51,500 km^2, and overflowed into the Snake River. A number of other basins also overflowed during wet periods, but they formed a series of interconnected lakes rather than a single large lake. The most extensive of these was in the southwestern Great Basin, where Mono, Owens, China, Searles, and Panamint lakes were connected by a series of overflow channels that reached into Death Valley (Smith and Street-Perrott 1983). A number of shorter, interconnected lake systems, such as the Malheur Lake–Harney Lake

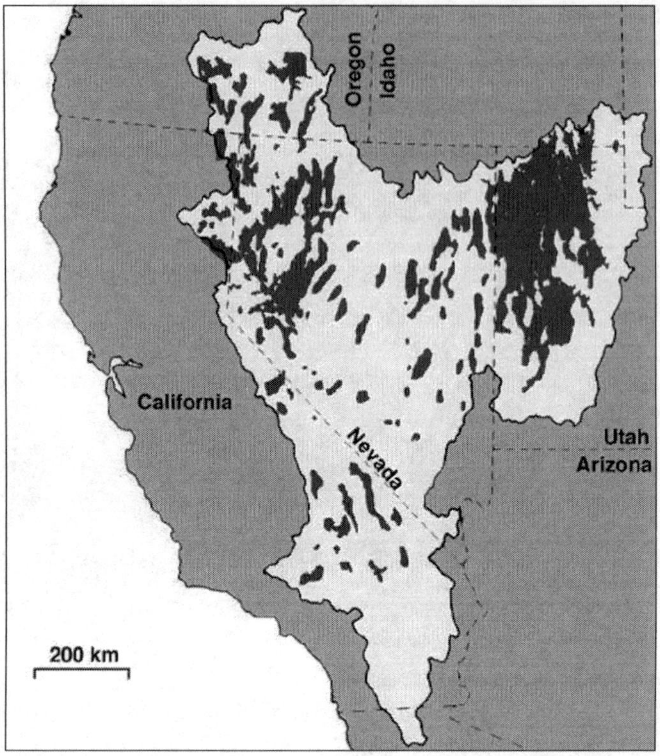

FIGURE 1.4. Pluvial lakes of the Great Basin (from Madsen et al. 2002).

chain in Oregon, were found in other parts of the Great Basin. Most of the basins in the central and southern Great Basin contained only smaller individual lakes, even at the Last Glacial Maximum, due to the small size and limited elevations of surrounding mountains.

Environmentally, the Great Basin is usually defined as arid to semiarid (Antevs 1948), but the basin-and-range topography engenders a climatic regime that may be better described as "hemiarid" (Currey 1991), a term I think captures the basic difference between the wet upland environments of the higher basin mountains and the arid environments of the basin valleys. In human terms, this difference is tempered somewhat by the closed-basin morphology that traps runoff from mountain precipitation in terminal valley lakes. This capture of mountain precipitation, much of it falling in the form of winter snow that gradually releases water as it melts throughout the rest of the year, creates a situation in which relatively lush riverine and lacustrine ecosystems, relying on what is essentially outside water, are found within relatively barren desert valley ecosystems supported only by limited local rainfall (25–40 cm/year at best). The contrast in productivity between the mountain-fed marshes and the surrounding valley floors and foothills is extreme in terms of both the plants and animals they support per unit area. In the simplest human terms, this means that spatial and seasonal distribution of foragers in the Great Basin tends to mirror the seasonal and spatial distribution of water.

A second critical aspect of Great Basin environments is the seasonality imposed by its middle-latitude setting. The Great Basin is characterized by warm summers and cold, sometimes harsh winters. Growing seasons are relatively short, and the amount and availability of food resources, particularly plants, are extremely limited during the winter. As a result, people had to either collect and store resources for winter consumption or situate themselves in the few locations where resources were available during the winter. Such localities were relatively few because of the aridity of Great Basin valleys and the deep winter snows blanketing most of the higher elevations.

Two additional, but related, environmental factors that structured prehistoric Great Basin foraging societies accompany these basic parameters of surface morphology and seasonality. First, the number and distribution of large herbivores in Great Basin environments were limited compared with many other areas of North America, both historically and throughout the late Quaternary. Because of the basin-and-range morphology of the Great Basin, habitats upon which many grazing animals depend are small and widely scattered, and large prey animals such as antelope (historically) and camels (prehistorically) were limited in number in any one location (e.g., Steward 1938). Second, the disparity in the distribution of water makes many of the habitats on which Great Basin peoples relied very sensitive to even short-term climatic changes. Resources away from stream, lake, and spring margin localities were limited and widely scattered, and when lacustrine and riverine habitats were reduced or eliminated by variations in climate, alternative food resources were difficult to acquire.

GREAT BASIN FORAGING STRATEGIES

Historically, Great Basin foragers employed one of two basic strategies in dealing with this morphologically and ecologically diverse hemiarid landscape.[2] A common strategy focused on the stream-fed, lake margin marsh environments common to most valley drainage systems. It involved a mobility pattern that is often referred to as the "limnosedentary" strategy (see also Fowler and Fowler 1990; Heizer 1967; Heizer and Napton 1970; Janetski 1986; Janetski and Madsen 1990; Madsen 1982; Thomas 1985; Zeanah 2004), but which I refer to simply as the lowland strategy (Madsen 2002). This strategy was most evident where large, stable marshes occur. Such marshes are situated primarily along the eastern and western margins of the Great Basin in the Lahontan and Bonneville drainages, where the extensive snowpack in the high and massive Sierra Nevada and Wasatch Mountains provides large amounts of water. However, similar marshes—such as the Ruby, Warner, and Malheur marsh areas—also occur adjacent to several of the large central and northwestern basin ranges.

People living in these large marsh areas were not sedentary in the strict sense; rather, they moved around within the marsh, collecting an array of seeds, tubers, and small animals throughout the year. Virtually all of these resources were relatively expensive to procure (in terms of time spent for the number of calories gained) but had the advantage of being both abundant and closely spaced (Madsen 2002; Zeanah 2004).

Raven (1993:8) has succinctly summarized this lowland marsh strategy, describing such foragers as living in "a fairly steady-state population within the marshes (but not at individual villages) throughout most of any hypothetical year."

> [G]overned chiefly by the exigencies of women's foraging, small scale adjustments in residence location were made frequently, but the sheer richness, diversity and predictability of marsh side productivity invited virtually continuous exploitation. Forays outside the immediate marsh precinct generally were short term (i.e., day trips) and for the most part were resource-specific. Foodstuffs were stored as they were collected, usually within shifting habitation sites; stores were tapped as needed by groups overwintering in the vicinity of the marshes. Only occasionally did groups leave the valley floor, as when the marsh simply could not supply a necessary resource (such as toolstone) or when the uplands offered seasonal resource opportunities more energetically profitable (such as the nearly annual pinyon harvest). During such episodes, various upland resources (such as large game) might be taken, and some surplus returned to marshland base camps; life on the valley floor, however, otherwise relied little on mountain resources.

Alternatively, a more mobile upland strategy was used when valley marsh ecosystems were too small or unreliable to provide storable surpluses sufficient for overwintering, forcing foragers to move themselves to a series of foraging areas, transport a variety of resources to a central area, or (more likely) both. In this strategy, mountain resources in both midslope and alpine areas were important components of the annual round as people moved through a series of foraging localities. However, unless one or more upland resource types, such as pine nuts, were particularly abundant in any given year, sites in or near valley marshes continued to be the focus of overwintering stays. This was particularly true where large game animals congregated around the fringes of the marshes during the winter months.

Because the transportation of bulk resources was relatively expensive (Jones and Madsen 1989; Metcalfe and Barlow 1992; Rhode 1990), the preferred alternative (that is, the most economically viable) was usually the sedentary lowland marsh strategy (Kelly 1995; Zeanah 2004, Zeanah et al. 1995). These stable lowland and more mobile upland strategies were not fixed and dichotomous, but flexible and varied, with permutations of both found through space and time in the Great Basin. Moreover, people could, and did, switch strategies as circumstances demanded, and such changes in circumstances were often related to changes in the size and/or productivity of the marsh systems. There are a number of ethnographic examples of these two basic adaptive strategies, but some of the best are those for the Kaibab Paiute (Fowler 1982; Kelly 1964),

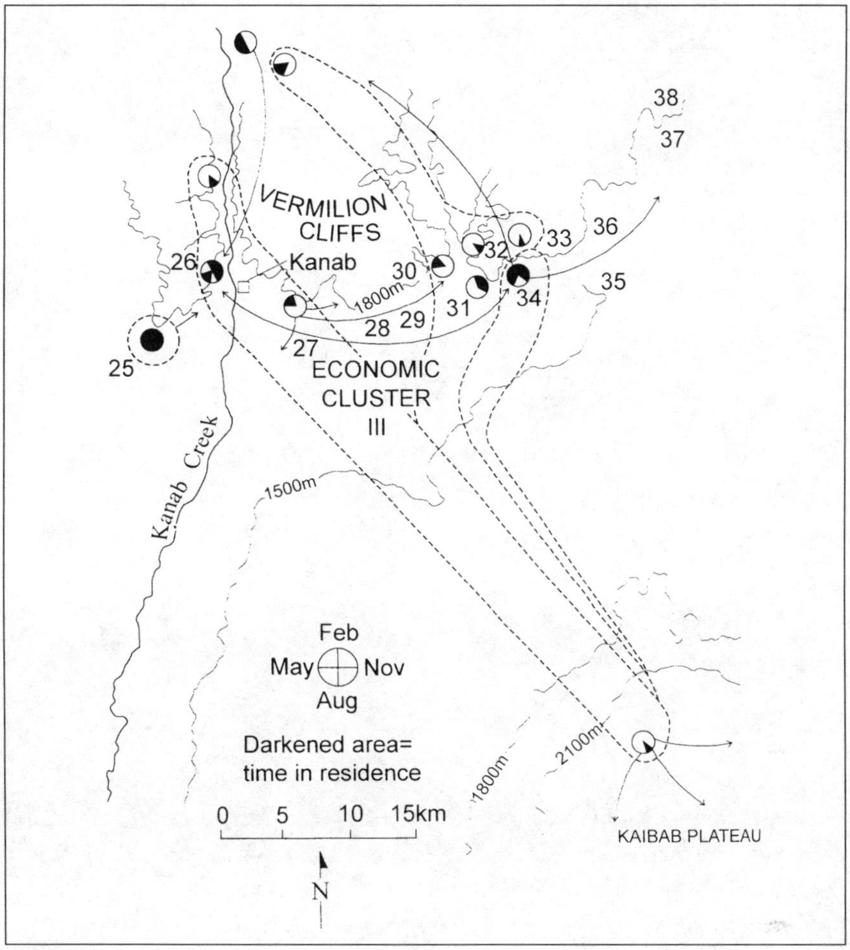

FIGURE 1.5. Kaibab Paiute "economic cluster III" (from Fowler 1982:127; Kelly 1964), representative of the Great Basin upland foraging strategy. Numbers (25–38) represent individual Southern Paiute camps associated with spring localities identified by Isabel Kelly.

illustrating the upland pattern, and the Northern Paiute "Cattail Eaters" of the Carson Sink (Fowler 1992; Madsen 2002; Zeanah 2004) and the Utes of Utah Lake (Janetski 1986) illustrating the lowland pattern. Fowler (1982, following Kelly 1964) has described the seasonal movement of several interrelated Kaibab Paiute households in what she calls an "economic cluster" (Figure 1.5). These foragers lived in an environmentally diverse region and had equally diverse foraging strategies. Not only did these economic clusters have high intercluster variability, but even within an economically interdependent group, individual camps composed of multiple households ranged from almost completely sedentary to highly mobile. Spring activities were oriented toward dispersed root procurement—of *Peteria* in the case of one camp, and *Tsii* in the case of another. Summer activities were oriented toward the collection of a variety of seeds—such as *Balsamorrhiza*, *Chenopodium*, *Epichapes*, *Mentzelia*, and *Urosepses*—in a variety of contexts. Fall activities revolved around the abundance of pinyon. In good years, trips for collecting pine nuts were made; in bad years, other resources, such as agave or *Artemisia* seeds, were taken. Winter locations

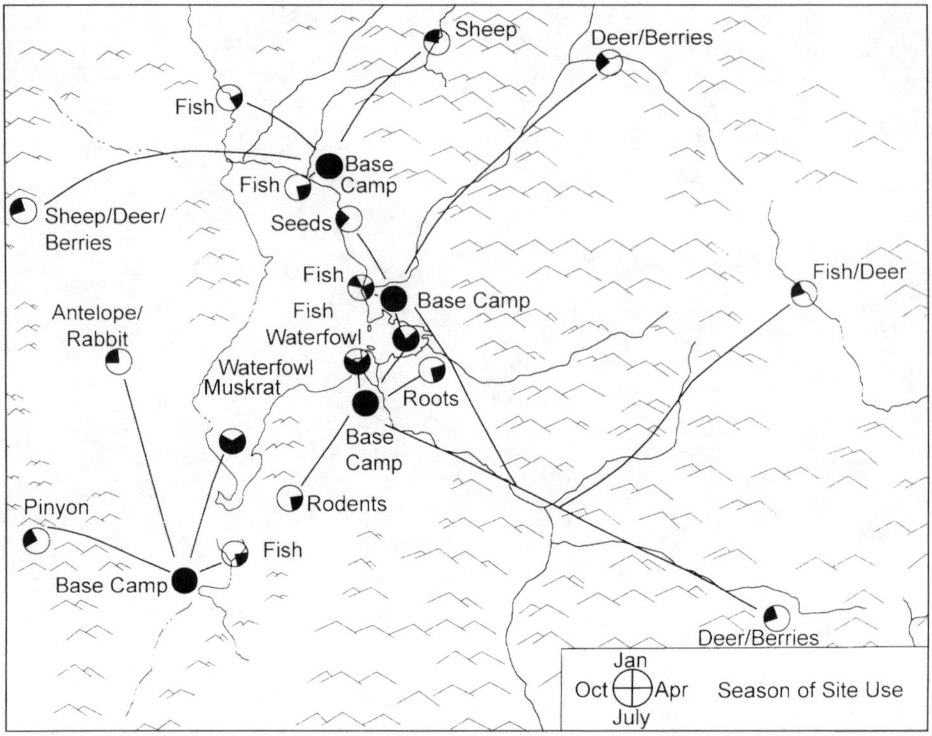

FIGURE 1.6. Settlement pattern of the Utah Valley Utes, representative of the Great Basin lowland foraging strategy.

were apparently dictated by non-food factors, possibly because all year-round activities took place within a 10–15 km radius. Stored resources were apparently cached near the point of collection and transported as needed.

On the other hand, Fowler (1992:212) suggests that the Cattail Eaters of the Stillwater and Carson Sink marshes, like Utes around Utah Lake (Figure 1.6), had a much narrower focus:

> The picture...appears to be one of considerable security, a tendency to remain in groups of perhaps 20 people or so, and certainly to focus on the richness of the marsh. *Toiwaitu*, "cattail place," was their home, their *Tibiwa*, and they seemingly wanted no other. They spent most of the year there, working out from small to medium-sized villages of cattail houses. If they moved away, it was only for short periods and then they returned.

The settlement strategy revolved around women's collecting activities (Madsen 2002; Zeanah 2004). In the early spring this involved foraging for cattail and bulrush roots, waterfowl eggs, and a variety of greens. In late spring they broke up into smaller family groups to collect a variety of greens and roots in various locations in the nearby foothills and/or to fish the spring spawning runs in the Truckee River. Summer collecting took place principally in the vicinity of the numerous shallow lakes or sloughs along the lower Carson River. Resources included small suckers and chubs, cattail pollen, and bulrush seeds. In the fall,

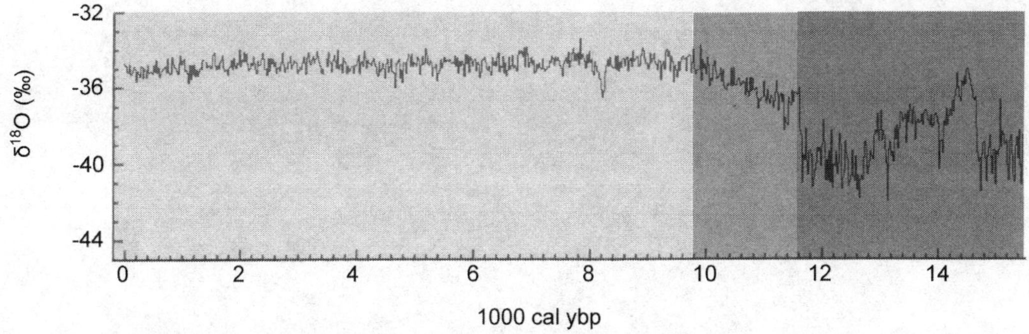

FIGURE 1.7. GISP2 oxygen isotope record used as a proxy for northern hemisphere temperatures (modified from Stuiver and Grootes 2000).

both marsh resources, such as cattail rhizomes, and upland resources, such as pine nuts, were collected. In the rare years when pinyon was highly productive, families would winter in the pinyon groves, but most years were spent back in the marshes. Stored resources were heavily utilized, but some persistent seeds were collected, and cattail and bulrush rhizomes were available well into winter.

Men's collection activities were determined primarily by those of the women, with the men coming and going from a central home base (Fowler 1992; Madsen 2002). DeQuille (1963:71), for example, describes one hunting party that left the women, children, and old folks in the principal base camps along the lower Carson River and traveled 50–70 km to the Clan Alpine Mountains to trap wood rats. This pattern was consistent throughout the prehistoric record for the area; skeletal remains indicate that males had a much higher incidence of skeletal pathologies associated with running and walking, suggesting that women were much more sedentary (Larson et al. 1995).

GREAT BASIN CLIMATE CHANGE DURING THE PLEISTOCENE/HOLOCENE TRANSITION

Because of the link between adaptive strategies and large valley marsh systems, major changes in the size of Great Basin lakes during the Pleistocene/Holocene transition had a dramatic effect on the concurrent Paleoarchaic/Archaic shift. A number of studies (e.g., Benson 1999; Benson et al. 2002; Madsen et al. 2001; Negrini 2002; Oviatt 1997; Oviatt et al. 2005; Zic et al. 2002) suggest that Great Basin environmental transitions reflect the globally linked climate changes identified in the Greenland ice core records (e.g., Mayewski et al. 1993). Assuming that to be the case, the GISP2 $\delta^{18}O$ record reflecting northern hemisphere temperatures is a good proxy for changes in Great Basin lake levels and the number, size, and productivity of associated marsh systems (Figure 1.7). Beginning about 15,000 calendar years ago (~13,000 ^{14}C BP), coeval with the earliest records of Paleoarchaic foragers in the Great Basin (e.g., Graf et al. 2004), temperatures warmed rapidly, and the large pluvial lakes began drying with equal rapidity. By ~13,000 cal BP (11,000 ^{14}C BP), lake levels

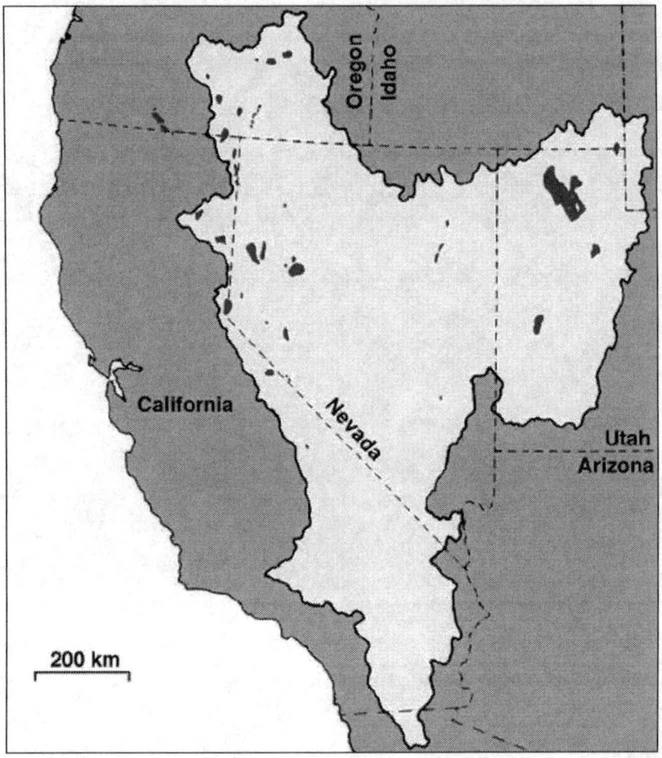

FIGURE 1.8. Modern Great Basin lakes (from Madsen et al. 2002).

approached and, perhaps, reached their modern levels (Oviatt et al. 2005) (Figure 1.8). During the Younger Dryas, ~13,000–11,600 cal BP (11,000–10,100 ^{14}C BP), lakes rebounded, with the surface areas of the larger lakes, such as Lahontan and Bonneville, increasing markedly, and the dry playas in many of the smaller basins flooding to at least a shallow depth. After ~11,600 cal BP (10,100 ^{14}C BP), temperatures again increased and, shortly after 10,000 cal BP (8900 ^{14}C BP), reached their Holocene average. During this early Holocene period, Great Basin lakes again approached their modern levels, or perhaps even lower, with the rate at which this happened depending largely on the size of the drainage basin.

Due to the peculiar morphology of the Great Basin, with steep mountain slopes rising rather precipitously from relatively flat valley floors, there may have been larger marsh systems at the ends of basin streams and rivers during the periods immediately preceding and following the Younger Dryas than during the Younger Dryas itself. When basin lakes completely flood valley floors, they leave only a narrow band of terrain suitable for marsh ecosystems, as opposed to the much larger areas available when water covers only part of the broad, flat basin floors. Where the size and height of surrounding mountains were sufficiently great, some Younger Dryas–age lakes may have actually reduced the size of local marsh ecosystems available to Paleoarchaic foragers.

In terms of Great Basin biotic communities, the period from ~15,000 to 10,000 years ago was one of dramatic change as plant communities previously restricted to lower elevations and/or to southern latitudes began to move north and upslope (Wigand and Rhode 2002). These transformations created a series of non-analog communities unlike many of those found today. In the northwestern Great Basin, during the warmer, but perhaps wetter, period before the Younger Dryas, sagebrush initially expanded rapidly, as did Utah juniper, and whitebark pine made a brief appearance. In the eastern Great Basin limber pine covered the lower slopes of the interior mountains, and Rocky Mountain juniper and sagebrush spread across valley floors. Toward the end of the Younger Dryas, however, saltbush and other thermophilous brush communities began to extend across valley floors, and limber pine and juniper started to retreat upslope. After the Younger Dryas, in the eastern Great Basin, Rocky Mountain juniper almost completely replaced limber and bristlecone pine on the lower mountain slopes, and pinyon pine began to migrate north and northwest into the central and northern Great Basin. The continuing transformation of Great Basin biotic communities lagged behind changes in temperature and lake levels, and they did not begin to approach those similar to the present until well into the Holocene. In the eastern Great Basin, for example, Rocky Mountain juniper continued to be a major element of low elevation woodland communities, and pinyon pine did not reach its modern distribution until well after ~7000 cal BP (~6100 ^{14}C BP). In the northwestern Great Basin, pinyon took even longer to appear and did not reach its modern limits until after ~2000 cal BP (~2000 ^{14}C BP). In terms of the Paleoarchaic/Archaic transition, what this means is that it was associated with a continuing, gradual transformation of Great Basin biotic communities and did not involve a simple, dramatic shift to modern conditions.

THE PALEOARCHAIC

From almost the beginning of archaeological research in the Great Basin, it was recognized that the earliest inhabitants of the region lived along the shores of now extinct late Pleistocene lakes (e.g., Campbell et al. 1937; Willig et al. 1988). Indeed, in some parts of the Great Basin, this lacustrine focus came to be known as the "Western Pluvial Lakes Tradition" (Bedwell 1973). With few exceptions, however, these sites consist of surface finds containing only lithic remains, and the lakeside adaptation is defined almost entirely by regionally unique lanceolate and stemmed biface forms known collectively as the "Western Stemmed Tradition" (Bryan 1980). This tradition (Figure 1.9) includes styles known as Haskett, Windust, Cougar Mountain, Parman, Lake Mohave, Silver Lake, and, my personal favorite, the Dugway Stubby, with these and other less easily categorized forms known more generally as the "Great Basin Stemmed Series" (GBSS) (Tuohy and Layton 1977). Often included with the GBSS is a crescent-shaped biface form unique to the Great Basin (see Schmitt et al., this volume, Figure 6.6). These styles date to between 11,200 and 7000 ^{14}C BP (13,100–7900 cal BP) (Beck and Jones 1997; Jones and Beck 1999) and thus overlap Clovis

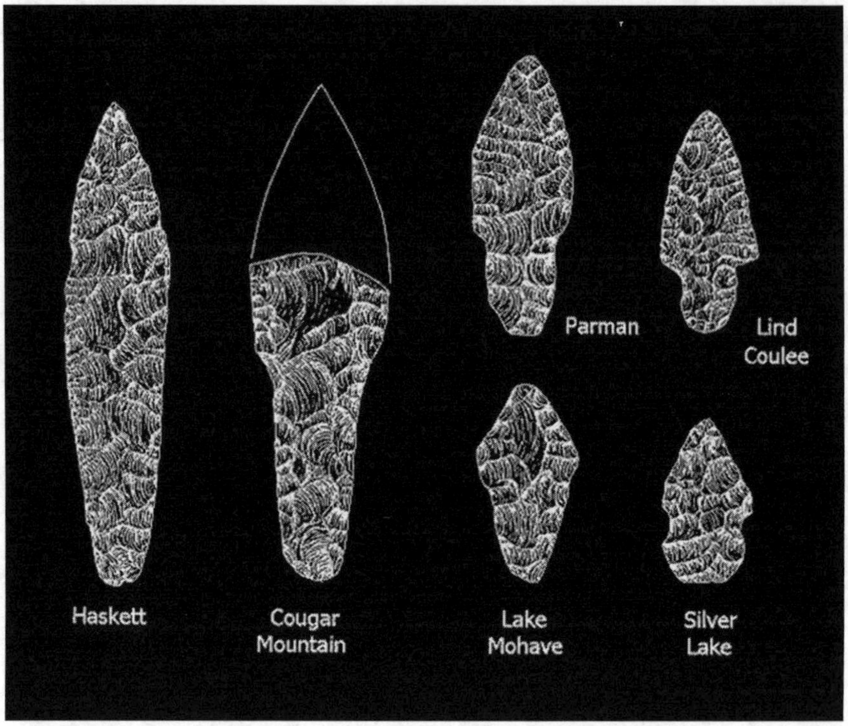

FIGURE 1.9. Common biface styles in the Great Basin Stemmed series (from Beck et al. 2004b).

biface forms on the early end of the sequence and Archaic dart points on the later end. In the Great Basin, Clovis forms tend to be found in the same lakeside settings, are often found with stemmed forms, and must be considered part of the overall Paleoarchaic lithic tradition (Taylor 2003). The exact chronological relationship between Clovis and GBSS remains unclear, however, and GBSS may predate, be contemporaneous with, or postdate Clovis. Beck et al. (2004b; see also Beck and Jones, this volume) suggest that Clovis in the northern Great Basin and on the Columbia Plateau may be somewhat younger than elsewhere, thus falling in about the middle of the prolonged GBSS tradition.

This tradition spans a period dating to between about 12,500 and 8500 ^{14}C BP and thus generally corresponds to what is known elsewhere in North America as the Paleoindian period (e.g., Jennings 1989). In the Great Basin, however, the term is not widely used, primarily because what little subsistence data are available suggest that these peoples were essentially broad-spectrum foragers who did not rely on Pleistocene megafauna and who were more similar to foragers of the following Archaic period (e.g., Beck and Jones 1997; Elston and Zeanah 2002; Jones et al. 2003). As a result, terms such as *Initial Archaic, Prearchaic,* and *Paleoarchaic* are often substituted to emphasize this broad adaptive focus while maintaining a chronological separation from the period that followed. *Paleoarchaic* is probably in widest use, and I employ the term here.

Briefly (see Beck and Jones 1997; Elston 1982; Elston and Zeanah 2002; Jones and Beck 1999; Jones et al. 2003; Pinson 1999; and Willig et al. 1988 for more detailed reviews), the Paleoarchaic foragers of the Great Basin are thought to have been characterized by small, mobile foraging groups tethered to marsh ecosystems associated with terminal Pleistocene lakes. Although a variety of lithic studies suggest a focus on large-game hunting (e.g., Amick 1997), this conjecture is not supported by the limited available subsistence data (primarily faunal remains) (e.g., Grayson 1993; Pinson 1999). In the western Great Basin, Paleoarchaic sites contain a variety of small mammals, birds, and fish (e.g., Grayson 1988), as do sites in the northwestern Great Basin (e.g., Bedwell 1973; Connolly 1999; Pinson 1999) and eastern Great Basin (e.g., Jennings 1957; Grayson 1988; see also Hockett, this volume). "Large game" were included in the mix, but the taxa involved, such as antelope, are similar to those taken during later Archaic periods (see also Jenkins, this volume; Henrikson and Long, this volume). Most illustrative of this orientation are the fecal remains associated with the only known Great Basin Paleoarchaic burial, from Spirit Cave in the Lahontan Basin. These remains contained small fish bones (probably Tui Chub [*Gila* sp.] and Tahoe Sucker [*Catostomus* sp.]) and *Scirpus* seeds (Eiselt 1997b; Napton 1997), betraying a subsistence focus not unlike that found ethnographically in the same area nearly 9,500 ^{14}C years later.

But if these subsistence data suggest a broad-spectrum adaptation tethered to marsh ecosystems, then what separates Paleoarchaic from Archaic foragers, if indeed they can be separated at all? Basically, there is but a single factor: the appearance of ground stone by about 8500 ^{14}C BP (9500 cal BP) (Rhode et al. 2006; see also Rhode and Louderback, this volume) and the processing of hard, small seeds that, because of their high processing costs, have relatively low return rates. These return rates are lower than those associated with small mammals, fish, rhizomes, and other taxa associated with the Paleoarchaic subsistence focus (e.g., Simms 1987), suggesting that the diet breadth of Archaic foragers was even broader than that of the Paleoarchaic. That is, the Paleoarchaic to Archaic transition in the Great Basin consisted essentially of a shift from broad-spectrum foraging to very broad-spectrum foraging, with the addition of movement to previously unoccupied areas where these lower-ranked resources were found. This appears to have occurred gradually between about 9500–8500 ^{14}C BP (10,700–9500 cal BP), concurrent with the final drying of many terminal Pleistocene lakes and the elimination of their associated marsh ecosystems.

A second feature that may separate the earlier from later foragers is a shift in mobility associated with this elimination of many of the smaller Great Basin marsh ecosystems. A variety of studies (e.g., Beck et al. 2002; Graf 2001; Jones et al. 2003) hypothesize that Paleoarchaic foragers were "highly mobile," with widely dispersed small groups moving long distances between increasingly smaller lacustrine habitats. However, much of this hypothesis is based on the distribution of obsidian sources (Figure 1.10), and it is not clear whether this may reflect the movements of Paleoarchaic groups as a whole or that of male hunters. If, as Elston and Zeanah (2002) suggest, male hunting parties were

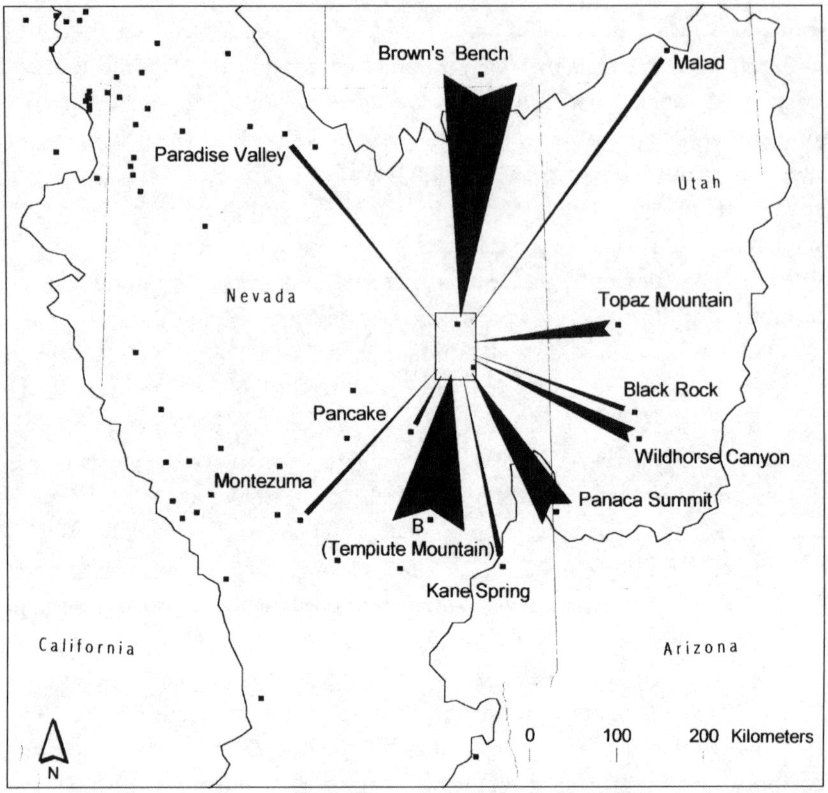

FIGURE 1.10. Obsidian sources at central Great Basin sites. Size of arrows represents approximate proportion in assemblages (from Jones et al. 2003).

coming and going from marsh ecosystem base camps occupied by women and the rest of the foraging group, then it is likely the size of these marsh habitats determined overall mobility. That is, toolstone data may reflect the frequent long-distance movement of Paleoarchaic men, not that of Paleoarchaic groups as a whole.

It is not yet possible to distinguish between these two models with the information at hand. The model employed by Jones et al. (2003) to explain the toolstone data suggests large-scale north-south movement of entire groups throughout a seasonal round (Figure 1.11), characterized by relatively short-term stays at a series of residential bases. A viable alternative is that the main group may have moved relatively short distances two or three times a year, with male parties traveling long distances to toolstone sources from these more permanent camps (Figure 1.12). Such toolstone procurement may have been embedded in hunting forays. Given the probable low population densities, some toolstone exchange may also have occurred at "jamborees" where multiple groups met to find mates, share information, and create support networks.

If the model proposed by Elston and Zeanah (2002) is valid (Figure 1.13), and I think it is, then it has a number of implications. One of the principal implications is that the model should play out differently in different situations. More precisely, patch-choice models (e.g., MacAruthur and Pianka 1966) predict that the group as a whole should move from one marsh ecosystem patch

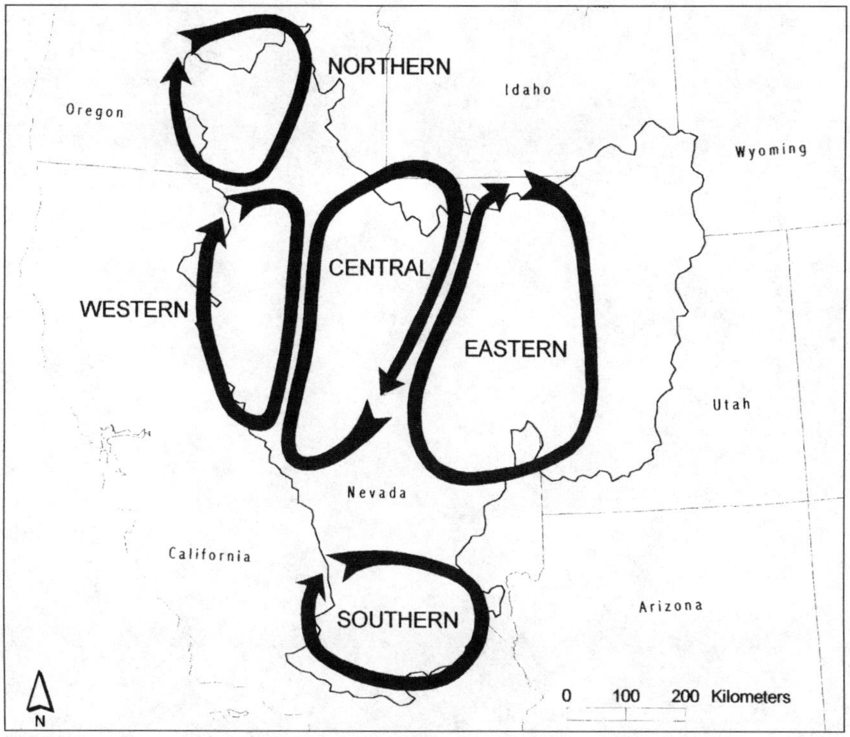

FIGURE 1.11. Possible toolstone conveyance zones in the Great Basin proposed by Jones et al. (2003) to explain toolstone source data.

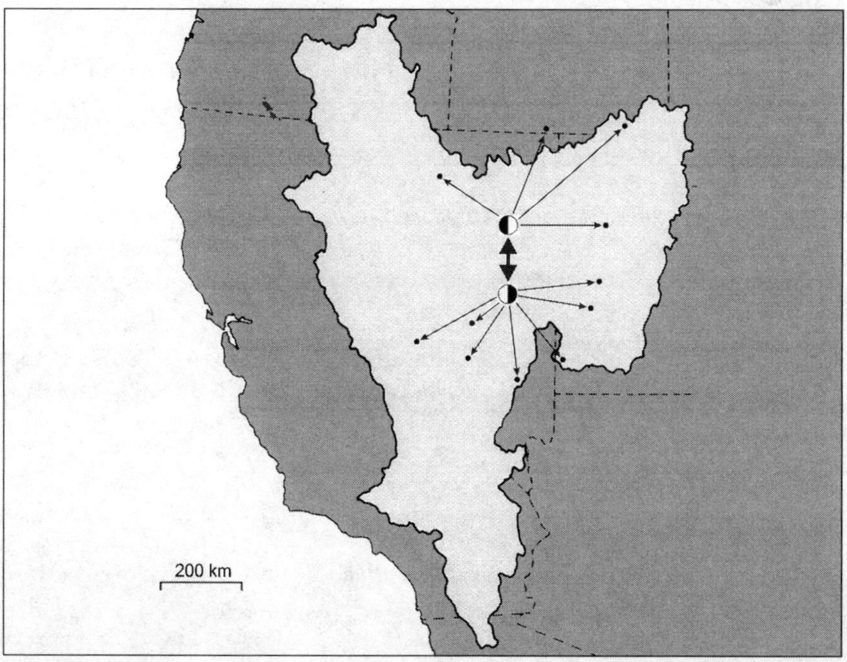

FIGURE 1.12. An alternative model of group fragmentation and movement that may explain the toolstone distributions in Figure 1.10. Small black circles represent toolstone sources traveled to by male parties; large half-shaded circles represent relatively permanent camps occupied by the rest of the group.

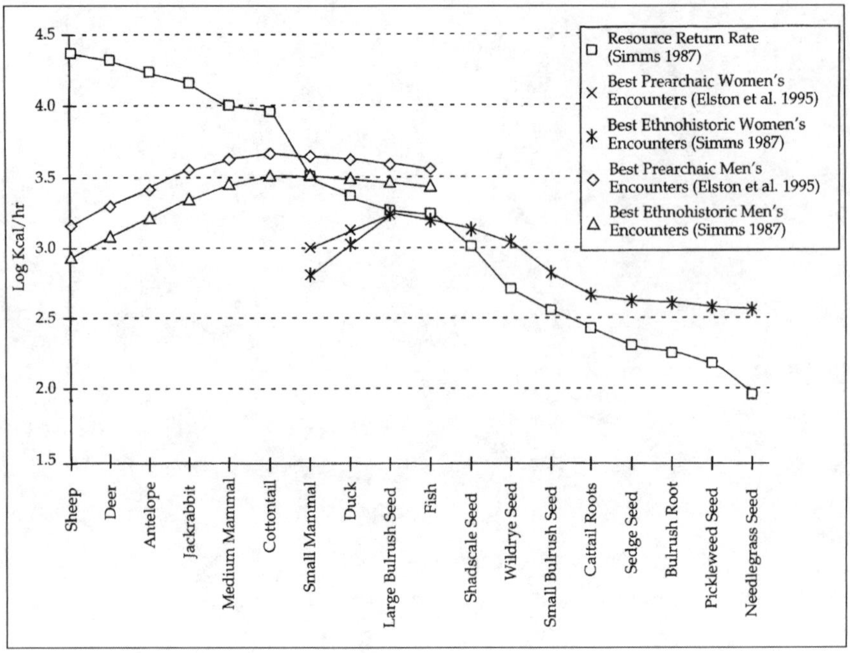

FIGURE 1.13. Estimated fall return rates for resources collected by male and female Great Basin foragers (from Elston and Zeanah 2002).

to another only when return rates in the patch they occupy begin to drop below the average for all other patches. This, in turn, means that the larger the patch, the longer it will take for foragers to reduce the overall return rate. One would expect Paleoarchaic mobility patterns to range from highly mobile long-distance movement where marsh habitats were small and widely scattered, to more sedentary and short-distance movement where marsh ecosystems were large and productive.

When one looks closely at Paleoarchaic material from the east-central Great Basin, variations in the patterns of mobility seem to reflect precisely these varied expectations. In the central Great Basin—where marsh systems dating to the Pleistocene/Holocene transition were small, relatively isolated, and widely scattered—movement was apparently around a number of short-term residential camps spread out primarily along the north-south oriented valley systems. In the eastern Great Basin, on the other hand, where the marsh ecosystem of the Old River Bed delta may have covered nearly 700 km^2, mobility seems to have been limited to short-distance movement within the large deltaic system itself (e.g., Oviatt et al. 2003; see also Duke and Young, this volume; Schmitt et al., this volume).

TRANSITION

The transition to the Archaic in the Great Basin appears to have been sudden in some locations. Where marsh habitats were small and the "lowland" adaptation quickly became unviable, Paleoarchaic foragers appear to have disappeared quickly after the Younger Dryas, ~10,000 ^{14}C BP (11,500 cal BP). Archaic foragers also appear suddenly in upland areas after ~8000 ^{14}C BP (8900 cal BP) in locations where Paleoarchaic foragers had been uncommon. With the development of ground stone seed processing, the "upland" adaptive strategy became a viable alternative, and areas of the Great Basin where many of the lower-ranked resources in the increasingly broad diet occur became suitable for foraging. In short, the transition appears dramatic, with Paleoarchaic foragers disappearing suddenly in some areas, while in others Archaic foragers seem to appear equally quickly.

Such a scenario may be more apparent than real, however. When one looks closely at the occupational sequences in large marsh ecosystems occupied continuously throughout the Pleistocene/Holocene transition, change seems to have occurred much more gradually. In the Old River Bed delta in the eastern Great Basin, for example, seriation of biface forms at sites dating to between ~11,000–8700 ^{14}C BP (12,900–9600 cal BP) indicates a gradual replacement of Paleoarchaic forms by those characteristic of the Archaic (see Schmitt et al., this volume, Figure 6.5). What this suggests, in turn, is that lowland adaptive strategies found ethnographically in the Great Basin appear to have been adopted to a greater or lesser degree through the entire history of human occupation. Although climate change may have altered the number and size of available marsh habitats, they continued to be used when and where they occurred. What the innovations of the Early Archaic provided was an additional strategy suitable for foraging in other habitats. This created an increased flexibility that allowed Great Basin foragers to operate successfully over more varied territories and in spite of the vagaries of environmental fluctuations.

This shift may also have altered mobility patterns in unexpected ways. As the early Holocene broad-spectrum diets of the Paleoarchaic became even broader, the diet breadth of male and female foragers began to overlap more extensively. As they did so, this likely altered the relative mobility of men and women. If the Paleoarchaic foragers were characterized by the long-distance movement of men from relatively stable camps occupied by women, the addition of an upland strategy with its broader diet meant that in some areas women and the main foraging group may have moved more frequently, but men likely traveled smaller distances to and from these shorter-term camps.

SUMMARY

In the Great Basin, the early Holocene transition from Paleoarchaic to Archaic foraging strategies is characterized primarily by the expansion of the number and kinds of habitats that foraging groups exploited, the addition of seed processing to the resource base, and a change in mobility by those groups focused on the exploitation of upland ecosystems. Despite these dramatic changes, the transition is also characterized by continuity in the way wetland communities at the ends of Great Basin rivers and major spring systems were exploited.

This lowland pattern involves a relatively sedentary occupation of large marsh areas by women, children, and the elderly, with able men coming and going in a highly mobile fashion to and from these processing areas. The transition was essentially an additive process. The basic late Pleistocene/early Holocene foraging strategies employed when lakes were high continued to be used in a reduced number of locations, but with the addition of an upland foraging pattern involving the movement of entire family groups through a series of residential bases situated in a variety of ecosystems.

NOTES

1. The following section is drawn largely from Madsen et. al. 2002.
2. Much of the following section is taken from Madsen 2002.

ACKNOWLEDGMENTS

Most of what is summarized here is from the work of several generations of Great Basin anthropologists, to whom I owe a great debt of gratitude. This benefited from ongoing, sometimes lively, discussions with Don Grayson, Charlotte Beck, Tom Jones, Jack Oviatt, Dave Rhode, and Dave Schmitt. I also thank Dave Schmitt and Evelyn Seelinger for editorial review, and Sam Gardner for assistance with image production.

PART II

Chronological and Contextual Studies

2

Early Paleoarchaic Point Morphology and Chronology

Charlotte Beck and George T. Jones

THE OCCURRENCE OF FLUTED POINTS IN THE GREAT BASIN HAS BEEN of interest to archaeologists of that region for nearly 80 years. The prevailing thought has always been, and continues to be, that these points are the western expression of the Clovis complex, believed by many to represent the earliest colonization of the continent. As we have argued for a number of years, however, neither inference has ever been demonstrated (e.g., Beck and Jones 1997; Beck et al. 2004a; Beck et al. 2004b). Only four radiocarbon dates have ever been reported for Great Basin Fluted points, and all are younger than dates for Clovis by at least 500 radiocarbon years (Table 2.1). Consequently, most Great Basin archaeologists discount them based on arguments concerning poor association (Danger Cave, Sunshine Locality), stratigraphic mixture (Connley Cave 5), or simply dating error (Henwood). For example, the date of 8470 ± 370 (AA-648) ^{14}C BP (9005–10,118 cal BP) reported by Douglas et al. (1988; see also Warren 1990; Warren and Phagan 1988) from the Henwood site in the Mojave Desert has been virtually ignored because it is so much later than those for Clovis, and for fluted points of any form across North America. In his discussion of fluted points in the Great Basin, Grayson (1993:237) suggests three possible explanations for the date: (1) the date may be incorrect; (2) the point may have been carried to the site by later peoples; or (3) fluted points may date later in the Great Basin than elsewhere in North America. The first two explanations are certainly possible, even probable, but what about the third? It is interesting that most archaeologists refuse to even consider the possibility that Great Basin Fluted points may date to a later period than Clovis on the Plains and in the Southwest (which is where all of the radiocarbon dates are from); why is this?

Fluted points in eastern North America date as late as 10,000 ^{14}C BP (11,400 cal BP) and perhaps later.[1] In fact, there are no dates in the East that are as old as those for Clovis on the Plains. In addition, numerous forms—such as Cumberland, Simpson, and Beaver Lake in the Southeast—are believed to occur successively in time (see Anderson et al. 1996:10, Figure 1.2). On the Plains and in the

TABLE 2.1. Radiocarbon Dates Reported to be Associated with Fluted Points in the Great Basin

Site	Uncalibrated Radiocarbon Date (BP)	Lab Number	Calibrated One-Sigma Range (rounded)[a]	Reference
Danger Cave, UT	10,080 ± 130	Beta-19333	11,400–11,960	Madsen and Rhode 1990
Sunshine Locality, NV	10,320 ± 50	Beta-86205	12,040–12,350	Jones et al. 1996
Henwood Site, CA	8470 ± 370	AA-648	9005–10,118	Douglas et al. 1988
Connley Cave 5, OR[b]	9540 ± 260	Gak-1744	10,514–11,204	Bedwell 1973
	7430 ± 140	Gak-2135	8057–8386	

[a] Dates were calibrated using latest version of CALIB (Stuiver et al. 2006).
[b] It is debated whether this point is fluted (see Beck et al. 2004a).

Southwest, Clovis is followed in time by Folsom, which lasts until 10,200 ^{14}C BP (11,850 cal BP and 11,970 cal BP) (Haynes 1992; Taylor et al. 1996). There is no reason to assume that such a succession of forms was not the case in the Great Basin as well. Even if Clovis is represented in the Great Basin, and we believe it is, this complex did not appear simultaneously throughout North America. It had to begin somewhere and spread from that place of origin. In addition, it was likely followed by other, later fluted forms in the Great Basin, as it was on the Plains and in the East.

The problem remains, of course, in the virtual absence of radiocarbon dates, how to evaluate if (1) fluted points appear later in the Great Basin than Clovis on the Plains, and (2) after its arrival, the Clovis fluted form was succeeded by at least one non-Clovis fluted form. We have discussed the first of these questions in detail elsewhere (e.g., Beck et al. 2004b) and thus focus here on the second, using an approach relatively new to archaeology: cladistics. But first we summarize the argument presented in Beck et al. 2004b concerning the appearance of the Clovis complex in the Great Basin.

CLOVIS ON THE PLAINS

Focusing on Clovis sites for which the radiocarbon dates are deemed by Taylor et al. (1996) to be reliable (Figure 2.1), we note two trends. First, the ages of Clovis sites are younger as we move west and north from Texas (Figure 2.2). This spatial trend is also evident in the distribution of 19 of the 20 Clovis caches identified by Kilby and Huckell (2003, n.d.), but the cache distribution continues across the mountains onto the Columbia Plateau (Figure 2.3).

The temporal and spatial patterns of the radiocarbon dates suggest a rapid spread of Clovis technology from the southern Plains northwest, along the Rocky Mountain front, and west, south of the Rockies (Figure 2.4). The latter route would provide an entry into the Great Basin from the south. A northern entry point is suggested by the presence of the Fenn, Simon, and East Wenatchee Clovis caches on the Columbia Plateau. The only date associated

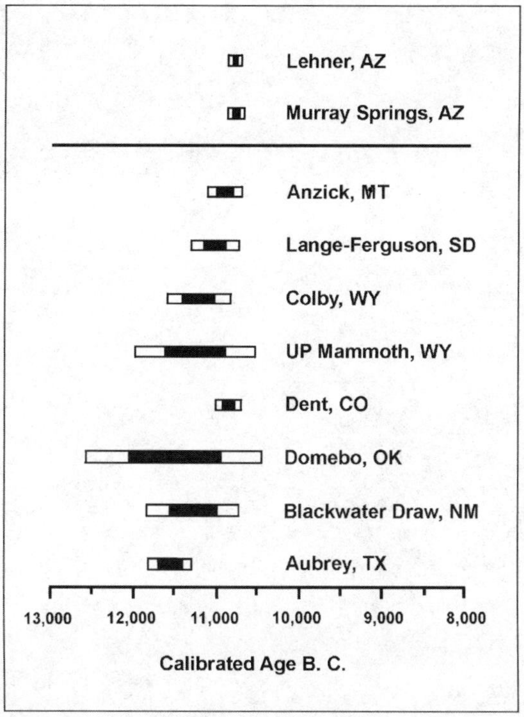

FIGURE 2.1. Graphic representation of the radiocarbon dates that Taylor et al. (1996) found to be reliable.

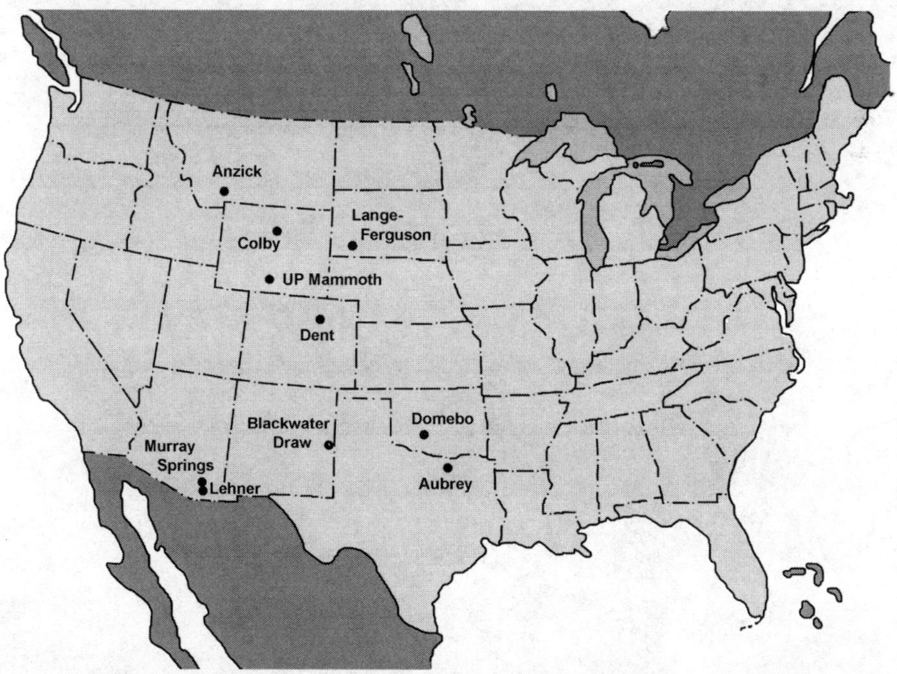

FIGURE 2.2. The distribution of Clovis sites from which the reliable dates come (Taylor et al. 1996).

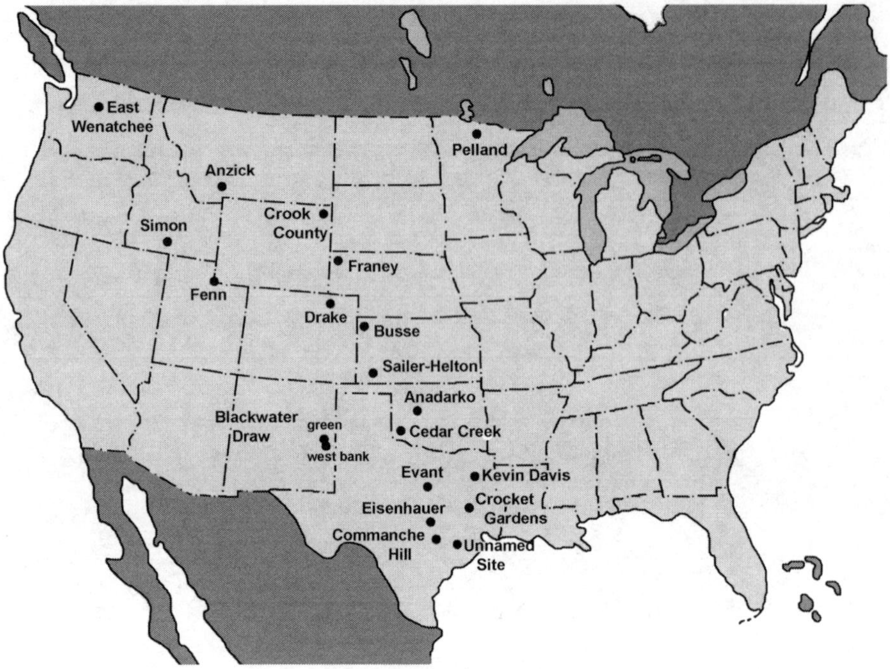

FIGURE 2.3. The distribution of Clovis caches (after Kilby and Huckell 2003, n.d.). "Green" and "west bank" are two caches found at the Blackwater Draw Clovis site.

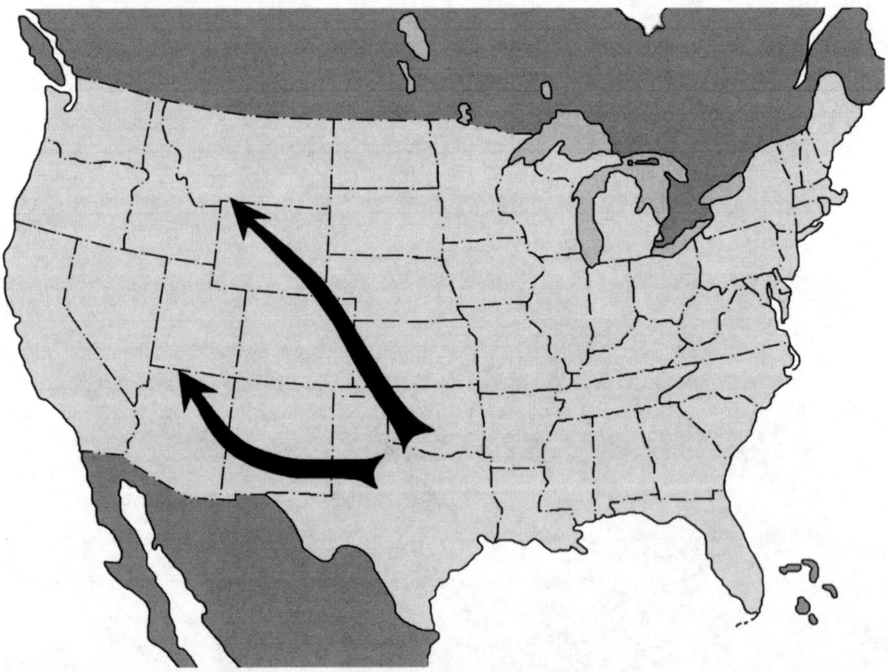

FIGURE 2.4. Two hypothetical routes of entry for Clovis technology into the Great Basin.

with any of the northwestern caches is a maximum date of ca. 11,250 ^{14}C BP (13,150 cal BP) for the East Wenatchee cache in Washington State (based on the dating of volcanic pumice contained in silicious crusts cemented to five artifacts from the cache by Mehringer and Foit 1990). This date suggests that Clovis technology was present on the Columbia Plateau about the same time or slightly later than it was on the northern Plains. Whether Clovis people entered from the south or north, Great Basin Fluted points should date later than Clovis in the southern and northern Plains, and as we noted above, this is certainly the case for the few dates that exist.

We also argue that the presence or absence of Clovis blades might give insight into the spatial expansion of Clovis technology and the relative timing of that expansion. Collins (1999b:14) states that in the southern Plains Clovis artifacts were produced through two reductive strategies: one resulting in bifacial projectile points and the other in prismatic blades. Clovis blades are large, ranging between 50 and 160 mm in length, and have "small platforms, flat bulbs, smooth interiors, and marked curvature" (Figure 2.5) (Collins and Lohse 2004:176). Collins and Lohse (2004:176) argue that "this constellation of traits is so distinctive that such blades are almost as diagnostic as are Clovis projectile points."

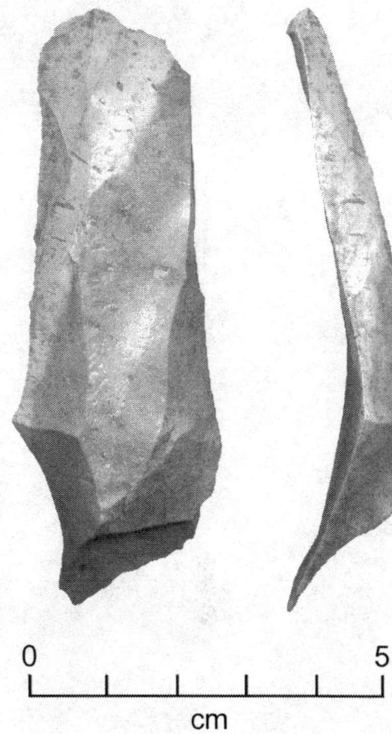

FIGURE 2.5. Example of a Clovis blade (from the Sunshine Locality in eastern Nevada) (Beck and Jones n.d.).

Figure 2.6 shows the locations of Clovis blades across North America. As is evident from this figure, blades are most numerous in the southern Plains, which also happens to be the location of the earliest radiocarbon date associated with Clovis (at the Aubrey site), but they grow less common beyond this area. Blades have been reported from several sites in the eastern U.S., but Collins (1999b:148) argues that only at Stanfield Worley in Alabama, Wells Creek in Tennessee, and Adams in Kentucky are they "the product of a robust blade technology." The Pelland cache in Minnesota, which contains nine blades, is also believed to be of Clovis affiliation (Collins 1999a; Kilby and Huckell 2003, n.d.).

To the west, 13 blades occur in the Murray Springs assemblage, and two scrapers collected from Lehner also appear to be made on blades. In addition, numerous blades and blade cores associated with 25 Clovis points were reported by Sanchez (2001) from a large surface site (El Bajio) in northern Sonora. In the Great Basin, one Clovis blade was found in the Pine Grove Hills of western Nevada (David Rhode, personal communication), at least three were collected at China Lake (Davis 1978), and one was recovered from the alluvial gravels at the Sunshine Locality (Beck and Jones n.d.; Beck et al. 2004b). Two obsidian blade cores were found by a private collector, one from northern Utah and the other from southeastern Idaho (Dennis Stanford, personal communication 2004). From the Columbia Plateau, Clovis blades have been reported from only one location, the East Wenatchee cache in Washington, in which there are five

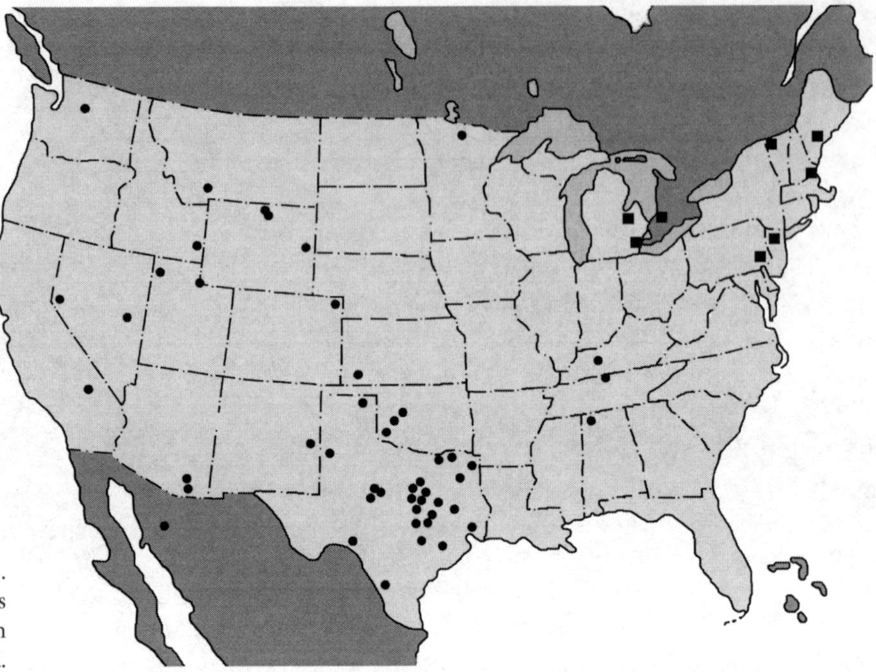

FIGURE 2.6. Locations of Clovis blades across North America.

(Gramly 1993). Even if blades have not been systematically reported, as Collins suggests, the pattern is clear: Clovis blades are numerous in the southern Plains and possibly in some parts of the Southeast; they diminish in number to the west, north, and northeast. A similar pattern is evident in the Clovis caches.

We have already pointed out that the spatial distribution of the Clovis caches follows that of the dated Clovis sites (Figures 2.2 and 2.3), moving primarily northwest along the Rocky Mountains and then onto the Columbia Plateau. Interestingly, the content of these caches changes as well from south to north. Caches to the south are either completely comprised of blades and blade cores or contain significant numbers of these artifacts in addition to other tools (Collins 1999a; Kilby and Huckell 2003, n.d.). Bifaces are rare in the southern Plains caches, and those that are represented are Clovis preforms (Collins 1999b). The caches to the north, however, contain very few or no blades, but they contain numerous bifaces, which vary more in form than those in the southern caches, including relatively large bifacial flake cores and bifacial fluted point blanks in addition to fluted point preforms. In fact, some of these caches, such as those from East Wenatchee, Simon, and Fenn, are extraordinary in their content of large, exquisitely formed fluted points and bifaces that have "almost no counterparts in the rest of the Clovis sub-regions" (Collins 1999b:31).

It appears that as people carrying Clovis technology moved out of the southern Plains (and/or perhaps the Southeast), bifaces replaced blades as the primary component of the toolkit. The radiocarbon record suggests that this replacement occurred during the Clovis time period (11,500–10,700 ^{14}C BP),

TABLE 2.2. Statistical Comparison of Clovis and Great Basin Fluted Points

VARIABLE	SEPARATE VARIANCE (t)	df	p	GBF (n)	CLOVIS (n)
Total Length	3.429	108.2	0.001	47	55
Maximum Width	-0.406	105.3	0.686	127	55
Maximum Thickness	4.419	98.9	< 0.001	170	55
Basal Indentation	-4.116	108.6	< 0.001	197	55
Basal Width	-1.988	101.4	0.049	201	55
BI/BW	-4.502	204.6	< 0.001	195	87
BW/MW	-0.975	193.8	0.331	112	87

with earlier assemblages, in general, containing blades but few bifaces, and later assemblages containing few blades but many bifaces. Therefore, Clovis assemblages in the Great Basin, which include very few blades, should date later than Clovis assemblages on the Plains.

THE GREAT BASIN RECORD

Once fluted point technology arrived in the Great Basin, it should have evolved morphologically over time, as it did in other areas, but not necessarily in the same direction. For instance, an examination of Figure 1.2 in Anderson et al. 1996:10 demonstrates such morphological evolution, from a Clovis-like form to forms that are increasingly constricted right above the base, eventually resulting in a pronounced "fish-tail" (see also Morrow and Morrow 1999). Fluting is decreasingly evident over the sequence, ending with Dalton, which is only sometimes fluted. In order to examine differences in morphology between Great Basin Fluted (GBF) points and Clovis points, we use a database originally created in 2002 by Amanda Taylor (see also Taylor 2003). Since the database was created, we have added 60 specimens, bringing the total to 239 points (20 of these are from Borax Lake in western California, outside the Great Basin). We made statistical comparisons between GBF points and 55 points from Plains Clovis sites and 32 points from the Fenn, Drake, Simon, and East Wenatchee Clovis caches (Table 2.2).

Cache points are, in general, larger than non-cache Clovis points, and thus statistical comparisons of individual attributes were made only with non-cache points. T-test results indicate significant differences in total length (although the sample size for GBF points is small), maximum thickness, basal width, and basal indentation. Specifically, GBF points are shorter and thinner but have a wider base and deeper basal concavities. Comparisons on two ratios, basal indentation/basal width and basal width/maximum width, between all Clovis points (cache and non-cache) and GBF points indicate that a significant difference exists only with respect to the former; GBF points exhibit a deeper basal concavity relative to the basal width.

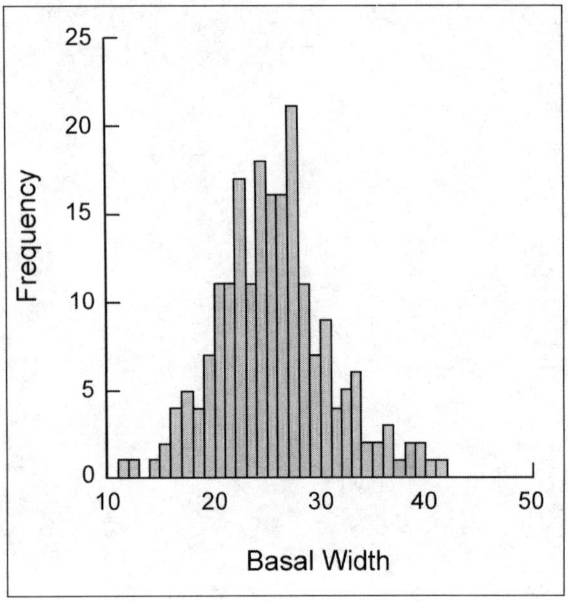

FIGURE 2.7. Distribution of basal width among Great Basin Fluted points.

In sum, these comparisons demonstrate that, overall, GBF points differ from Clovis points on some attributes. However, there is considerable variability in GBF points, as the histogram of basal width in Figure 2.7 indicates. This variability, we believe, may be, at least in part, temporal. Certainly a subset of GBF points matches the characteristics of Clovis points. Others exhibit consistent differences, and we believe these represent at least one later fluted form. For example, an examination of 17 fluted points from the Sunshine Locality in eastern Nevada revealed that these points are much more uniform in their morphology, and most show significant differences from Clovis points on a majority of both quantitative and qualitative attributes (Beck and Jones n.d.; Beck et al. 2004b).

Great Basin archaeologists have been searching for at least 50 years for fluted points in undisturbed, datable, stratigraphic contexts but have not found them. Without a lot more work and a significant change in fortunes, we should not expect to easily solve the question of the timing and affiliation of fluted points through conventional approaches. This does not, however, prevent us from generating hypotheses about what the differences in morphologies might indicate. Change in form over time is something archaeologists have been dealing with for a long time (Lyman and O'Brien 2000; Lyman et al. 1997). Such changes are highly constrained in that new forms usually represent modifications of old ones, leading to what culture historians call "tool traditions" and what selectionist archaeologists have termed "intellectual traditions" (O'Brien et al. 2001:1117). If we assume that different fluted point morphologies are related in a "genetic" sense, then from the standpoint of the GBF point record, Clovis is presumably ancestral and left descendent forms there, much as we find in other regions with forms like Folsom. To generate ideas about the nature of these changes, we turn to cladistics, a systematic approach for the study of evolving lineages.

CLADISTICS

One of the hallmarks of the culture history paradigm is the space-time chart, which displays similarities in material culture over space and time.[2] Similarities in both space and time were attributed to cultural contact, the first being primarily synchronic (horizon) and the second being primarily diachronic (tradition). In both cases change is portrayed as being continuous and gradual, as typified by Ford's (1952, 1962) well-known ceramic sequences in the Southeast,

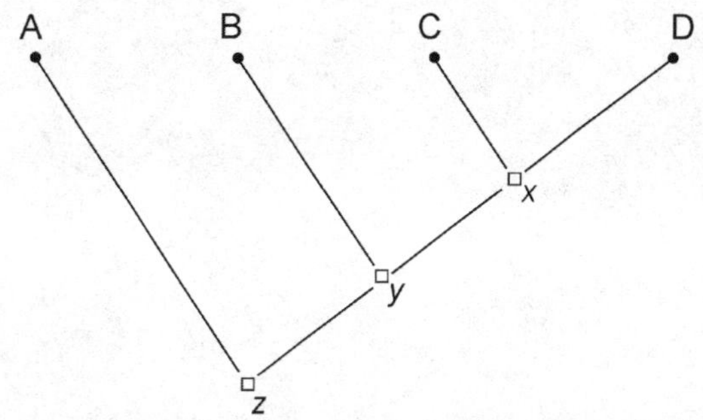

FIGURE 2.8. Hypothetical cladogram (after O'Brien et al. 2001:1119).

based on changes in both form and design. In evolutionary terms, similarity due to cultural contact, be it in space or time, is known as homologous similarity (geneology); the alternative, analogous similarity, arises independently, in the absence of cultural contact (parallelism or convergence) (Dunnell 1980). It is with homologous similarity through time that we are concerned here.

Cladistics, initially proposed by Hennig (1950, 1966), is a method that employs homologous features for the study of phylogenies and is widely accepted in biology and paleobiology as the method for this purpose (e.g., Brooks and McLennan 1991; Kitching et al. 1998; Wiley et al. 1991). Cladistics has also been used to some extent in historical linguistics (e.g., Platnick and Cameron 1977; Ross 1989), but its application in archaeology has been quite recent (e.g., O'Brien et al. 2001; O'Brien and Lyman 2003). O'Brien et al. (2001) argue, however, that cladistics has great promise for constructing intellectual traditions, similar to the culture historian's tool traditions but based on a phylogenetic approach. The method produces a cladogram (Figure 2.8), similar in form to a phenogram produced by numerical taxonomy (e.g., Sokal and Sneath 1963) to summarize a pattern of trait distribution. For example, in the cladogram in Figure 2.8, C and D, the most closely related groups, form a clade. B is more distantly related to C and D than they are to each other, and thus the three form a more inclusive clade, and so on. As is the case for seriation, however, the cladogram is a statement only about similarity; it is up to the analyst to explain the pattern.

Cladistics differs from numerical taxonomy in that it employs only homologous traits, whereas numerical taxonomy uses traits regardless of whether they are homologous or analogous. Cladistics differs from other phylogenetic methods as well; the latter utilize shared ancestral traits (all homologous), while cladistics employs shared derived traits, a subset of homologous traits held in common only by sister taxa and their most recent common ancestor. To demonstrate, we use an example employed by O'Brien et al. (2001) focusing on a single trait, fluting, for which there are two character states: unfluted and fluted (Figure 2.9). In the first diagram Taxon A is unfluted and gives rise to two new taxa: 1, which is unfluted like A, and 2, which is fluted. "Fluted" is a derived character state, evolved from its ancestral character state, "unfluted." Next (Figure 2.9b), Taxon 2 gives rise to Taxa 3 and 4, both of which are fluted. Fluting is now a derived character state shared by two sister taxa (3 and 4) and their most recent common ancestor (Taxon 2). In the third diagram (Figure 2.9c), Taxon 4 gives rise to Taxa 5 and 6, both of which are also fluted. At this point, "fluted" is now an ancestral character state because it is shared by more than sister taxa and their immediate common ancestor.

In order to plot the evolutionary relationships among a set of taxa, we must distinguish between ancestral and derived character states, termed *character*

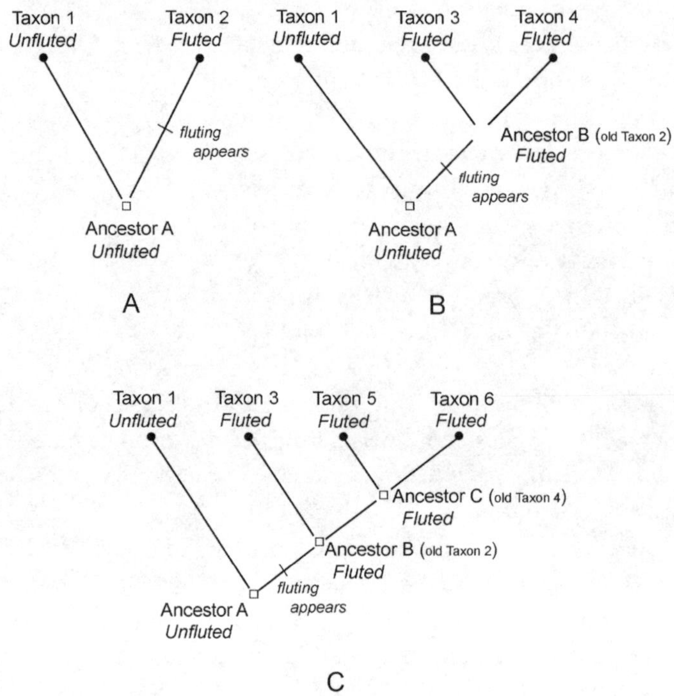

FIGURE 2.9. Phylogenetic trees showing the evolution of projectile point taxa (after O'Brien et al. 2001:1119).

polarity; because a character state can be both derived and ancestral, however, depending on the scale at which it is viewed (O'Brien and Lyman 2003:59), this can often be difficult. A commonly used approach is comparison with an outgroup, a taxon outside the taxa being investigated (the ingroup taxa) but believed to be related to it through an earlier common ancestor (O'Brien and Lyman 2003:62). The most common state of a trait among members of the outgroup is assumed to be ancestral relative to the ingroup taxa. "The more closely related the outgroup is to the ingroup taxa, the greater the number of characters or character states that might be clarified relative to polarity" (O'Brien and Lyman 2003:65). Of course, this assumes that we know the evolutionary sequence that we are attempting to demonstrate, creating a degree of circularity. Given that we really do not know the evolutionary sequence but are choosing an outgroup based on our "best guess" of ancestral relatedness, cladistical methods often yield multiple solutions (cladograms). These are alternative hypotheses about the actual history of cladogenesis or branching events in a lineage. The general rule of thumb, or decision rule, is that the solution that requires the fewest evolutionary events is probably the best. For example, it is more parsimonious to assume that a particular character state (like fluting) evolved only once, although, in reality, this is not likely to hold true in most cases. More likely, a cladogram will contain character states that appear in lines not related to a common ancestor. These are called *homoplasies*. A character state reversal, for example, is a type of homoplasy where one state, say unfluted, is replaced by another state, fluted, and then at some later point reverts back to

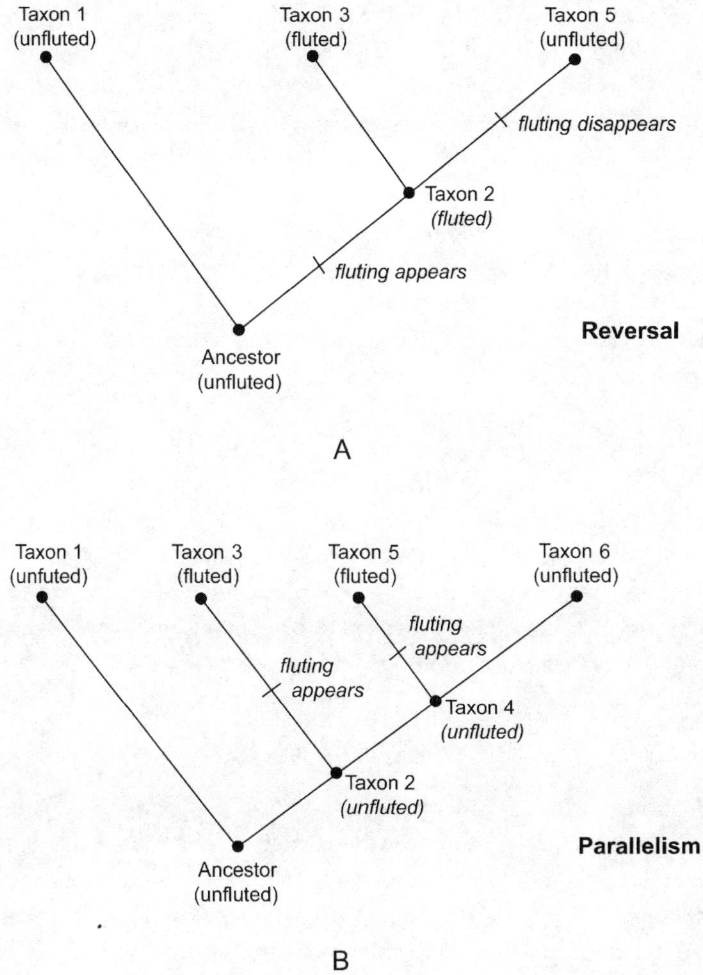

FIGURE 2.10. Phylogenetic trees showing two types of homoplasy.

its original, unfluted state (Figure 2.10a). A second type of homoplasy is parallelism, in which a new character state appears independently in two or more different branches (Figure 2.10b). Thus, in choosing among a number of different cladograms, the best (most parsimonious) solution is the one with the fewest steps and the fewest homoplasies.

The use of analytical parsimony, O'Brien et al. (2001:1121) suggest, helps to minimize the circularity inherent in the use of the outgroup. In addition, if several traits suggest the same phylogenetic sequence, we might have more confidence in the cladogram we have selected. Finally, we can use independent temporal information, such as stratigraphic sequence, that corroborates the phylogenetic sequence.

Once we have selected a solution, however, how do we judge its overall robustness? That is, how accurate a representation is it of the true phylogeny? One test discussed by O'Brien et al. (2001; O'Brien and Lyman 2003) is a consistency index (CI), which rates the solution on the basis of the number of homoplasies relative to the number of characters on the cladogram (Farris 1989; Goloboff 1991). The CI ranges from 0 to 1.0, with 0 equaling complete homoplasy and 1.0 indicating none; this index is calculated by dividing the number of character states represented in the classes being analyzed (exclusive of the outgroup class) by the number of character states on the cladogram (O'Brien et al. 2001:1124). For instance, in the fluted example, unfluted is the ancestral (outgroup) state, and thus the number of character states in the actual analysis is one: fluted. In all of the diagrams in Figure 2.9 the CI is 1.0: the number of character states used in the analysis (1) divided by the number of character states in the cladogram (1). In the cladogram in Figure 2.10a, however, the CI is 0.50, because, due to the reversal, there are now two character states in the cladogram.

As O'Brien and Lyman (2003) point out, however, there are several problems inherent in this index. For example, where there are only two alternative states of a character (e.g., unfluted, fluted) as opposed to multiple states (e.g., fluted, stemmed, notched), it is impossible for the index to equal zero. In addition, the index value is not independent of the number of taxa being analyzed. Several approaches have been proposed to overcome some of these problems; for a complete discussion, see O'Brien and Lyman 2003:72–73.

CLADISTIC ANALYSIS OF GREAT BASIN FLUTED POINTS

In their application of cladistics to fluted points in the Southeast, O'Brien et al. (2001:1126) selected eight traits that they believed were most likely to change over time "as a result of [cultural] transmission and thus create a strong phylogenetic signal" (Table 2.3, Figure 2.11). Their sample contained only complete fluted points, and thus they were able to use traits based on metrics, such as a length-to-width ratio and maximum blade width location. The Great Basin sample that we use here contains few complete points; for example, out of 239 GBF points, there are length measurements for only 47. Consequently, we were unable to employ several of the traits used by O'Brien et al. In addition, their analysis is of eastern fluted points, which are distinctive with respect to fluted points in other parts of North America in that most have a prominent "fish-tail" shape to their base. Therefore, the two characters O'Brien et al. used to measure the degree of "fish-tailing"—constriction ratio and outer tang angle—are not relevant to either Clovis or GBF points. For the present analysis, then, we developed a paradigmatic classification[3] based on six traits measurable on a larger number of GBF and Clovis points (Table 2.4, Figure 2.11), which produces 960 possible classes. Even so, only 137 of 326 (87 Clovis and 239 GBF) points could be identified with classes; the remaining points are missing data on one or more traits. These 137 points identify with 72 classes, most of which have fewer than three members. For our analysis we used only those classes that contain three or more members, for a total of 12 classes containing 67 points (Table 2.5).

Before proceeding with our analysis, we had to determine which class to use as an outgroup. If we subdivide the Clovis points into southern and northern

TABLE 2.3. Paradigmatic Classification Used by O'Brien et al. (2001) in Their Cladistic Analysis of Eastern Fluted Points

A. Location of Maximum Blade Width	E. Outer Tang Angle (degrees)
1. proximal quarter	1. 93–115
2. secondmost proximal quarter	2. 88–92
3. secondmost distal quarter	3. 81–87
4. distal quarter	4. 66–88
	5. 51–65
	6. < 50
B. Base Shape	F. Tang-tip Shape
1. arc-shaped	1. pointed
2. normal curve	2. round
3. triangular	3. blunt
4. Folsomoid	
C. Basal Indentation Ratio[a]	G. Fluting
1. no basal indentation	1. absent
2. shallow (0.90–0.99)	2. present
3. deep (0.80–0.89)	
D. Constriction Ratio[b]	H. Length/Width Ratio
1. 1.00	1. 1.00–1.99
2. 0.90–0.99	2. 2.00–2.99
3. 0.80–0.89	3. 3.00–3.99
4. 0.70–0.79	4. 4.00–4.99
5. 0.60–0.69	5. 5.00–5.99
6. 0.50–0.59	6. ≥ 6.00

Source: O'Brien et al. 2001:1126; O'Brien and Lyman 2003:152.
[a] Medial length/total length.
[b] Minimum blade width/maximum blade width.

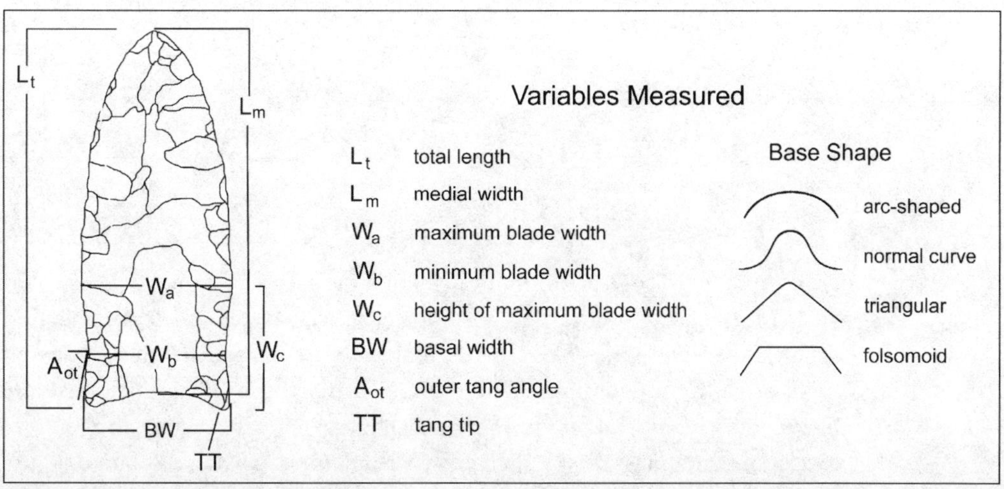

FIGURE 2.11. Variables measured by O'Brien et al. (2001; O'Brien and Lyman 2003) as well as those measured in the present research.

TABLE 2.4. Paradigmatic Classification Used in the Cladistic Analysis

A. Base Shape	D. BI/BW Ratio
1. arced	1. 0–0.11
2. normal curve	2. > 0.11–0.19
3. Folsomoid	3. > 0.19–0.27
4. straight	4. > 0.27
B. Flute Face	E. BW/MW Ratio
1. both sides	1. < 0.81
2. single side	2. 0.81–0.98
	3. > 0.98
C. Number Flute Scars	F. Basal Width
1. single	1. < 20.0 mm
2. multiple	2. 20.0–24.0 mm
	3. > 24.0–30.0 mm
	4. > 30.0–35.0 mm
	5. > 35.0 mm

TABLE 2.5. Classes Used in the Cladistic Analysis

CLASS NUMBER	CLASS DESIGNATION	CACHE POINTS	NON-CACHE CLOVIS POINTS	GBF POINTS[a]	TOTAL
1	111124	1	—	3	4
2	111223	—	2	5	7
3	111233	—	—	3	3
4	111322	—	1	2	3
5	111323	—	1	3	4
6	112123	2	3	5	10
7	112124	10	—	2	12
8	112222	—	2	2	4
9	112223	1	4	6	11
10	112224	2	—	1	3
11	112233	—	—	3	3
12	112324	—	—	3	3

[a] GBF = Great Basin Fluted.

groups, the most logical outgroup would be the southern group because Clovis is oldest in the southern Plains. However, we have complete information for only 22 Clovis points from this region, and they fall into 13 different classes, none of which predominate. The number of northern points for which we have complete information is not so much larger (n = 28 points falling into 16 classes), but one class predominates here, containing 10 of the 28 points (Figure 2.12). Therefore, we selected this class as our outgroup for this initial exercise.

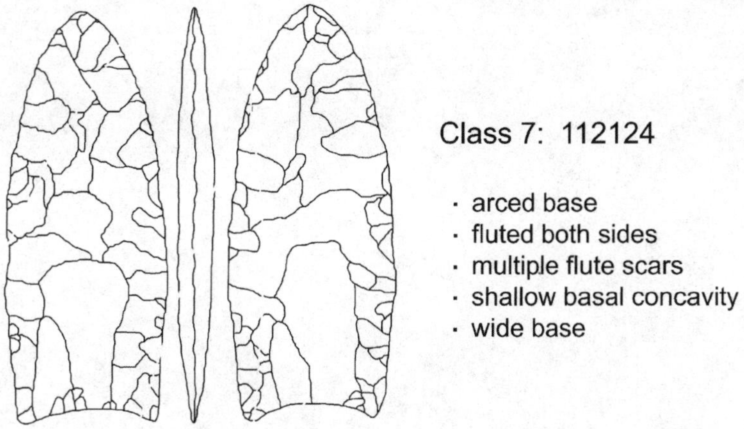

FIGURE 2.12. The class used as an outgroup in the cladistic analysis of Great Basin Fluted points.

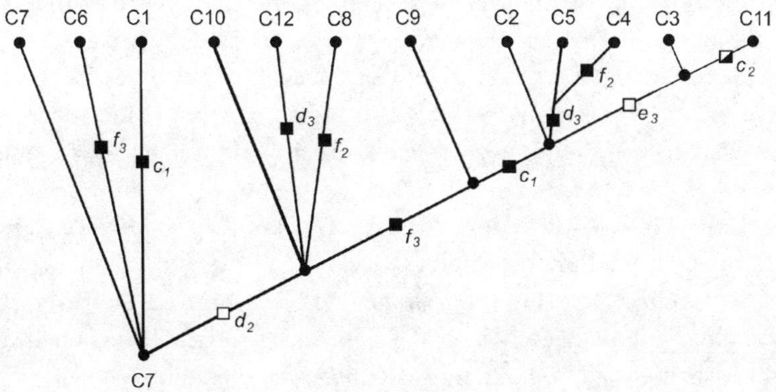

FIGURE 2.13. Phylogenetic tree (cladogram) of 12 projectile point classes.

Using the PHYLIP program (Felsenstein 2004), our analysis produced a single solution (Figure 2.13). To provide some sense of the significance of this result, when other classes in the analysis were substituted as outgroups, they yielded as many as 55 equally parsimonious solutions. In the cladogram in Figure 2.13, there are 11 character state changes, represented by the small boxes. The letters adjacent to each represent the trait, and the subscript represents the state. Open boxes indicate changes that occur only once in the cladogram, while solid boxes represent those that occur more than once; that is, these are parallel changes in two or more branches. The half-shaded box represents a reversal; that is, a return to the ancestral state. The CI for this cladogram is 0.55, as compared with O'Brien et al.'s (2001) value of 0.59. O'Brien et al. (2001) state that 0.59 is a relatively high value compared with those commonly resulting for cladograms produced in biology and paleobiology, which suggests our result of 0.55 is relatively robust.

The first changes that occur are in Traits C (number of flute scars), D (depth of basal concavity), and F (basal width). The change in flute scars is from multiple to single in Class 1, and the change in basal width is a narrowing in Class 6. The change from shallow to deeper basal concavity produces Class 10, which after another change in concavity depth (even deeper) leads to Class 12, and an

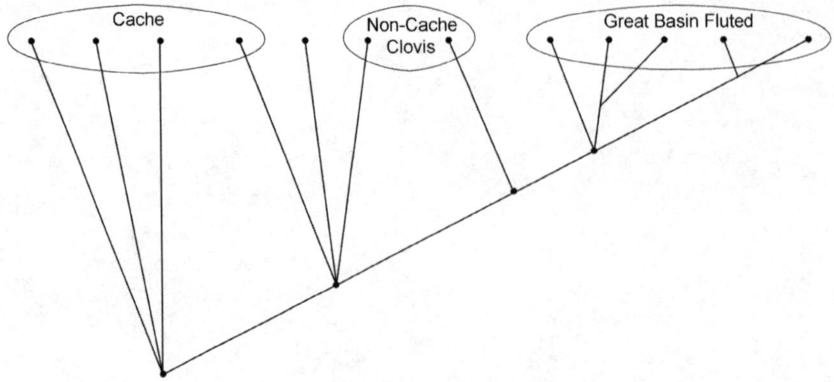

FIGURE 2.14. Pattern of projectile point occurrence in the cladogram.

increasing narrowing of the basal width to Classes 8 and 9. As can be seen from the cladogram, the picture gets more complex with the latest branches. Overall there are five cases of homoplasy, four cases of parallelism, and one case of reversal. For example, Trait C (number of flute scars) changes from 2 (multiple scars) to 1 (single scar) in Clade 1 and again between Clades 3 and 4. The reversal occurs at the outermost branch, from single flute scars back to multiple scars.

As in paleontology, where an incomplete fossil record prevents thorough testing of phylogenetic hypotheses, the absence of stratified sequences of fluted points in the Great Basin limits our evaluation of this exercise. However, we can look to other criteria to evaluate this result. One criterion often applied is geographic consistency. Do clades form sensible spatial patterns, showing a "center of gravity," minimal fragmentation, or spatial discontinuities, and are they relatively non-overlapping? We have already noted that fluted points, including Plains and Great Basin Clovis and their derivatives, are highly variable morphologically. We may surmise from this that geographic consistency will be low, and this seems to be the case, as O'Brien et al. (2001; O'Brien and Lyman 2003) also found to be true among fluted points in the Southeast. Nevertheless, we found some measure of consistency in the solution presented (Figure 2.14) in that there is a tendency for fluted points from cache contexts to cluster around early ancestors, while non-cache Clovis points are distributed on branches at the center of the cladogram. Although GBF points occur throughout the cladogram (see Table 2.5), 42 percent occur in the last four classes, and Classes 3 and 11 on the last two branches contain only GBF points. This is what we would expect if some of the points represent Clovis and others represent different, presumably later, fluted forms.

Spatially there also appears to be a pattern. We subdivided the region into four zones: northern, western, eastern, and southern (Figure 2.15). Table 2.6 shows the proportion of fluted points, exclusive of Clovis on the Plains and in the Southwest, that occur in each class in the cladogram, with classes arranged in the order that they appear. For example, Class 7 is the outgroup and appears first in the cladogram, while Class 11 appears at the opposite end. Keeping in

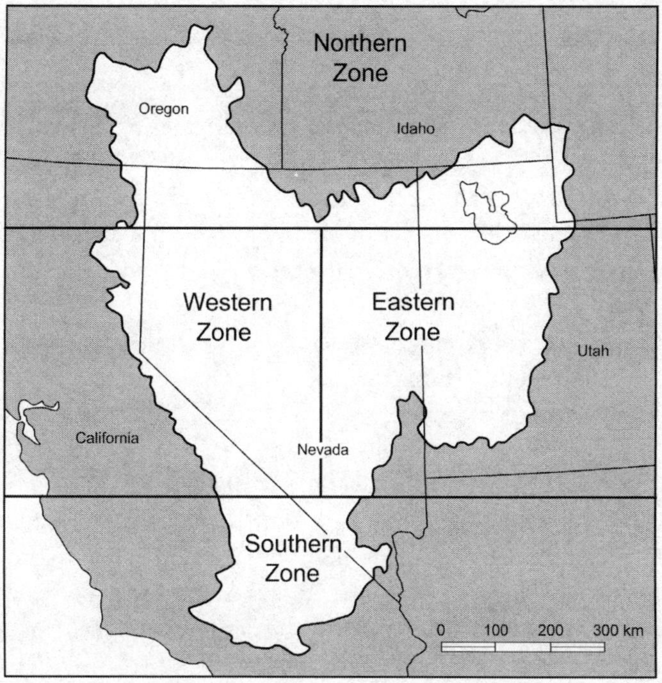

FIGURE 2.15. Subdivision of the Great Basin into four spatial zones: northern, western, eastern, and southern.

TABLE 2.6. Representation of Classes Across Areas of the Great Basin in the Order They Appear in the Cladogram

	CLASS											
	7	6	1	10	12	8	9	2	5	4	3	11
Northern	.92	.43	.50	1.00	.50	1.00	.50	.40	.33	.50	.33	1.00
Western	.08	.28	50	—	.50	—	.50	.40	.33	.50	.33	—
Eastern	—	.28	—	—	—	—	—	—	.33	—	—	—
Southern	—	—	—	—	—	—	—	.20	—	—	.33	—
Southwestern Clovis Sites[a]		x				x	x	x			x	

Note: Figure 2.15 shows the four spatial zones of the Great Basin.
[a] Lehner, Naco, and Blackwater Draw.

mind that the sample size relative to the area encompassed within the Great Basin is extremely small, we point to a pattern of migration from north to south, with movement primarily into the western zone and finally into the southern zone. Table 2.5 also indicates in which classes Clovis points from the southwestern sites (Lehner, Naco, and Blackwater Draw) also occur. Interestingly, in all but one case (Class 6), they are all in the second half of the cladogram, where we begin to see GBF points in the southern zone. We hesitate to make too much of this pattern, but it is intriguing nonetheless and perhaps should be investigated further.

CONCLUSION

As we pointed out earlier in our discussion of cladistics, a cladogram does not specify phylogeny; it is only a summary of trait distribution. It must be considered a hypothesis of phylogeny, one that must be tested. For the moment, the best we can do is evaluate the results against geographic criteria as suggested above and examine if the sequence of evolutionary events suggested in the cladogram makes sense. Several morphological trends are evident here:

- A narrowing of the base
- An increase in the basal concavity depth
- A change from multiple to single flute scars
- A change to more parallel sides

Focusing on the first, the solution gives the expected trend of decreasing point size, here monitored by basal width. Regarding the second, the result conforms to what has been described as a shift in the preparation of the base for detachment of the flute flake. A deeper concavity is consistent with preparation of an isolated platform along the base and use of a punch technique. This change of technique is evident in a fluted point preform recovered from the Sunshine Locality that was broken during the fluting process (Figure 2.16).

A change from multiple to single flute scars is a trend that is observable in all later fluted forms, in the eastern United States and on the Plains. Finally, while later fluted forms in eastern North America are characterized by a constriction above the base yielding a "fish-tail" effect, Folsom points become parallel-sided, which appears to be the trend of later GBF point forms.

We do not pretend to have accounted for the large degree of variation in GBF points, but we believe that cladistic analysis can help identify trends in trait changes over time, and our effort is a first in that direction. We hope that with additional data on both Clovis and GBF points, we may be able to better identify different morphological (and temporal) forms.

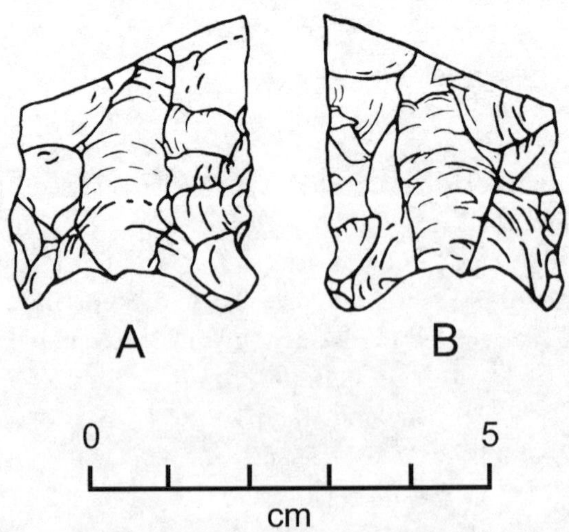

FIGURE 2.16. Fluted point preform from the Sunshine Locality in eastern Nevada (Beck and Jones n.d.).

NOTES

1. The ^{14}C date of 10,000 BP actually calibrates to four different calendar years (11,370 BP, 11,400 BP, 11,460 BP, and 11,575 BP), but the most likely according to Stuiver et al. (2006) is 11,400 BP.
2. Our discussion of cladistics relies heavily on O'Brien et al. 2001 and O'Brien and Lyman 2003.
3. See Dunnell 1971 and 1986 for a discussion of the strengths of paradigmatic classification, and O'Brien et al. 2001 and O'Brien and Lyman 2003 as to its specific use in cladistical analysis.

ACKNOWLEDGMENTS

We thank Lee Lyman, Mike O'Brien, and Don Grayson for their helpful comments on an earlier version of this chapter.

3

In Pursuit of Humans and Extinct Megafauna in the Northern Great Basin

Results of the Kelvin's Cave Excavations

L. Suzann Henrikson and Montana M. Long

THE CURRENT RESEARCH FOCUS OF THE UNIVERSITY OF OREGON Museum of Natural and Cultural History is to examine archaeological sites in the northern Great Basin that have been reported to contain evidence for an association between humans and extinct megafauna. Our objectives are to reexamine existing collections and conduct independent research to evaluate the claims of previous researchers (Jenkins et al. 2004). This paper specifically addresses our recent investigations at Kelvin's Cave, located on the eastern Snake River Plain in southern Idaho (Figure 3.1).

Previous efforts to investigate the terminal Pleistocene archaeological record in Idaho have failed to produce definitive evidence regarding the timing of human entry into the region. Although early dates were recovered from Wilson Butte Cave in the late 1950s (Gruhn 1961), the scant evidence of human occupation at the site has since been dismissed by most in the archaeological community (Cannon and Meltzer 2004).

Fluted points have been recovered from the Snake River Plain; however, most are isolated surface finds (Woods and Titmus 1985). Those recovered in a subsurface context include the Simon Clovis Cache (Butler 1963) and the "Folsom" points from Owl Cave (Miller and Dort 1978; Miller 1982). Unfortunately, no faunal remains were recovered with the Simon Cache, and according to Grayson and Meltzer (2002:331), an unequivocal association between the stone tools and extinct megafauna from Owl Cave has yet to be demonstrated.

Kelvin's Cave was originally investigated in 1988 by staff from the Herrett Center at the College of Southern Idaho. Although the researchers were initially interested in the cave for its Holocene-aged deposits, the discovery of Pleistocene megafauna from the cave's lowest levels piqued their interest, especially when waste flakes were recovered from the same depths.

Immediately following the 1988 excavations, the site was publicized as

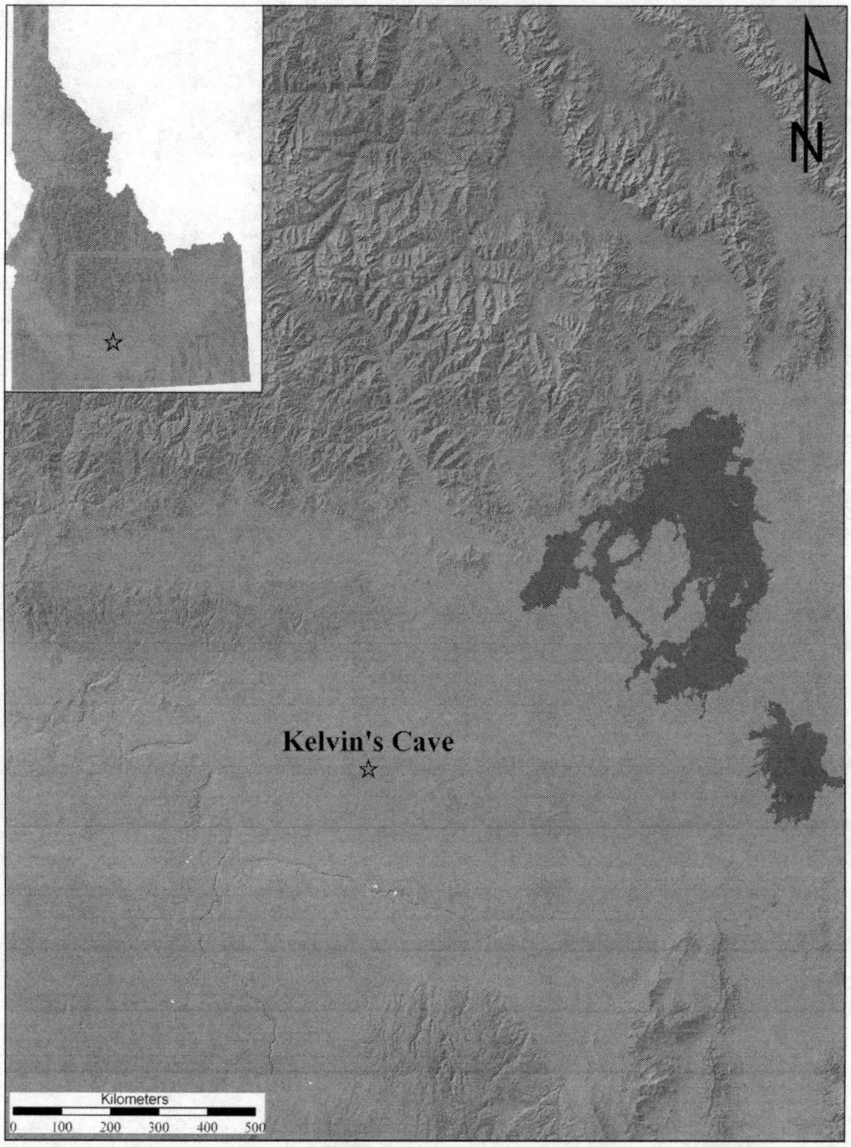

FIGURE 3.1. Location of Kelvin's Cave on Idaho's Snake River Plain.

possessing "the potential to help solve the question of who the first people were in North America" (Woods 1988). Unfortunately, the excavations were a volunteer effort, and only limited analyses were conducted after completion of the fieldwork. Although a preliminary report was published (Meatte et al. 1989), it lacked adequate information for an independent review of the research. In 1998, Bureau of Land Management (BLM) archaeologist Lisa Cresswell inherited the situation and pursued additional funding. Money became available only after recent vandalism drew much needed attention to the site. Cresswell subsequently asked the University of Oregon to revisit the cave.

FIGURE 3.2. Kelvin's Cave during the 2004 excavations.

ENVIRONMENTAL SETTING

The eastern Snake River Plain is a region of low topographic relief bordered by the Basin and Range Province to the south and the Northern Rocky Mountain Province to the north. The landforms of the plain were created primarily by basalt flows from low shield volcanoes and fissure eruptions ranging from Pliocene to Holocene in age (Greeley and King 1977; Kuntz et al. 1992). These features include rolling basalt pressure ridges, shallow basins, swales, knolls, buttes, fissures, and lava tube systems. With elevations ranging from 4,500 to 6,000 ft asl, the region receives less than 10 in. of annual precipitation, primarily as snow and early spring rains (Butler 1978). The native vegetation includes big sagebrush, three-tip sagebrush, bitterbrush, horsebrush, rabbitbrush, and snakeweed as the dominant shrubs. Common native grasses include bluebunch wheatgrass, junegrass, ricegrass, poa, and needle and thread grass (Franzen 1980). Mammals commonly exploited by the historical inhabitants include bison, pronghorn antelope, mule deer, elk, blacktailed jackrabbit, cottontail, and pygmy rabbit. Common birds include blue grouse and sage grouse.

2004 EXCAVATIONS

Kelvin's Cave is part of a collapsed lava tube system associated with a Pleistocene-aged basalt flow from a nearby shield volcano. The cave is open at both ends, with a northwest-southeast orientation, and contains roughly 2 m of aeolian fill interspersed with roof fall (Figure 3.2). Excavations in 1988 focused on the southeast-facing entrance of the cave, where the cave ceiling provided greater clearance. The 2004 fieldwork expanded the previous excavations and opened units in the west end of the cave (Figure 3.3). Although occupational debris dating from the middle and late Holocene was recovered from the cave, the nature of these deposits will be discussed in another venue.

Our research objective in 2004 was to reexamine the deposits reported in 1988 to contain the remains of megafauna and cultural debris (Figure 3.4). These deposits, originally referred to as Stratigraphic Unit 3, consist of a zone of angular basalt rubble enveloped in fine silty sand with small fragments of

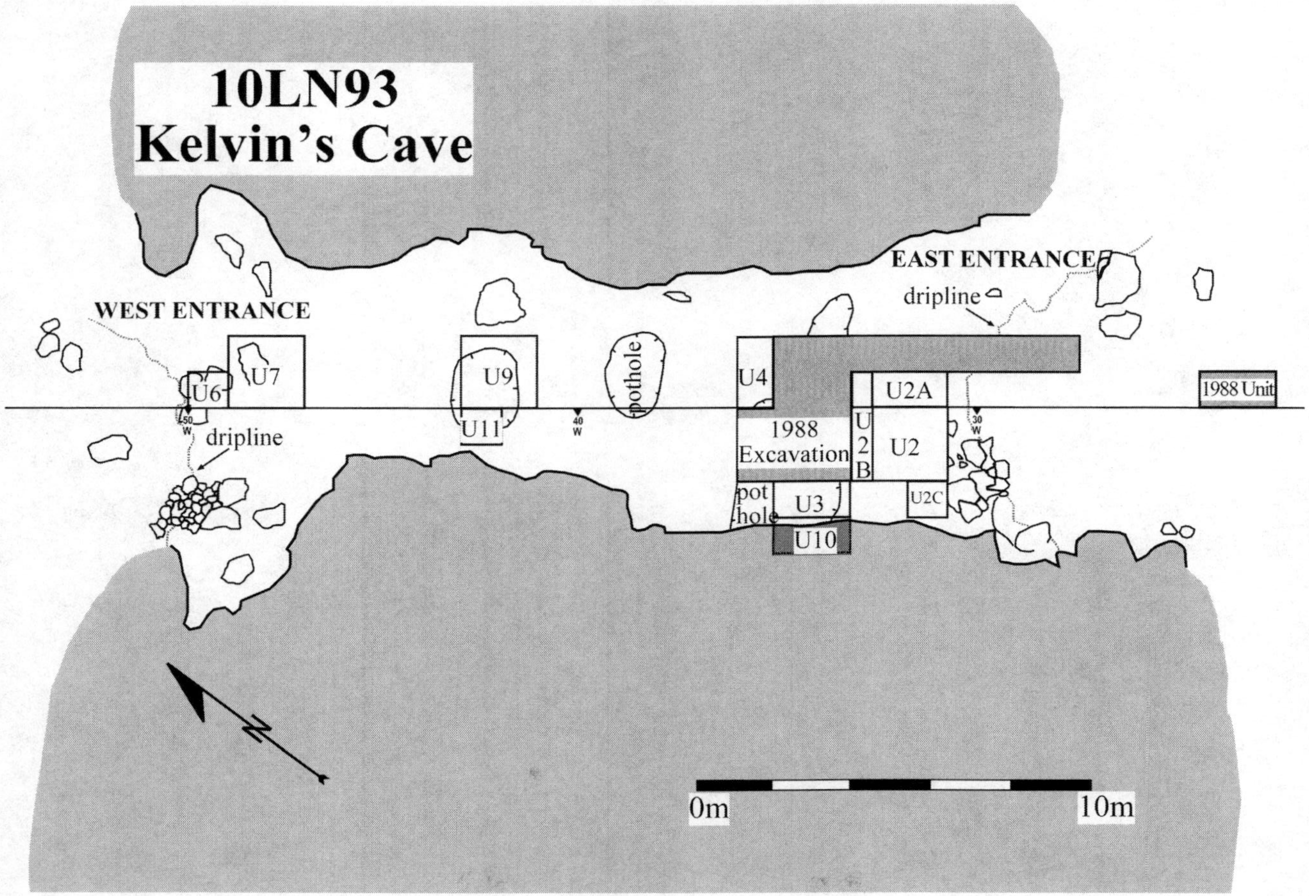

FIGURE 3.3. Plan view of Kelvin's Cave (adapted from Meatte et al. 1989), showing locations of 1988 and 2004 excavation units.

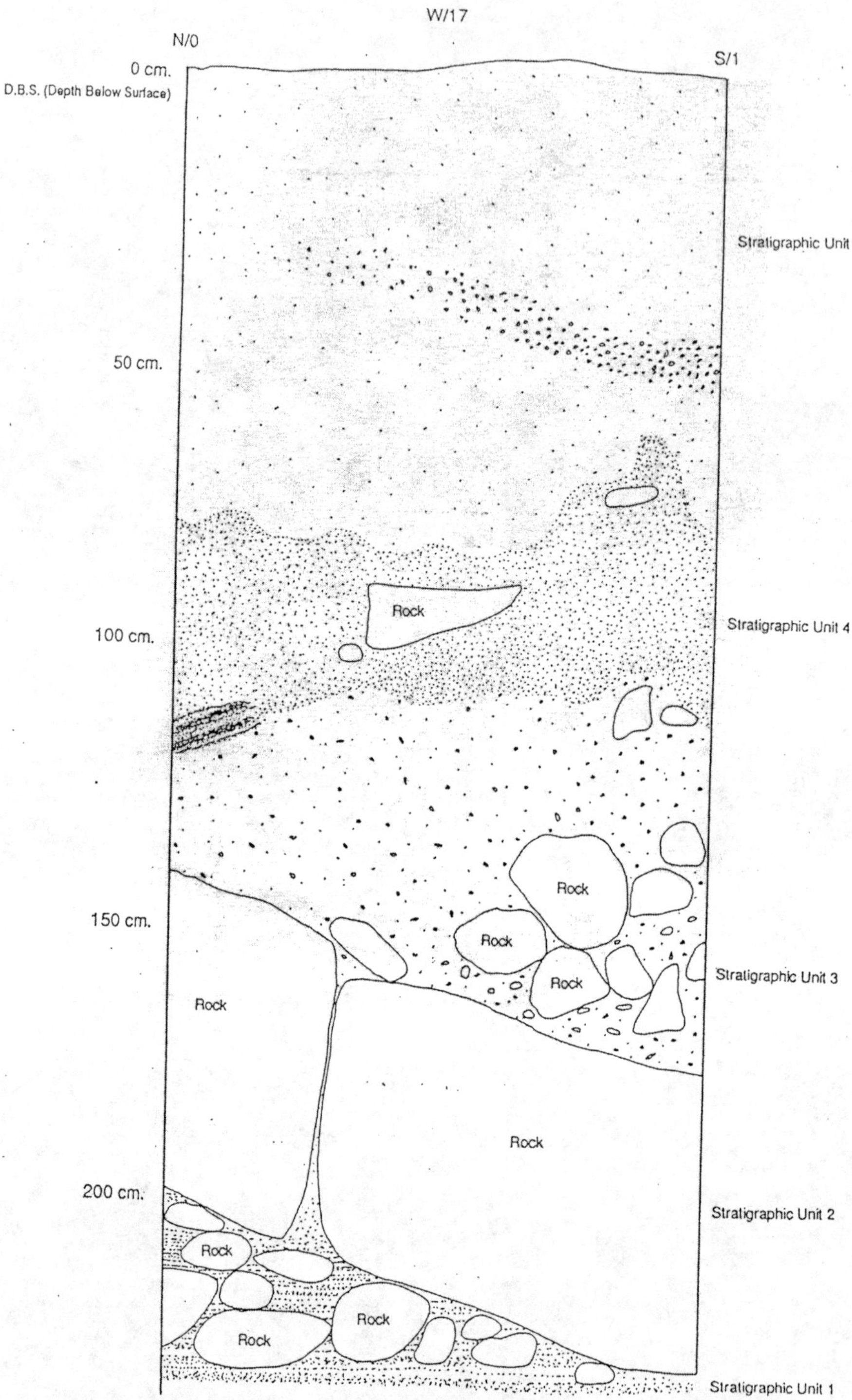

FIGURE 3.4. Wall profile from 1988 showing location of Stratigraphic Unit 3, defined here as the rubble stratum (taken from Meatte et al. 1989).

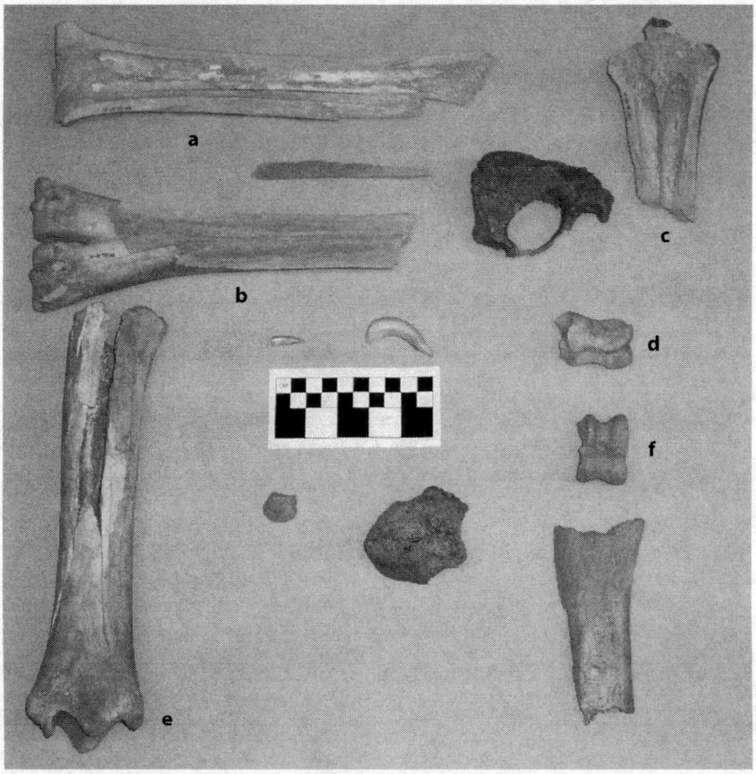

FIGURE 3.5. Megafauna remains from the 1988 excavations, including *a–c*, *Camelops hesternus* (camel); *d*, *Bootherium bombifrons* (musk ox); *e*, *Equus* (horse) sp.; and *f*, bovid.

decomposing basalt matrix. This rubble stratum, ranging from 40 to 60 cm in thickness, rests below roughly 80 cm of aeolian sands and silts, and above enormous roof fall boulders resting on the floor of the cave. The generally small size of the basalt cobbles and boulders contained in the rubble stratum is more suggestive of roof spall associated with seasonal freeze/thaw activity than a single cataclysmic geologic event (Miller and Dort 1978).

Faunal Assemblage from the Rubble Stratum

Diagnostic megafauna remains recovered from the rubble stratum in 1988 include a metatarsal identified as *Camelops hesternus* (camel), a *Bootherium bombifrons* (musk ox) phalanx, and most of a right tibia assigned to the genus *Equus* (horse) (Figure 3.5). During the 2004 excavations, we recovered five additional specimens from the rubble stratum that could be identified to taxa. Faunal identifications were made using the comparative collection of Rancholebrean specimens at the Idaho Museum of Natural History with assistance from paleontologist Mary Thompson.

Specific elements recovered in 2004 include a distal radius and sacrum identified as *Bootherium bombifrons* (musk ox), an astragalus assigned to *Equus* sp. (horse), a cuboid identified as *Camelops hesternus* (camel), a proximal radius identified as *Bison antiquus* (extinct bison), and two bovid phalanges (Figure 3.6). In addition, nondiagnostic large mammal remains, microfauna, and

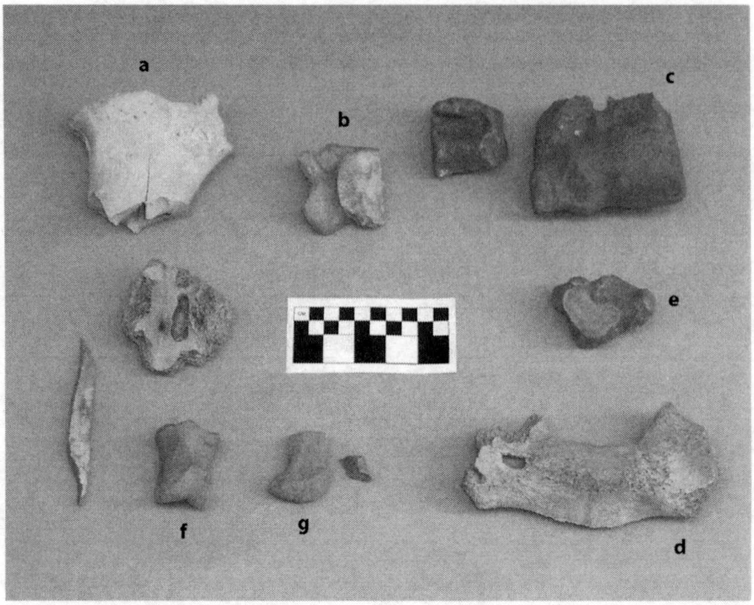

FIGURE 3.6. Megafauna remains from the 2004 excavations, including *a, Bison antiquus* (extinct bison); *b, Equus* (horse) sp.; *c, d, Bootherium bombifrons* (musk ox); *e, Camelops hesternus* (camel); and *f, g,* bovid.

FIGURE 3.7. Green fractured *Bison antiquus* (extinct bison) radius in the lower portion of the rubble stratum in Unit 4.

highly fragmented bone were also recovered. Although the exact provenience of the megafauna within the rubble zone was not discussed in the 1989 preliminary report, it was apparent during the 2004 excavations that these remains are located in the lowest levels of this stratum (Figure 3.7).

While no stone tools or diagnostic projectile points were found in the rubble stratum during either the 1988 or 2004 excavations at Kelvin's Cave, waste flakes were recovered from this zone. Of primary interest here is the lithic debris recovered in association with the extinct fauna. A total of 47 tertiary flakes were

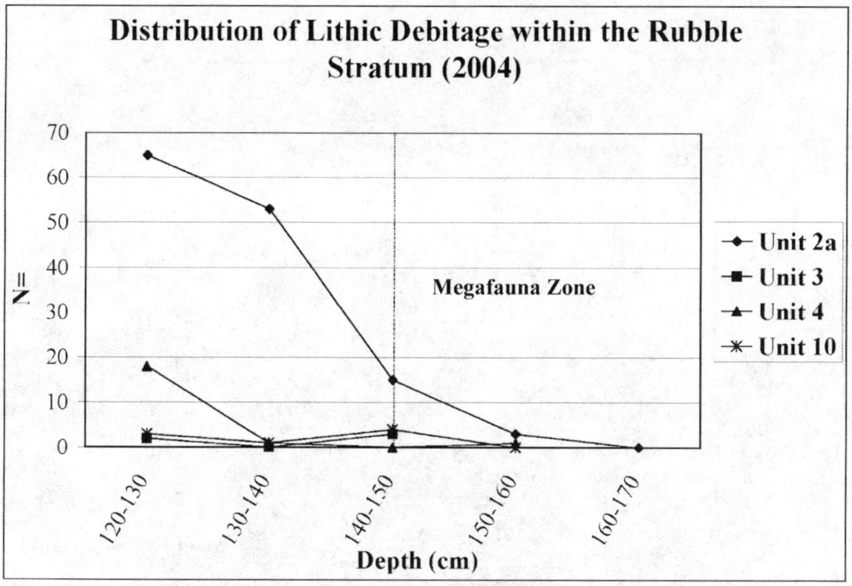

FIGURE 3.8.
Graph showing the distribution of debitage recovered from the rubble stratum in 2004.

recovered from the base of the rubble stratum during the 2004 excavations (Figure 3.8). These consist of 32 flakes from Unit 2A, 10 from Unit 3, 1 from Unit 4, and 4 from Unit 10. The bulk of the debitage from Unit 2A is obsidian, while cryptocrystalline silicate flakes occur more frequently in the other units. The flakes range from roughly 0.5 to 3 cm in diameter.

DISCUSSION

The 2004 investigations produced very similar results to those reported for the 1988 research. Debitage was found in the same levels as Pleistocene megafauna. However, any interpretations regarding the nature of this assemblage must involve a critical analysis of the potential processes that may have contributed to its formation. During the 2004 excavations, it was quite apparent that a variety of taphonomic processes had likely compromised the stratigraphic integrity of the cultural deposits in Kelvin's Cave.

Bioturbation

The extent of bioturbation in the cave is clearly evident in the profile of Unit 4 (Figure 3.9). Krotovina were encountered throughout the upper levels of this unit, and although every attempt was made to excavate rodent burrows separately, the distribution of modern debris in Unit 4 reflects the magnitude of disturbance associated with burrowing animals (Figure 3.10). However, the frequency of krotovina appears to decrease with depth in all of the excavation units. While this decrease may be associated with an impasse created by the rubble stratum, we cannot assume that this zone was immune to rodent or badger activity.

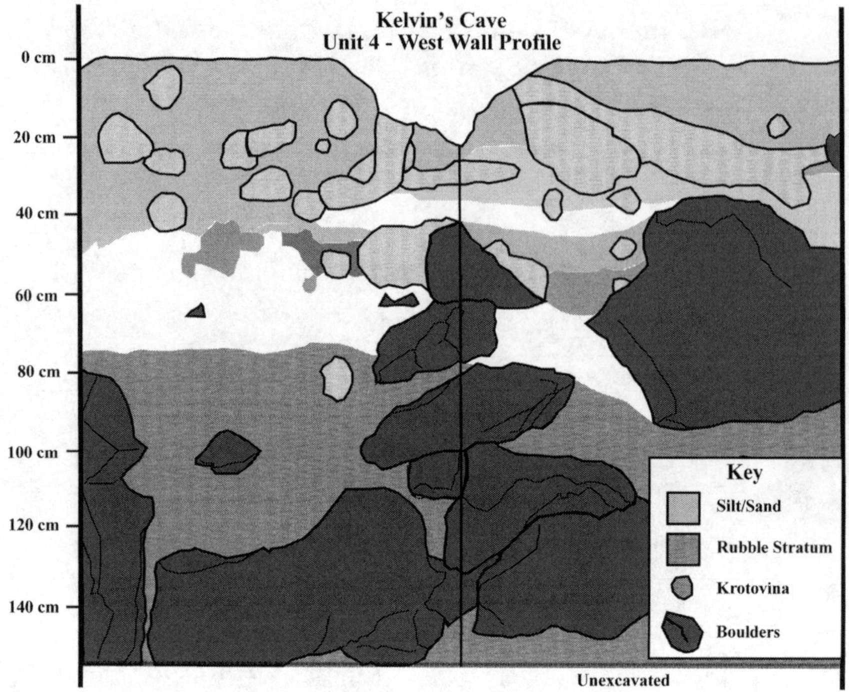

FIGURE 3.9. West wall profile of Unit 4 showing prevalence of krotovina in the upper and middle levels of the unit.

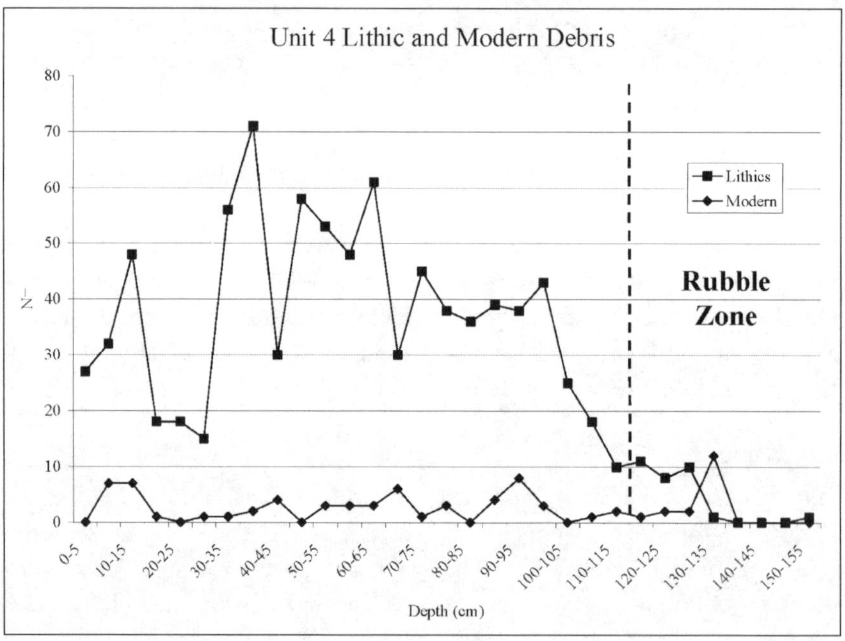

FIGURE 3.10. Distribution of lithics and modern debris within the rubble stratum in Unit 4.

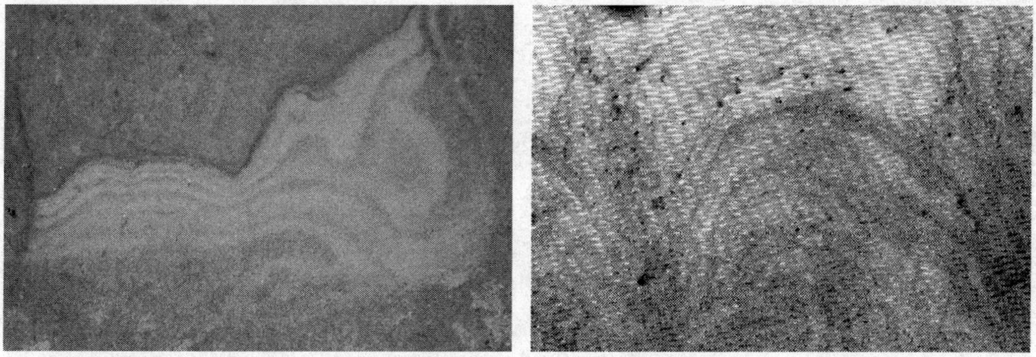

FIGURE 3.11. Involutions in Kelvin's Cave (*left*) and Owl Cave (*right;* taken from Dort 1968) resulting from long-term seasonal freeze/thaw regimes on the Snake River Plain.

Cryoturbation

In addition to bioturbation, the 2004 investigations revealed elaborate contortions of sediment in the profiles and floors of excavation units in the east entrance. These features are extremely similar to those documented in the sediments of Owl Cave (Figure 3.11). According to Dort (1968:34), these "involutions" are likely the result of sediment deformation occurring as a result of continuous seasonal freezing and thawing. Because lava tube ceilings are notoriously porous, winter precipitation can often introduce considerable amounts of water into cave sediments. It is reasonably certain that the involutions in Kelvin's Cave were formed through processes similar to those documented in Owl Cave, and the potential effect of these features on the displacement of cultural debris needs to be considered.

Vertical Displacement

Effects of seasonal freezing and thawing may include vertical displacement or dispersal of cultural material. Vertical displacement of artifacts has been demonstrated to occur under a variety of depositional and post-depositional conditions (Gifford-Gonzalez et al. 1985; Villa 1982). In her analysis of Old World sites, Villa found that conjoinable artifacts representing a single cultural occupation were distributed over 20–30 cm of sediment and even dispersed through different geologic layers. Therefore, prior to making any grand declarations about the cultural debris and extinct fauna resting at the base of the rubble stratum, it is imperative to disprove vertical displacement as a contributing factor in the formation of the archaeological record at Kelvin's Cave.

Analysis of Pleistocene Fauna Assemblage

As Grayson and Meltzer (2002) have stated, the recovery of extinct mammals and artifacts from the same stratigraphic layer does not necessarily imply that humans were subsisting on Pleistocene megafauna. Even if vertical displacement and bioturbation can be eliminated as explanations for the co-occurrence of debitage and bones of extinct mammals in Kelvin's Cave, these bones must also exhibit characteristics that clearly demonstrate they are the products of

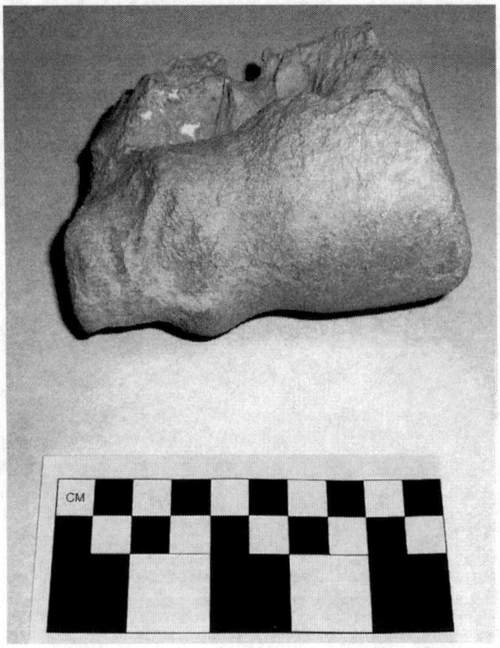

FIGURE 3.12. *Bootherium bombifrons* (musk ox) specimen with evidence of carnivore furrowing.

human subsistence behavior. Unfortunately, the extinct megafauna remains recovered in both 1988 and 2004 do not provide definitive evidence of cultural modification. No butcher marks could be identified on any of the specimens, and none exhibit characteristics suggestive of burning or charring.

In fact, the nature of the megafauna assemblage from Kelvin's Cave shares greater affinity with bone assemblages produced by carnivores, as indicated by Binford's (1981) analysis of faunal material left by wolves in rockshelters and caves in Alaska. Faunal elements recovered from these dens consist primarily of dense lower limb bones such as astragalus, tarsals, tibia, phalanges, and sesamoids. When compared to the composition of the Kelvin's Cave assemblage, the similarities are conspicuous. However, most of the Kelvin's Cave specimens do not exhibit punctures, pits, scoring, or other features typical of carnivore gnawing. Only one element, the *B. bombifrons* distal radius (Figure 3.12), shows linear grooves identified as canine furrows (Binford 1981:48).

One proximal radius recovered from the rubble stratum in Unit 4 (Figure 3.13) exhibits spiral fractures on the diaphysis reminiscent of breakage patterns associated with human marrow extraction. While Binford (1981) has argued that such fractures are not typical of carnivore breakage, Haynes (2002:135–148) documented similar fractures on elephant long bone subjected to trampling and emphasized that such patterns are commonly misinterpreted as cultural.

Radiocarbon Dates

Eight faunal specimens recovered from Kelvin's Cave were submitted to Beta Analytic for accelerator mass spectrometry (AMS) dating (Table 3.1). These include megafauna remains from the lowest portions of the rubble stratum (a camel tarsal from the 1988 excavations; a proximal radius from Unit 4; a long bone fragment, vertebrae, sacrum, and astragalus from Unit 2A; and a distal radius from Unit 7) and an indeterminate fragment of mammal bone from immediately above the rubble stratum (see Table 3.1). In most cases, sufficient collagen could be extracted in the dating process without requiring destruction of the entire specimen. As indicated in Figure 3.14, the faunal assemblage recovered from the base of the rubble zone ranges from 14,000 to 20,000 cal BP. The youngest specimens include the green fractured *B. antiquus* proximal radius from Unit 4 and the *B. bombifrons* distal radius from Unit 7 (see Figure 3.6). There also appears to be some overlap in the dates associated with the older specimens, most of which were recovered from Unit 2A. The early dates generated from the extinct megafauna in Kelvin's Cave are significantly

FIGURE 3.13. Green fractured *Bison antiquus* (extinct bison) radius recovered from Unit 4.

TABLE 3.1. Radiocarbon Dates from Kelvin's Cave

SAMPLE	SPECIMEN	^{14}C AGE BP	SAMPLE #	δ ^{13}C	CAL BP
10LN937B17	*B. bombifrons* humerus	12,250 ± 60	Beta-194532	-19.3	14,320–14,110
10LN934D143	*B. antiquus* radius	13,320 ± 60	Beta-195673	-19.9	16,430–15,680
10LN932AA	Large mammal	14,210 ± 50	Beta-194531	-18.8	17,380–16,670
10LN932AA150	*B. bombifrons* sacrum	16,260 ± 80	Beta-195670	-19.4	19,890–18,980
10LN932AA155	*Equus* sp. astragalus	16,340 ± 90	Beta-195671	-20.0	20,000–18,980
10LN932AA156	Large mammal	16,350 ± 90	Beta-195672	-19.5	20,020–18,990
10LN93C-14#2(1988)	*C. hesternus* tarsal	16,360 ± 120	Beta-43435	—	19,834–19,261
10LN93U3QD15	Large mammal	5790 ± 0	Beta-207836	-18.7	6690–6460

older than those from the elephant bone in Owl Cave, which produced dates of between 12,800 and 10,900 ^{14}C BP (Miller and Dort 1978).

The indeterminate mammal bone collected from directly above the rubble stratum in Unit 3 produced a radiocarbon date of 5790 ± 50 (Beta-207836)^{14}C BP (6460–6690 cal BP). This date confirms that the rubble stratum is terminal Pleistocene/early Holocene in age. Some initial information regarding the climatic conditions on the eastern Snake River Plain for the period following the Last Glacial Maximum (LGM) has been generated from the analysis of pollen and macrofloral remains extracted from a Unit 4 column sample (Figure 3.15). While *Artemisia* pollen dominates all levels sampled, the lowest levels of the cave, representing roughly 20,000 to 12,000 years ago, contain lower frequencies of *Artemisia* and grasses along with pollen from higher-elevation taxa,

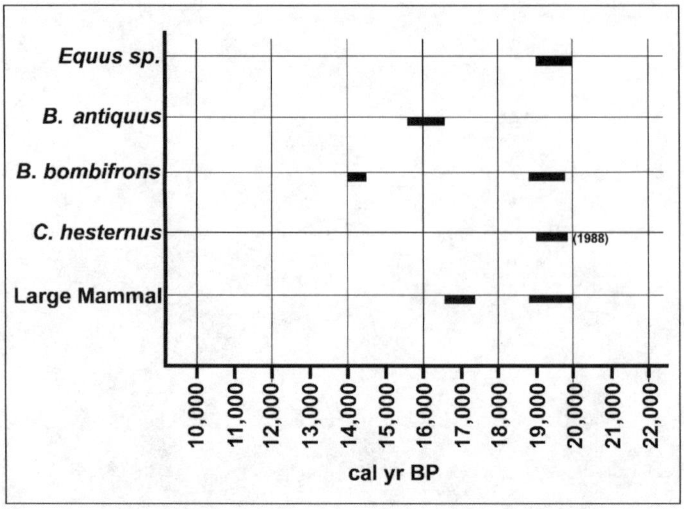

FIGURE 3.14. Distribution of calibrated radiocarbon dates from Pleistocene megafauna specimens within Kelvin's Cave (1988 and 2004).

including members of the rose family, mountain mahogany, and chokecherry (Cummings and Puseman 2005).

The results of the pollen analysis suggest an increase in available moisture on the eastern Snake River Plain immediately following the LGM. Other caves and rockshelters in the region also exhibit depositional features suggestive of cool, moist conditions well into the early Holocene. Swanson (1972:55) attributed the massive Period II rock fall in the Birch Creek rockshelters to cooler temperatures between 11,950 and 7150 ^{14}C BP, and Gruhn (1961:28) argued that the water-borne sand characterizing Wilson Butte Cave's Stratum 3 indicates a "much more moist climate" prior to 6850 ^{14}C BP.

The ice wedge features in Owl Cave, all of which were confined to sediments beneath Mazama ash, have been attributed to periods of severe cold (Dort 1968), leading Butler (1968:9) to surmise that "the cold-climate aspect of the Late Pleistocene may have continued well into the early Postglacial period of the northern Intermontane region."

Although explanations involving freeze/thaw have been overly used to account for rock fall episodes (Stein and Farrand 2001), the formation of the rubble stratum in Kelvin's Cave, as indicated by sedimentation in other regional caves, is likely the result of cooler, wetter climatic conditions during the terminal Pleistocene/early Holocene. Other processes responsible for rock fall, such as earthquakes and hydration spalling, are less likely candidates in this case due to the length of time associated with the formation of the rubble stratum.

CONCLUSION

The 2004 University of Oregon investigations were conducted to evaluate earlier claims regarding the nature of the association between the archaeological material and extinct megafauna at Kelvin's Cave. Based on our investigations

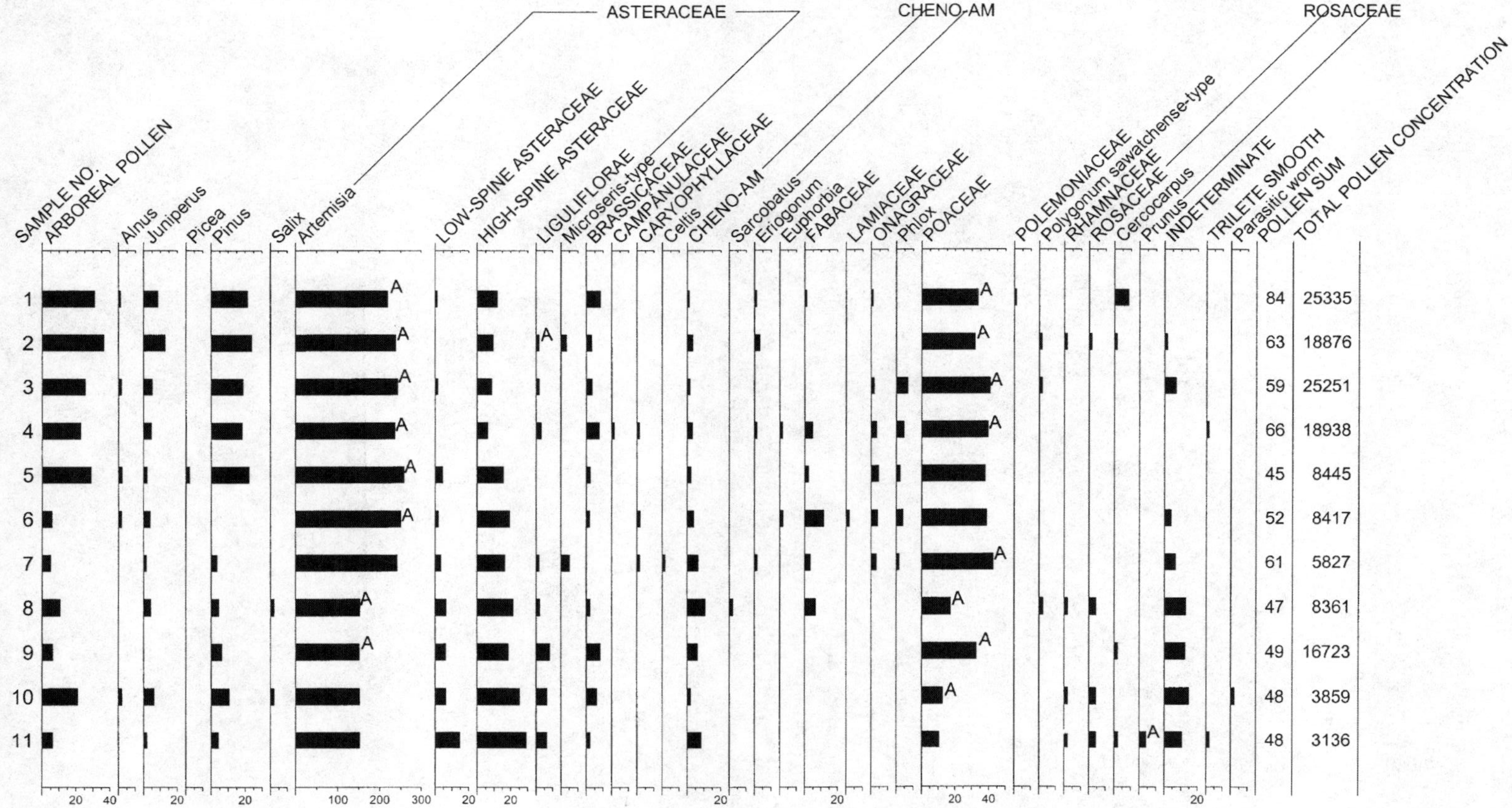

FIGURE 3.15. Pollen diagram (from Cummings and Puseman 2005) displaying relative proportions of pollen present in Kelvin's Cave from 17,000 ^{14}C BP to the present.

thus far, there is no conclusive evidence to suggest that humans were present at the site during the terminal Pleistocene, much less hunting extinct megafauna in the region. The faunal assemblage at Kelvin's Cave exhibits characteristics more indicative of a carnivore den than human subsistence activities. This is a reasonable assumption, given that large carnivores also roamed the Snake River Plain during the terminal Pleistocene (Jefferson et al. 2002).

No stone tools were found in association with the megafauna remains, and based on bioturbation, cryoturbation, and roof fall episodes evident in the sediments, vertical displacement appears to be the most plausible explanation for the presence of waste flakes within the rubble stratum (see Figure 3.8). However, although a terminal Pleistocene/early Holocene occupation of the cave is considered unlikely at this juncture, experiments need to be conducted to identify and quantify the specific processes responsible for the migration of debitage through rocky, compacted aeolian sediments.

While our preliminary analyses do not indicate an early human presence at Kelvin's Cave, these findings do not preclude the possibility of encountering early human occupations in buried contexts on the eastern Snake River Plain. The University of Oregon plans to continue investigations in southern Idaho, specifically in hopes of working with the original researchers at Owl Cave, where an unequivocal association of fluted points and extinct megafauna has yet to be demonstrated. Perhaps additional fieldwork and a more thorough reporting of the existing data can resolve the ambiguities surrounding this archaeological assemblage and provide critical data regarding the nature of human subsistence patterns during the terminal Pleistocene.

4

Distribution and Dating of Cultural and Paleontological Remains at the Paisley Five Mile Point Caves in the Northern Great Basin

An Early Assessment

Dennis L. Jenkins

THE PAISLEY FIVE MILE POINT CAVES (35LK3400) ARE LOCATED AT the southeast end of the Summer Lake basin near the town of Paisley in south-central Oregon (Figure 4.1). Accompanied by Ernst Antevs, Luther Cressman was taken to the site in 1937 by Walter Perry, a local resident who had been alerted to its existence by a woman who had been digging in the caves for artifacts. Upon seeing the evidence for cultural materials predating the eruption of Mount Mazama, Antevs and Cressman recognized the extreme importance of the locality to northern Great Basin prehistory and began making plans for its excavation the following year (Cressman 1988). Cressman returned with a field crew to excavate exploratory trenches in Caves 1, 2, and 3 in 1938 (Figure 4.2). Returning in 1939, he removed the majority of deposits from the main chambers of these same caves. He briefly returned a third time in 1940 to broaden the excavated area in Cave 3 where he had discovered Pleistocene faunal remains together with obsidian artifacts the year before (Cressman 1940).

Initially, Cressman's greatest interest in the excavation of the Paisley Caves was to establish proof of pre-Mazama (6850 ^{14}C BP [7600 cal BP]) human occupations in the region (Cressman 1942). Consequently, he dug the caves in three basic stratigraphic units: roughly 1 m of sediment above the Mount Mazama tephra, as much as 30 cm of tephra, and 1 m of sediment below the tephra. Photographs, field notes, and excavation-damaged artifacts suggest that deposits were predominantly removed with pick and shovel. Only those artifacts and faunal remains found in situ were recorded with precise vertical and horizontal provenience.

In Cave 3, Cressman discovered a U-shaped living floor, cleared of stones and outlined with boulders, well below the Mazama tephra layer. The remains of late Pleistocene camel, bison, horse, and waterfowl had apparently been

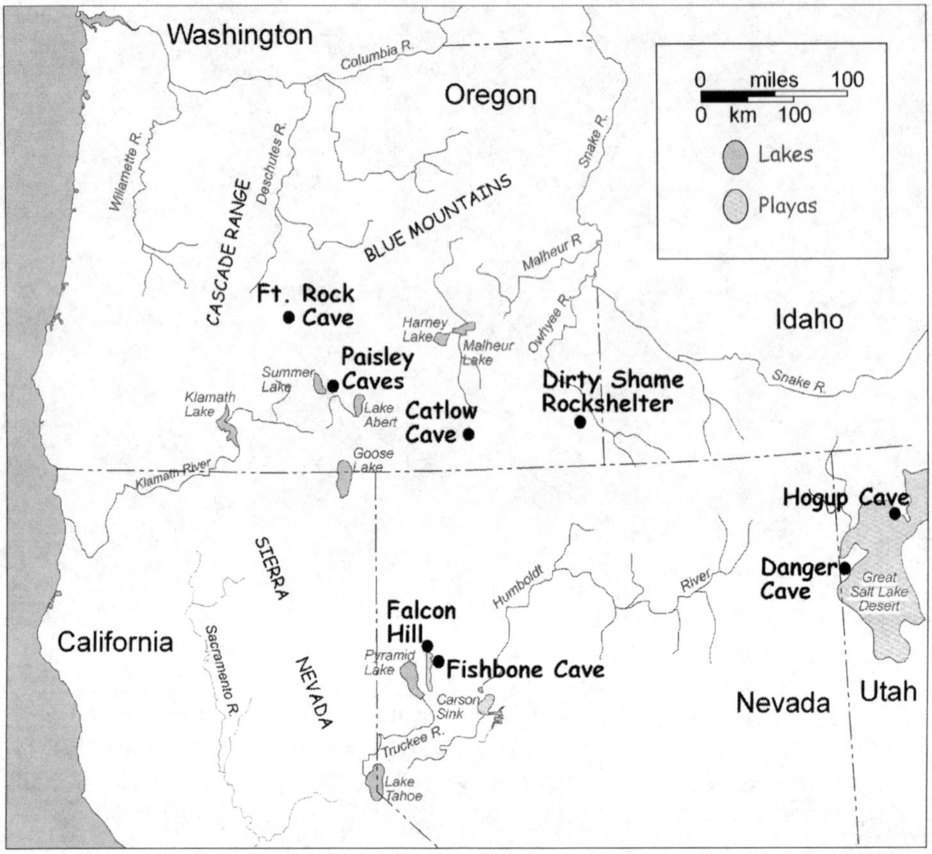

FIGURE 4.1. Location of the Paisley Five Mile Point Caves in the northern Great Basin, south-central Oregon.

tossed out of this area, piling up around it, particularly near the rear wall. A small assemblage of obsidian artifacts was recovered from the living floor and the surrounding area. The bone discard pattern combined with the charring of some of the specimens suggested to Cressman that the megafauna bones represented food remains left by the occupants of the cave (Cressman 1942:93).

While Cressman believed this evidence demonstrated the association of cultural materials with extinct Pleistocene fauna at the Paisley Caves (Cressman 1986:121, 1966:41, 1942; Cressman and Williams 1940), Heizer and Baumhoff (1970:5), among others (cf. Krieger 1944), questioned his interpretations on the grounds that he had failed to provide adequate quantification and precise provenience data for the artifacts and megafaunal remains. Still, they acknowledged that they were "inclined to accept Paisley Five-Mile Point Cave No. 3 as the strongest candidate for contemporaneity of man and extinct animals in the Great Basin."

A strong candidate it might be, but Jesse Jennings's (1986:115) abrupt conclusion on the matter—"no situation has been found where extinct mammals, stone tools or weapons, or evidence of fire, are credibly reported in good association [in the Great Basin]"—remains the standard after another two decades of research on the topic. With this statement, unabashedly included

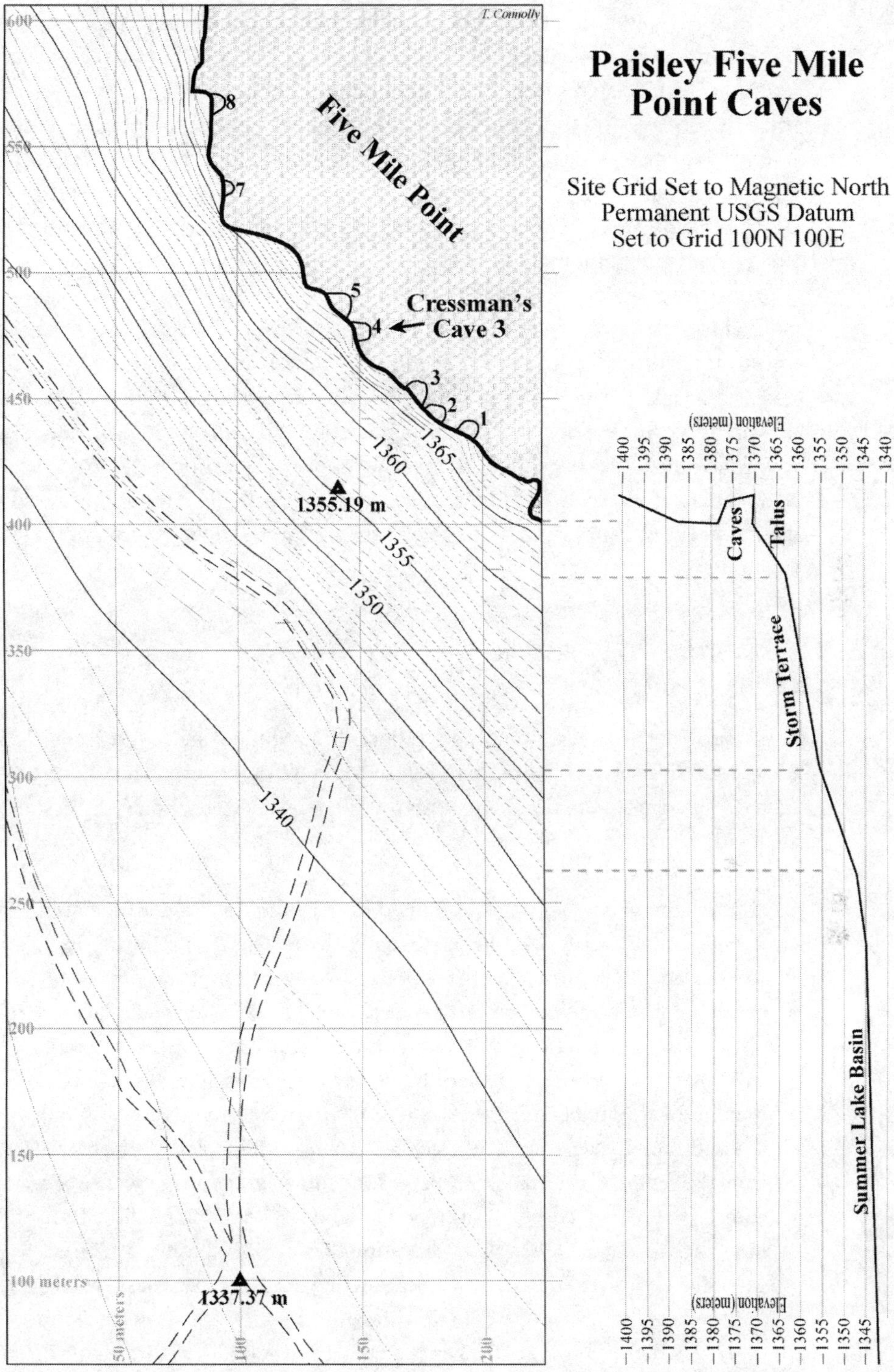

Figure 4.2. Map of the Paisley Five Mile Point Caves site (35LK3400).

in the same volume that Cressman (1986:121) repeated his assertion that he had proven human/megafauna contemporaneity in the northern Great Basin, Jennings implied that the association of artifacts and megafauna remains in the Paisley Caves was spurious. Considering these very different interpretations of the Paisley Caves data, and the lack of irrefutable evidence supporting late Pleistocene human/megafauna interactions in the Great Basin, a reinvestigation of the Paisley Caves seemed critical to establishing the timing of the first human occupations in the region.

The University of Oregon archaeological field school conducted excavations at the Paisley Caves during the summers of 2002 and 2003 with the goal of testing Cressman's theory of Pleistocene human/megafauna contemporaneity at that site. This chapter reports on the current status of investigations into the distribution and dating of the cultural and paleontological remains found there. Specifically, it will address the following questions: (1) Do cultural and megafaunal remains occur in horizontal, vertical, and stratigraphic association in the Paisley Caves? And (2) are the cultural and megafaunal remains the same age?

CRESSMAN'S DESCRIPTIONS OF STRATA AND CULTURAL/MEGAFAUNAL ASSOCIATIONS

Cressman (1942:93) described the stratigraphic sequence he found in Paisley Cave 3. Of Strata 6 and 7 (numbered from top down), the only strata that contained Pleistocene faunal remains, and the "floor" they were apparently associated with, he writes:

> Stratum 6 was dry, soft, and contained rather numerous rodent remains. It ran from less than .1 m. in thickness at the rear to .25 m. in the front, where it abutted against the large rock [Figure 4.3]. Stratum 7 was predominantly sand and gravel from the old beach, along the level of which the shelter was formed. It had a depth of approximately .25 m. resting on large water smoothed rocks forming the floor of the cave. Strata 6 and 7 contained the remains of an extinct fauna, worked obsidian, and ash lenses with both fine ash and thumb-sized bits of charcoal.... The cave floor sloped up sharply from the center on both sides. The large rock shown in the profile was apparently a part of the bedrock.... It was very smooth from water action. Running back on either side to the wall a number of rocks, all water smoothed, extended up into the dry stratum overlying the sand. Some of these seemed to have been in place as cave floor, while others apparently were placed in the spaces between some of the larger ones. At any rate, it was in this roughly U-shaped space that the faunal remains and evidence of human occupation were found. The mixture of bones and ash in this small space tells a story of these hunters bringing a part of their kill to this shelter, where it was cooked and the refuse thrown back over the rock or against the wall to pile up midden-like.

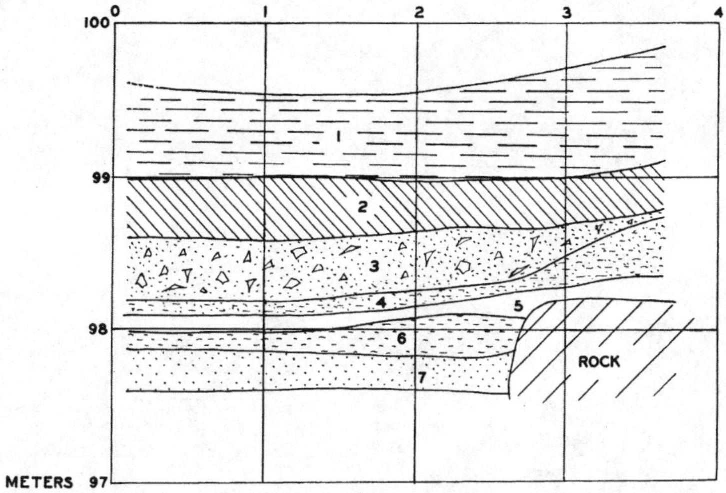

FIGURE 4.3. Profile of strata in Paisley Cave 3, from rear of cave to front.

UNIVERSITY OF OREGON ARCHAEOLOGICAL FIELD SCHOOL INVESTIGATIONS

University of Oregon field school excavations were conducted in Caves 1, 2, 4 (Cressman's Cave 3), and 5, generally revealing dry, well-stratified sediments wherever undisturbed deposits were encountered. Sediments were removed by 5 cm levels in cultural deposits and 10 cm levels in culturally sterile deposits. Levels were dug within strata to a mean depth of 250 cm. All materials were screened through ⅛" wire mesh. Students were closely supervised at all times, with special attention paid to recovering paleontological and cultural materials in situ, noting stratigraphic and spatial associations, and recording evidence for possible stratigraphic mixing.

Though extinct faunal remains have been recovered from three of the four investigated caves, Cave 1 being the exception, it is only in Cave 5 that sufficient samples of artifacts and extinct faunal remains have been recovered in close proximity to warrant the distributional analysis offered here. Data collected from the other caves offer good supportive contextual information—datable cultural features and artifacts stratigraphically sealed between thin strata—as well as isolated elements of extinct fauna. Consequently, here the attention is focused on the distribution of megafaunal remains in Cave 5 assemblages. However, dating for the site is based on data collected from all of the caves investigated by the field school.

EXCAVATIONS AND STRATIGRAPHY

The strata in Cave 5 are predominantly composed of organic debris (rodent feces, vegetation, and bone), exfoliating volcanic breccia and basalt from the cliff formation, volcanic tephra deposited during the climactic eruption of Mount Mazama, and aeolian silt and sand transported from exposed lake bottom as Pleistocene Winter Lake retreated, and the Holocene floor of the Summer Lake

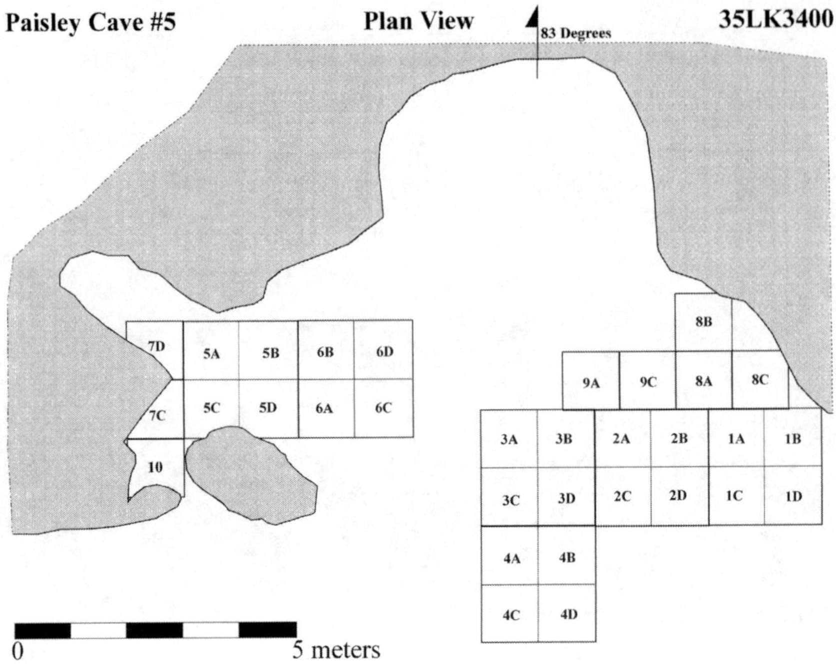

FIGURE 4.4. University of Oregon field school excavations in Paisley Cave 5.

basin was exposed. Organic remains, reflecting the occupations of people and the nesting behaviors of wood rats and various avian species, occur predominantly in the more sheltered northern and eastern sections of the cave. Preservation of cordage, netting, thin twisted threads, processed sinew, hide, bone, coprolites, and plant material was exceptional in these sections of the cave. However, the main chamber in Cave 5, in particular those deposits nearest the cave walls, had been extensively vandalized prior to University of Oregon excavations. Consequently, investigations were focused near the mouth of the cave in two excavation blocks (Figure 4.4).

The first, located at the south end, was situated on the interior slope of the berm formed by the redeposition of interior deposits near the drip line during the vandalism of the cave. Here, rocky gravel deposits predominate in younger (post–Mount Mazama) strata. The older (pre-Mazama) strata are generally more fine-grained, composed of thin layers of silt and sand showing evidence of pedogenic development, separated by thin layers of sandy gravels composed of roof spalls (Davis 2006). Deposits nearest the drip line remain wet throughout the summers, and the composting effects of moisture prevail. Bone and perishable artifacts are uncommon in these strata, and seriously degraded when they are present.

The second block excavation, located at the north end of the cave in a small alcove, was situated between a trench dug by vandals along the rear (east) wall and a large boulder lying on the surface to the west across the mouth of the alcove. Compact rodent feces covering the surface of this sheltered area gave clear evidence of the undisturbed nature of the deposits preserved in the "pocket" behind this boulder, although surface deposits along the east side of the block sloped steeply to the east and southeast toward the vandal trench along the wall of the cave. Subsequent excavations revealed a series of very thin (2–3 cm) layers of alternating fine-silty/coarse-sandy deposits of dusty, acrid rat feces, vegetation, bone, feathers, coprolites, and artifacts. These deposits were divided into 10 basic lithostratigraphic units (numbered from the bottom up) by Loren G. Davis (2006; Figure 4.5A):

LU1a: Moderately sorted mix of fine sediments (mainly fine sand to silt) with occasional subangular to angular pebble and cobble clasts of igneous rock, very consolidated (firm to hard), common macrobotanical remains and *Neotoma* sp. fecal pellets, matrix supported, abrupt irregular boundary as deposit overlies basal boulder deposit, unit dips to the southeast.

LU1b: Poorly sorted mix of angular to subangular fine pebble to coarse cobble igneous clasts and fine interstitial sediments (mainly fine sand to silt), weakly consolidated (loose to friable), *Neotoma* sp. fecal pellets, some macrobotanical organics, clast supported, abrupt irregular boundary as deposit overlies basal boulder deposit, unit dips to the southeast.

LU2: Moderately sorted mix of fine sediments (mainly fine sand to silt) with occasional subangular to angular pebble and cobble clasts of igneous rock, consolidated (firm), common macrobotanical remains and *Neotoma* sp. fecal pellets, matrix supported, abrupt wavy boundary, unit dips to the southeast.

LU3: Thin (ca. 2–4 cm thick) layer of whitish volcanic tephra, suspected as representing Mazama set O eruption, abrupt wavy boundary, unit dips to the southeast.

LU4: Moderately sorted mix of fine sediments (mainly fine sand to silt) with occasional subangular to angular pebble and cobble clasts of igneous rock that increases notably to the south, consolidated (firm), common macrobotanical remains and *Neotoma* sp. fecal pellets, matrix supported, abrupt wavy boundary, unit dips to the southeast and interfingers with LU5.

LU5: Poorly sorted mix of angular to subangular fine pebble to fine boulder igneous clasts and fine interstitial sediments (mainly fine sand to silt), weakly consolidated (loose to friable), rare macrobotanical organics, clast supported, abrupt wavy boundary, unit dips to the southeast and interfingers with LU4 and LU6 to the north.

LU6: Moderately sorted mix of fine sediments (mainly fine sand to silt) with occasional subangular to angular pebble and cobble clasts of igneous rock

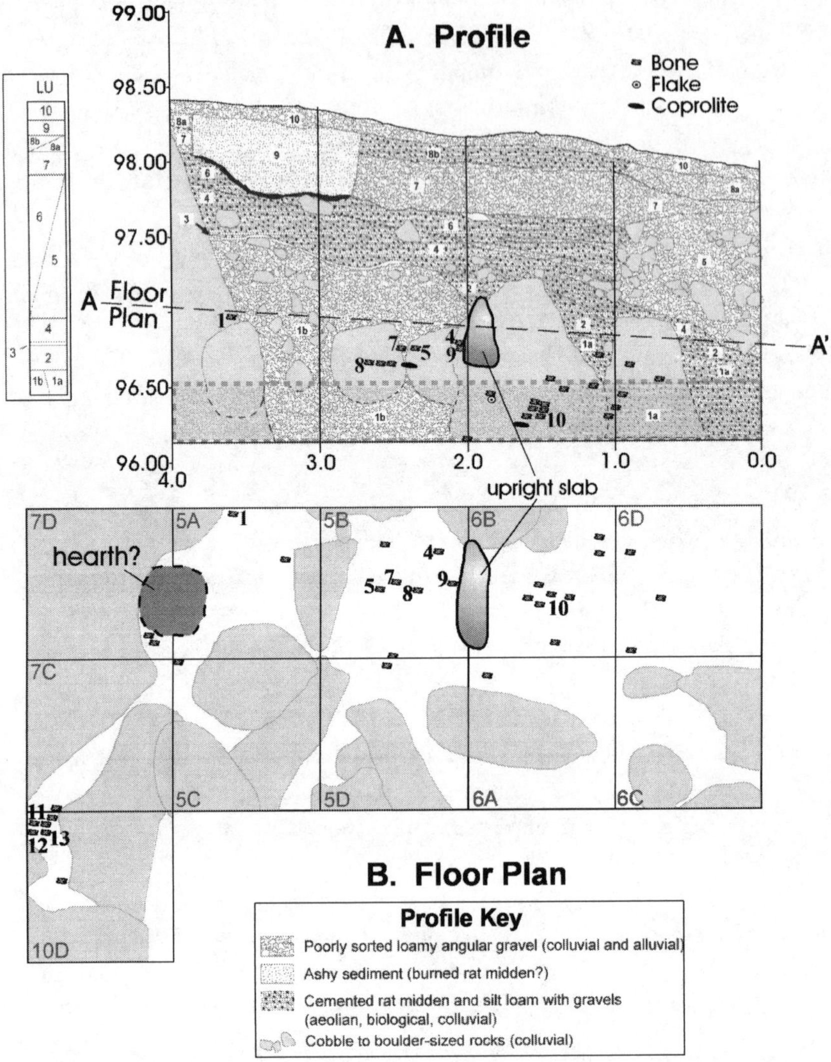

FIGURE 4.5. Stratigraphic profile of the east wall of Units 5 and 6 showing relationship of in situ bone to strata, and floor plan of Units 5, 6, 7, and 10 showing horizontal distribution of in situ bone.

that increases notably to the south, consolidated (firm), common macrobotanical remains and *Neotoma* sp. fecal pellets, matrix supported, abrupt wavy boundary, unit dips to the southeast and interfingers with LU5.

LU7: Poorly sorted mix of angular to subangular fine pebbles and coarse cobble-sized igneous clasts and fine interstitial sediments (mainly medium sand to silt), moderately consolidated (friable), clast supported, occasional macrobotanical organics and *Neotoma* sp. fecal pellets appear in discrete layers or dispersed in matrix, abrupt wavy boundary, unit dips to the southeast.

LU8a: Moderately sorted mix of fine sediments (mainly fine sand to silt) with occasional subangular to angular pebble and cobble clasts of igneous rock, consolidated (firm), common macrobotanical remains and *Neotoma* sp. fecal pellets, matrix supported, abrupt wavy boundary, unit dips to the southeast and interfingers with 8b to the south.

LU8b: Poorly sorted mix of angular to subangular fine pebbles and coarse cobble-sized igneous clasts and fine interstitial sediments (mainly medium sand to silt), moderately consolidated (friable), clast supported, occasional macrobotanical organics and *Neotoma* sp. fecal pellets appear in discrete layers or dispersed in matrix, abrupt wavy boundary, unit dips to the southeast.

LU9: Zone of burned organics and ash accumulation with layer of charcoal at base. Prior to burning, deposit may have appeared as a lateral extension of LU7 and LU8a, as seen to the south. Unit has abrupt wavy boundaries on sides and base.

LU10: Poorly sorted mix of angular to subangular fine to medium pebbles with occasional cobble-sized igneous clasts and fine sediments (mainly medium sand to silt), poorly consolidated (loose to friable), matrix and clast supported, occasional macrobotanical organics and *Neotoma* sp. fecal pellets, abrupt wavy boundary, unit dips to the southeast.

All identifiable megafauna remains in Cave 5 were recovered from Strata LU1a and LU1b in Units 5, 6, 7, and 10 (Figure 4.5A). These two strata are predominantly differentiated by the quantity of subangular to angular pebble and cobble clasts of igneous rock, degree of poorly to moderately sorted mixes of fine sand and silt, and loose to consolidated (firm to extremely hard) compaction. All units dip to the southeast, thickening as the distance from the north wall of the cave increases, and most contain some measure of wood rat midden materials (rodent fecal pellets, coprolites, bones, feathers, seeds, and twigs). Capping these units, the Mazama tephra lens (Stratum LU3) is thin (1–2 cm) but easily discernible in the northeast corner of the excavations.

The majority of identifiable megafauna remains were recovered from excavation Units 5 and 6 (Figure 4.5B). As noted above, the strata thicken and slope downward away from the walls in these units in the same manner as those described by Cressman for Cave 3. Consequently, Units 7 and 10, located along the north wall of the cave, contain strata that are substantially thinner than those of Units 5 and 6. Their strata are also more compacted with crystallized rat urine, suggesting that moisture (condensation or melting snow along the walls resulting in dripping amberat) may have had some compacting effect on the deposits. A 15 cm thick deposit of extremely hard wood rat feces below the Mazama tephra spread across Unit 7 and the north end of Unit 5. This deposit was so hard that it had to be chipped apart with a pick. While there were no rodent holes passing through this lens, there was evidence that rodents had dug through softer sediments skirting around its edges to reach the lowest deposits

(Strata LU1a, LU1b, and LU2) of the cave. This allowed oxygen to reach these deposits and set the stage for smoldering fires to destroy significant portions of them. Rodent holes filled with white ash—as well as broader, more diffuse, and amorphous charred features—were particularly evident in Unit 5. One of these smoldering fires destroyed a large area of organic sediment near the modern surface in the northeast corner of Unit 5A (LU9 in Figure 4.5A) and then followed a rodent hole down through the flammable deposits to near the floor of the cave. Perishable artifacts were destroyed by such fires, and bones were charred by them. It is therefore very important to evaluate the distribution of both cultural and faunal remains when attributing the charring of bone from these deposits to cultural activity.

Also, while camelid (probably *Camelops* sp. [James Martin, personal communication]) and horse bones were recovered from Units 7 and 10 (Figure 4.5B), the situation of most specimens in narrow crevices between boulders on the cave floor at the base of the cave walls suggest—along with evidence of rodent gnawing—that they could have been accumulated there due to wood rat scavenging in the cave. In Units 5 and 6, camel and horse bones were recovered from late Pleistocene deposits (Strata LU1a and LU1b) that were generally thicker, looser, and often separated by up to 60 cm of culturally sterile early Holocene deposits from the overlying middle Holocene cultural deposits.

DISTRIBUTION AND DATING OF MEGAFAUNA BONES AND CULTURAL MATERIALS

Faunal remains recovered in situ from late Pleistocene deposits—generally large game (camel, horse, bison, and artiodactyls) but also birds (swans, geese, herons, seagulls, and sage hens), fish, and small mammals—tended to be located in the east half of the excavations toward the rear of the cave. Smaller clusters of large bone were encountered near a possible hearth (Feature 5/3 described below) and in a boulder crevice in the floor along the protected north wall of the cave (Figure 4.5B). The more exposed west side of the excavations contained little bone but more sand, rock, and gravel that had accumulated at the cave mouth as the cliff face weathered.

A relatively thin upright rock slab that appears to have been culturally emplaced (possibly as part of an encircling wall such as Cressman reported for the living floor in Cave 3) separated the majority of in situ bone in Units 5 and 6 into northern and southern clusters (Figure 4.5B). The main bone cluster north of this slab was discovered in the floor of the cave between elevations of 96.65 and 96.80 m (as measured from the permanent block datum set at 99.00 m in the large boulder on the west edge of the excavation block) in a pocket of LU1b deposits covered by an elongated boulder. These faunal remains included a large camel vertebra, a camel astragalus, a horse phalanx, a camel mandible, a sheep mandible, and a horse hoof. None of the identifiable bone in this group comes from prime meat-bearing portions of the carcass. Preliminary analysis indicates there is evidence of moderate large carnivore and rodent gnawing. What percentage of the assemblage was first contributed by humans and later scavenged by carnivores either accompanying them (domesticated dogs) or occupying the caves after they left (wolves, coyotes, bears, and large cats) is currently unknown. In fact, while there are spiral green fractures and conchoidal impact depressions on some of the large mammal bones, suggesting possible human modification (O'Connor 2000:43), the possibility that the megafaunal assemblage accumulated naturally, rather than culturally, cannot be discounted at this stage.

Tiny threads of tightly twisted grass, sinew, and possibly hemp or hair were recovered from the deposits in and around the megafauna remains in the pocket of LU1b sediment. However, a tiny piece (5 × 6 mm) of tightly woven light brown fabric was recovered from the adjacent Unit 5D at an elevation of ca. 96.70 m, raising the specter of rodent disturbance as a significant consideration in determining the possible association of megafauna remains with cultural materials nearby.

As mentioned above, rodent tunnels were occasionally followed through the course of excavations across stratigraphic boundaries from near surface deposits to the lowest strata of the cave. The possibility that modern threads, fabrics, paper, and cotton items could have been incorporated into these deposits as nest building materials cannot be discounted. In some cases, cultural materials were believed by the excavators to have been recovered from filled-in rodent nests. In fact, one of the human coprolites submitted for ^{14}C dating, a specimen recovered 10 cm above camel and horse bones in a boulder crevice in Unit 10, proved to be stratigraphically out of place with an age of only 4130 ± 40 (Beta-213427) ^{14}C BP (4770 cal BP), and a cotton thread recovered from Level 28 of Unit 2, at the south end of the cave, produced a statistically modern AMS date of 139.1 ± 0.2 pMC (Beta-221344).

The organic deposits of LU1a and LU1b were generally compact where exposed, but looser under boulders. The latter was possibly due to the preference of rodents for building nests under boulders, protecting them to some degree from moisture that may have occasionally penetrated interior cave deposits as drifting snow. Finally, moisture may have caused sediment compaction through composting; accumulation of salts, calcium carbonates, and silica; and crystallization of rodent urine, which permeates the deposits. However, it must be remembered that the preservation of perishable cultural materials (threads and coprolites, in particular) and soft tissue on bones of great antiquity establishes limits on the amount of moisture that could have penetrated these deposits without also destroying the perishables.

In Figure 4.6, the vertical distributions of all bone, cultural materials (36 artifacts and 10 lithic debitage), and identifiable megafaunal elements recovered from Unit 5B (1 × 1 m) are compared as the percentage of class totals in 5 cm levels. The megafauna class includes seven diagnostic camel and horse elements, predominantly joint and foot bones (Table 4.1: nos. 2–8). Thick-walled,

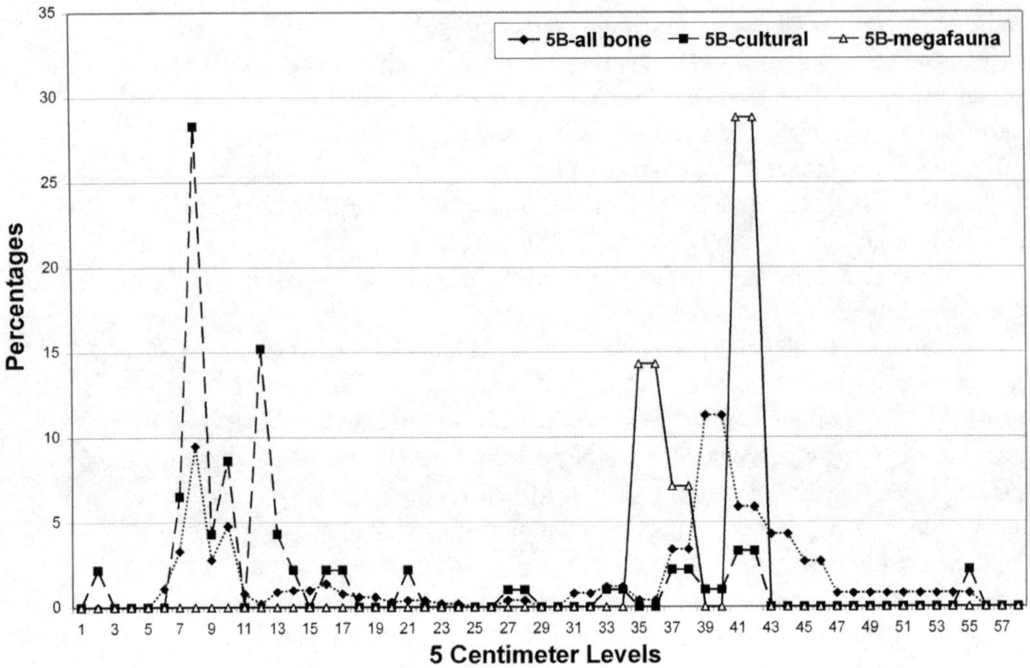

FIGURE 4.6. Distribution of cultural and faunal remains in Unit 5B.

nondiagnostic long bone fragments far outnumber diagnostic elements. However, because they lack genera identifications, these bone fragments are included in the "all bone" category.

Bone and cultural materials exhibit strong bimodal distributions, with the peaks separated by 60 cm of culturally sterile deposits. The upper peak represents roughly 1 m of late and middle Holocene sediment; the lower peak, roughly 80 cm of late Pleistocene and early Holocene sediment. All megafauna remains were recovered from the lower deposits (Strata LU1a and LU1b), and cultural materials were recovered stratigraphically below them. Most of these artifacts are tiny threads particularly susceptible to movement through the deposits by rodent activity. On the other hand, obsidian and cryptocrystalline silicate artifacts (flakes and bifaces) were also found stratigraphically below the identifiable megafauna remains, and the results of obsidian hydration (OH) studies from this unit and the units to the south suggest that that is where they originated (see discussion below). This interpretation may be further supported by the synchronous fluctuations in megafauna elements and cultural remains within the lowest deposits of this unit (see Levels 34–38 and 41–43 in Figure 4.6). The point here is that while there is the possibility of stratigraphic mixing in the lower cave deposits, there is also the distinct possibility that two separate late Pleistocene occupations occurred here.

Radiocarbon Dating

Radiocarbon dating of cultural features, coprolites, perishable artifacts, bone, and rat feces has provided 27 ^{14}C dates spanning >14,000 calibrated years at

TABLE 4.1. Distribution of Megafaunal Elements in Paisley Cave 5

UNIT	STRATUM	GENUS	ELEMENT	ELEVATION (BD)[a]	COMMENTS
1) 5A	1b	horse	phalange	97.03	burned, both sides
2) 5B	1b	camelid	pisiform	96.95	burned, one side
3) 5B	1b	horse	phalange	96.95	partially charred, 11,130 ± 40 (Beta-185942) ^{14}C (13,140 cal BP)
4) 5B	1b	horse	hoof	96.81	
5) 5B	1b	camelid	vertebra	96.76	
6) 5B	1b	horse	phalange	96.75	
7) 5B	1b	camelid	mandible	96.66	
8) 5B	1b	camelid	astragalus	96.65	12,300 ± 40 (Beta-172663) ^{14}C (14,290 cal BP)
9) 6B	1a	horse	hoof	96.81	
10) 6B	1a	horse	calcaneous	96.39	
11) 7C	1a	camelid	phalange	97.83	rodent gnawing, flesh still attached
12) 10D	1a	horse	phalange	97.83	
13) 10	1a	camelid	metatarsal	97.66	

Note: Locations of megafauna elements found in situ are identified in Figure 4.5 by the corresponding numbers in the first column.
[a] BD = block datum elevation, arbitrarily set at 99.00 m.

the Paisley Caves (Table 4.2). Cressman dated wood rat droppings from just below Mazama ash in Cave 3 during the 1950s (Preston et al. 1955). Connolly et al. (1998) and Connolly and Barker (2004) dated two pieces of basketry and a sandal fragment, respectively, from the Cressman collections. Connolly and Trieu (2006) dated the cotton thread recovered out of stratigraphic sequence from Unit 2. The remaining 23 dates are field school and DNA study related samples, 22 of which are stratigraphically consistent.

Radiocarbon dating of a horse phalange, a camelid astragalus, and three culturally processed (stripped and tightly twisted) threads of grass recovered from LU1b deposits in and around the pocket of megafauna bone north of the upright slab serves to illustrate the continued need for extreme caution when interpreting stratigraphic associations. Bone collagen from the horse phalange produced an AMS date of 11,130 ± 40 (Beta-185942) ^{14}C BP (13,140 cal BP). The camelid astragalus produced an AMS date on bone collagen of 12,300 ± 40 (Beta-172663) ^{14}C BP (ca. 14,290 cal BP), while the three grass thread specimens produced an AMS date of 10,550 ± 40 (Beta-171938) ^{14}C BP (12,750 cal BP). This situation demonstrates two important points: (1) there are late Pleistocene–aged threads suitable for use in sewing and weaving activities in these deposits, and (2) those dated so far are not the same age as the megafauna remains with which they were found (Table 4.2). The megafauna date from the Allerød between ca. 11,130 and 12,300 ^{14}C BP (ca. 13,140 and 14,290 cal BP). The perishable artifacts have so far dated from the Younger Dryas between 10,160 and 10,690 ^{14}C BP (ca. 11,860 to 12,750 cal BP).

TABLE 4.2. Radiocarbon Dates from the Paisley Five Mile Point Caves (35LK3400)

SAMPLE NUMBER	CONVENTIONAL ^{14}C AGE	CAL BP AT 2 SIGMA	MATERIAL	REFERENCES
Beta-221344	139.1 ± 0.2	Modern	Cotton thread	Previously unpublished
AA-19151	145 ± 50	300–0	*Scirpus* basket	Connolly et al. 1998
Beta-195907	1060 ± 40	1060–920	Cloth	Jenkins 2005
OxA-16377	1308 ± 28	1293–1178	Human coprolite	Previously unpublished
Beta-147424	2270 ± 50	2340–2310	*Scirpus* sandal	Connolly & Barker 2004
Beta-213427	4130 ± 40	4830–4520	Human coprolite	Previously unpublished
AA-19153	6560 ± 70	7560–7420	*Scirpus* basket	Connolly et al. 1998
OxA-16496	6608 ± 35[a]	7566–7436	Human coprolite	Previously unpublished
Beta-213428	6640 ± 40[a]	7580–7440	Human coprolite	Previously unpublished
Beta-191540	7600 ± 70	8510–8220	Sagebrush	Jenkins 2005
Y-109	7610 ± 120	8630–8180	Rodent feces	Preston et al. 1955
Beta-191539	7640 ± 50	8540–8360	Sagebrush	Jenkins 2005
Beta-213429	7860 ± 40	8720–8560	Human coprolite	Previously unpublished
Beta-213423	10,050 ± 50[b]	11,950–11,280	Human coprolite	Previously unpublished
OxA-16376	10,965 ± 50[b]	13,030–12,839	Human coprolite	Previously unpublished
Beta-182920	10,160 ± 60	12,320–11,440	P.E.T.[f]	Jenkins 2005
Beta-195908	10,290 ± 40	12,360–11,870	Sagebrush rope	Jenkins 2005
Beta-171938	10,550 ± 40	12,880–12,330	Grass threads	Jenkins et al. 2004
Beta-213425	10,690 ± 60	12,960–12,380	Human coprolite	Previously unpublished
Beta-185942	11,130 ± 40	13,190–12,990	Horse, phalange	Jenkins et al. 2004
OxA-16495	12,140 ± 70[c]	14,156–13,821	Human coprolite	Previously unpublished
Beta-216474	12,260 ± 60[c]	15,340–13,880	Human coprolite	Previously unpublished
Beta-213426	12,290 ± 60[d]	15360–14,100	Human coprolite	Previously unpublished
OxA-16497	12,345 ± 55[d]	14,695–14,052	Human coprolite	Previously unpublished
Beta-172663	12,300 ± 40	15,340–12,170	Camelid astrag.	Jenkins et al. 2004
OxA-16498	12,275 ± 55[e]	14,505–13,971	Human coprolite	Previously unpublished
Beta-213424	12,400 ± 60[e]	15,430–14,130	Human coprolite	Previously unpublished

[a–e] Matched pairs of ^{14}C dates on same samples.
[f] P.E.T. = processed edible tissue.

Radiocarbon-dated samples taken from the interiors of five human coprolites (as indicated by DNA and protein residue) add further insight and support for dual late Pleistocene and early Holocene occupations of the caves (Table 4.2). Mitochondrial DNA testing of the coprolites was conducted by three independent ancient DNA laboratories in Denmark, Sweden, and Germany. Genetic profiles were established for a total of 58 field school supervisors, students, and site visitors for comparative purposes. It is standard protocol for all laboratory personnel involved in DNA studies to have genetic profiles entered in the laboratory data bases to verify they have not contaminated the samples they process. No modern matches were found to the haplogroup Types A and B (Native American) mDNA identified in the dung samples. In other words, contamination of the samples by modern DNA has been ruled out (Willerslev et al. 2006).

Further verification that the coprolites are indeed human was sought through protein residue analysis. Human and bison protein residues (Parr 2006) were recovered from specimen 1294-5/6-B-50, which has been dated at 12,140 ± 70 (OxA-16495) ^{14}C BP and 12,260 ± 60 (Beta-216474)^{14}C BP (14,290 cal BP). In fact, three of the coprolites date to a mean of about 12,260 ^{14}C BP (14,280 cal BP), making them the same age as the camel astragalus (Allerød), and two date to about the same time as the Younger Dryas hearth, thread, and braided sagebrush rope (ca. 10,050–10,690 ^{14}C BP [ca. 11,560–12,830 cal BP]). These five coprolites are in proper stratigraphic and chronologic order, with the younger coming from above the stone slab covering the pocket of megafauna bones in Unit 5 and similar elevations in Unit 6, and the older samples being recovered from the lower deposits of Units 5, 6, and 7. The DNA and blood protein analyses prove the coprolites are human, and dating by two independent radiocarbon dating laboratories (Beta Analytic, Inc., and Oxford AMS Radiocarbon Dating Laboratory) prove they are as old as the megafauna. Thus the question of whether or not humans occupied the northern Great Basin before the megafaunal extinctions occurred has finally been resolved in the affirmative.

A small charred feature that may have been a hearth was located adjacent to the cave wall (Unit 7D/Unit 5A), approximately 2 m north of the upright slab (Figure 4.5B), at a unit excavation elevation of 96.85 to 97.20. Feature 5/3 was roughly 60 cm in diameter and 35 cm deep. This feature was bisected in 2002, revealing a deep bowl shape with charred rocks lining the base of the depression. Charred rat feces surrounded the sides and base of the feature (Stratum 1b). The lack of charring in the overlying rat feces suggests that the feature may have been ignited while the top of the depression was exposed to the atmosphere, as one would expect of a hearth but not of a smoldering rodent burrow. Still, the proximity of the LU9 charred feature near Feature 5/3 (the possible hearth) dictates extreme caution in the acceptance of this feature as cultural. It is for this reason that the feature has not yet been directly radiocarbon dated.

Large mammal bones, seldom encountered in noncultural deposits, were recovered in increased quantities in and around Feature 5/3. A charred horse phalange and two unidentifiable large game bone splinters were recovered along the west side of the feature (Figure 4.5B). A large burned bird bone was recovered from the feature, as were fragments of large mammal ribs and various other nondiagnostic bone fragments. Cultural remains recovered from around the feature include a tiny obsidian flake recovered at an elevation of 97.10 m and cordage made of vegetable fibers and possibly hair.

Bones recovered south of the upright slab tended to be found at somewhat greater depths (Figure 4.5A). This is probably due in large part to the nature of the southeast-dipping strata in the cave. However, these deposits also reach proportionately greater depth, so cultural components that are somewhat "compressed" in shallower deposits north of the upright slab could be more widely separated in deeper deposits south of the slab. Indeed, the results of OH analysis on flakes (n = 2) and bifaces (n = 2) recovered from Strata LU1a and LU1b in this area suggest that this could well be the case. However, to adequately evaluate the OH data, we must first understand the parameters of the OH process as it has occurred within the Paisley Caves.

Obsidian Hydration Dating

Obsidian hydration (OH) dating has been extensively applied with good results in the northern Great Basin (Jenkins 2000, 2004; Jenkins et al. 2004; Pettigrew and Hodges 1995; Skinner et al. 2004). OH is the process of molecular water adsorption by exposed obsidian surfaces (Friedman 1968, 1977; Friedman and Long 1976; Friedman and Smith 1960; Friedman and Trembour 1983). The predominance of obsidian and its use in the production of stone tools in the northern Great Basin (it generally composes 90 percent of cultural assemblages) make this dating technique widely applicable throughout the region.

Establishing reliable OH rates requires prior obsidian source characterization of the specimens. Individual volcanic sources have unique chemical compositions that affect hydration rate (Friedman and Long 1976; Friedman and Smith 1960; Friedman and Trembour 1983). Theoretically, each source hydrates at its own rate. However, regional clusters of obsidian sources, such as those in and around the greater Summer Lake–Chewaucan Basin, often hydrate at rates more similar to each other than to sources at greater distances (Hughes 1986).

OH offers a method of directly addressing the age of obsidian specimens. However, the acceptance of hydration ages must be tempered with the knowledge that the hydration rind thickness of any particular specimen is more than just a measure of its age. Hydration rinds form in response to many factors. Soil temperature and moisture content are two of the most important and highly variable of these. Hydration rinds form more quickly as ambient temperatures increase, and more slowly as they decrease (e.g., Ridings 1991, 1996). Thus, postdepositional histories of individual artifacts, as well as the depositional settings in which they reside, affect the cumulative rate of hydration for each specimen. Artifacts that remain on or near the surface hydrate more quickly than rapidly buried artifacts. It is not surprising, then, that most groups of OH measurements contain some specimens with incongruent measurements. It is generally the mean of a group of specimens in good association and context that is important, not the measurement of each individual specimen.

The current sample of 136 OH measurements at the Paisley Caves offers good evidence that this dating method works reasonably well. Tucker Hill obsidian, locally available in Pleistocene lake gravels near the site, accounts for 75 percent of the late Pleistocene lithic debitage and artifacts at the site. Obsidian source use patterns changed during the middle Holocene as sediment covered Pleistocene gravel deposits incorporating Tucker Hill obsidian nodules near the site. While Tucker Hill obsidian continued as a major stone tool source, apparently being brought to the site from the south in the form of completed stone tools, other nearby sources such as Coglan Buttes and Silver Lake/Sycan Marsh contributed proportionately more to site assemblages during the middle and late Holocene.

The Paisley Caves vary considerably in size, with interior dimensions ranging from a few meters to almost 10 m. During the summer, the sun reaches the mouth of most caves by about 1 P.M., rapidly increasing surface temperatures throughout the afternoon. However, deep in some caves are locations that get very little if any direct sunlight and thus remain relatively cool throughout the year. Interior cave deposits are generally exceedingly dry and dusty with silt blown in from the surface of Summer Lake, while exterior cave deposits experience repeated wetting and drying events and are predominantly composed of gravel, rock, and sand. The combination of these radically variant thermal and hydrological settings in close proximity within the site has resulted in widely variant OH rates (Jenkins and Skinner 2004). Obsidian is hydrating at much faster and variable rates at cave exteriors than it is in the dry, cool cave interior deposits. Further analysis involving the collection of soil temperatures both within and without the caves over a period of several years is currently being conducted to determine adequate OH rates for specimens recovered from different settings at the site.

Traditionally, OH dating has depended in large measure on the comparison of associated radiocarbon ages with OH measurements (see Stevenson et al. 2000 for a differing opinion). Development of a hydration rate for obsidian recovered from cool, dry interior settings at the Paisley Caves involved the comparison of hydration rims on 13 specimens (including one stemmed or foliate point base) associated with a stratigraphically sealed-off hearth at the back of Cave 2 (Figure 4.2). Charred, processed edible tissues (possible food) from this hearth produced an AMS date of 10,160 ± 60 (Beta-182920) ^{14}C BP (ca. 11,860 cal BP). An associated segment of braided sagebrush rope produced a confirming AMS age of 10,290 ± 40 (Beta-195908) ^{14}C BP (ca. 12,000 cal BP). The 13 associated OH measurements ranged from 4.8 to 5.3μ, producing a rounded off mean of 5.1μ. Using a hydration rate of 2.2μ^2 per thousand years, this mean indicates an age of roughly 11,800 cal BP for this deposit, fitting comfortably with OH measurements and ages of cave interior assemblages across the entire site (Figure 4.7).

In Cave 5, four obsidian specimens (two flakes, one biface, and a drill) recovered from Strata LU1a and LU1b in Units 5 (n = 1) and 6 (n = 3) had hydration rinds that ranged in thickness from 5.1μ to 6.1μ, producing a mean of 5.5μ and a mean hydration age of 13,750 cal BP. While individual OH measurements can be notoriously misleading as dating tools (for the reasons noted above), it is interesting to note that in this tiny collection the thinner (presumably younger) measurements occur stratigraphically well above the thicker (presumably older) specimens. A flake and a biface fragment recovered from elevations of 97.15 m (Unit 5A) and 97.25 m (Unit 6A), respectively, produced hydration measurements suggesting that they were deposited during the

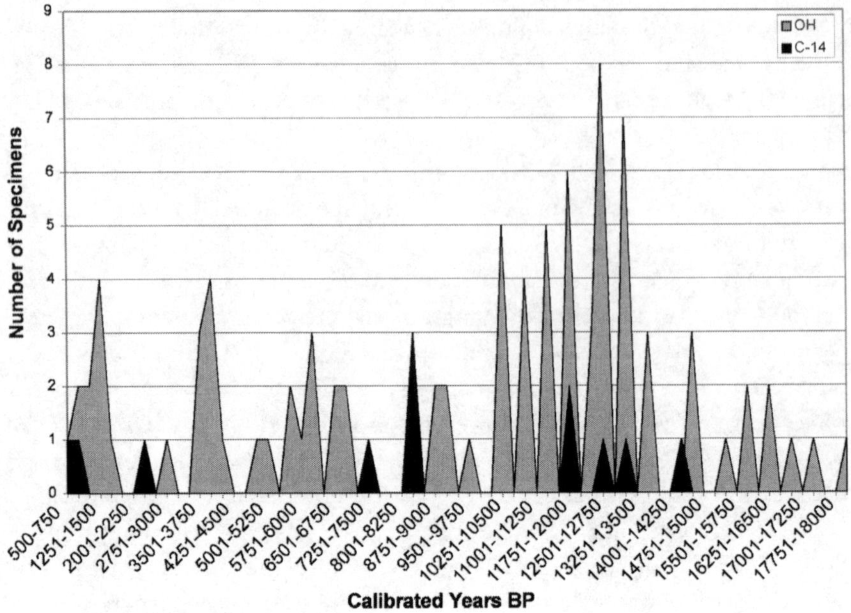

FIGURE 4.7. Comparison of radiocarbon and obsidian hydration dates.

Younger Dryas period (ca. 10,200–10,900 ^{14}C BP), while a flake and a drill fragment recovered at greater depths of 96.35 m (Unit 6B) and 96.05 m (Unit 6D), respectively, produced OH measurements suggesting that they were deposited substantially earlier (Table 4.3).

Here, OH dating (employing a site-specific hydration rate based on direct associations between a firmly ^{14}C dated cultural feature and 13 OH measurements in a sealed-off stratum in Cave 2) indicates hydration ages in Cave 5 that are older than the AMS dated perishable artifacts (10,550 ^{14}C BP [12,750 cal BP]) and consistent with the AMS dated megafauna bones and human coprolites (11,130 to 12,300 ^{14}C BP [13,140 to 14,340 cal BP]; Tables 4.2 and 4.3). Comparison of all of the acceptable OH ages obtained for cool, dry cave interior deposits with the radiocarbon dates from the same settings suggests that the most intense site occupation occurred between ca. 13,500 and 11,000 cal BP. However, an occupation during which a bison was consumed also occurred at about 12,260 ^{14}C BP (14,280 cal BP), and even older occupations may be indicated by OH dating results (Figure 4.7).

SUMMARY

Luther Cressman conducted three phases of excavations at the Paisley Five Mile Point Caves between 1938 and 1940 (Cressman 1940, 1942; Cressman and Williams 1940). Finding megafauna remains (camelid, horse, and bison) in apparent association with a few obsidian artifacts, he believed he had proven that humans had hunted megafauna in the northern Great Basin near the end of the late Pleistocene (Cressman 1966:41, 1986:122). However, the validity of his claims were challenged on various grounds (Heizer and Baumhoff 1970:5;

TABLE 4.3. Comparison of Obsidian Hydration and ^{14}C Dating in Units 5 and 6

Unit/ Strat.	Elev.	Item	Source	OH	OH Age	Unit/ Strat.	Elev.	^{14}C Age
5A/1b	97.15	flake	Tucker Hill	5.1μ	11,820 cal BP			
6A/1a	97.25	biface	Coglan Butte	5.3μ	12,770 cal BP	5B/1b	97–96.75	12,750 cal BP[a]
						5D/1b	96.90	12,830 cal BP[b]
						5B/1b	96.75	13,140 cal BP[c]
						5B/1b	96.65	14,290 cal BP[d]
						5D/1b	96.60	14,340 cal BP[b]
6B/1a	96.35	flake	Tucker Hill	5.7μ	14,770 cal BP	6B/1a	96.40	11,560 cal BP[b]
6D/1a	96.05	drill	Tucker Hill	6.1μ	16,910 cal BP	6B/1a	95.95	14,260 cal BP[b]

[a] Processed threads.
[b] Apparently human coprolite.
[c] Horse phalange.
[c] Camel astragalus.

Krieger 1944; Jennings 1986:115), and the issue of human/megafauna contemporaneity at the Paisley Caves has remained in question for two more decades. The University of Oregon's northern Great Basin archaeological field school returned to the site to resume investigations and resolve the issue.

This chapter addressed two questions related to the apparent association of cultural and megafaunal remains in the Paisley Caves. The first concerns whether cultural and megafaunal remains occur in horizontal, vertical, and stratigraphic association in the Paisley Caves, and the answer to this question is yes. Figures 4.5 and 4.6 clearly demonstrate that there is no separation—horizontal, vertical, or stratigraphic—of cultural materials from paleontological materials in Units 5 and 6. Coprolites, cordage, threads, sinew, bifaces, and lithic debitage are found above, among, and below camelid and horse remains in Units 5, 6, 7, and 10 in Cave 5. However, site formation processes such as floor clearing by humans, bone and coprolite scavenging by rodents and carnivores, and movement of artifacts through the site deposits by rodents appear to have affected their distributions to some degree.

While bone taphonomy is not the main thrust of this chapter—and is so important to the analysis of the site that it must remain essentially a separate topic of investigations—it provides an important consideration of site formation processes that cannot be entirely ignored here. The fact is, megafaunal remains identifiable to genus in Cave 5 consist predominantly of phalanges, metatarsals, various other foot bones, vertebrae, and mandibles (James Martin, personal communication). These elements are dense bones that do not contain marrow and did not support thick muscle (meat) at the time of death. Nondiagnostic thick-walled bone fragments of marrow and meat-bearing long bones commonly outnumber the identifiable remains reported more fully here, indicating that these elements were usually broken or masticated prior to burial. Virtually all large carnivores—including humans and domesticated dogs—break long bones to extract marrow. The differing mechanisms by which this process occurs (hammer and anvil breakage by humans, and chewing by carnivores) result in distinctive patterns of damage to the bone (O'Connor 2000; Miller 1979). Bone masticated by large canines chewing with their molars exhibits distinctive U-shaped grooving on the chewed ends from which the bone is systematically reduced, while rodents, chewing with chisel-like front teeth, leave much smaller, more symmetrical parallel grooves transverse to edges and ridges of bone. Preliminary analysis indicates that some of the large mammal remains at the Paisley Caves have been chewed by carnivores. Some of these bones also exhibit moderate to extreme degrees of checking and weathering, whereas others were buried while they still had blood and cartilage attached to them (all stages of weathering as defined by Behrensmeyer [1978]).

What percentage of the assemblage(s) in Strata LU1a and LU1b was first contributed by humans before being scavenged by carnivores (dogs, wolves, coyotes, bears, and large cats that occupied the caves when humans were not present) is currently unknown. In fact, while there are green fractures and conchoidal impact depressions on a few long bone fragments (suggestive of human breakage patterns), the possibility that the megafaunal assemblage accumulated naturally, rather than culturally, cannot be discounted at this stage. Thus, while the evidence currently suggests the association of megafauna and cultural remains in the site, the faunal evidence by itself is insufficient to unequivocally demonstrate the contemporaneity of the two. The most direct way to establish human/megafauna contemporaneity has been ^{14}C dating of human coprolites and megafauna elements to an overlapping period of time.

This brings us to the second question addressed here regarding whether the cultural and paleontological remains are the same age. The answer to this is also yes.

Hearths, rope, cloth, threads, sandals, and basketry provide 10 cultural radiocarbon dates ranging from 139.1 to 10,550 ^{14}C BP (< 200 to 12,750 cal BP), and human coprolites add an additional 14 dates ranging from 1308 to 12,400 ^{14}C BP (ca. 1275 to 14,360 cal BP). Partially charred horse (n = 1) and uncharred camelid (n = 1) bones have been dated to 11,130 and 12,300 ^{14}C BP (ca. 13,140 and 14,290 cal BP), respectively. Thus, among late Pleistocene dates currently available, mean radiocarbon date intercepts for cultural and megafaunal elements overlap between 11,130 ^{14}C BP (ca. 13,140 cal BP) and 12,300 ^{14}C BP (14,300 cal BP).

The results of DNA tests and ^{14}C dating (verified through replication of results by multiple independent laboratories) have now conclusively resolved the question of human/megafaunal contemporaneity in the Great Basin. These studies clearly indicate that human occupations in the Paisley Caves were contemporaneous with megafauna (camel, horse, and bison).

The suite of ^{14}C dates from the Paisley Caves offered the opportunity to compare OH measurements of 136 specimens (artifacts and debitage) from the cave deposits with calibrated radiocarbon dates spanning more than 14,000 years. Comparing the results of these two dating methods offers better insight to the total length and intermittency of human occupations at the site than ^{14}C dating does alone.

The proposed OH rate for dry/cool deposits within cave interiors was established on the basis of close association between 13 obsidian specimens recovered around a 12,000-year-old hearth sealed off by extremely dry, highly organic deposits at the back of Cave 2. Processed edible tissues and a segment of sagebrush rope associated with this hearth produced AMS dates of 10,160 ^{14}C BP (ca. 11,860 cal BP) and 10,290 ^{14}C BP (ca. 12,000 cal BP), respectively. The associated hydration measurements ranged from 4.8 to 5.3μ and produced a mean of 5.1μ. Application of a hydration rate of 2.2μ^2 per thousand years to this mean suggests an age of 11,800 years for these deposits, providing close correspondence between radiocarbon and OH dating in the dry/cool deposits of cave interiors for the site.

Application of this rate to OH measurements acquired from a biface, a drill, and two flakes recovered stratigraphically with (one biface and one flake) and below (a large drill/perforator and a flake) megafauna specimens in LU1a and LU1b deposits in Cave 5 produce suggested ages ranging from ca. 11,800 to 16,900 cal BP. While the sample is tiny, the distribution of these specimens in upper (97.25–97.15 m) and lower (96.35–96.05 m) elevation components (Table 4.2) corroborate the existence of two separate late Pleistocene occupations, the older dating between possibly 17,000 and 14,000 cal BP and the younger between ca. 13,000 and 11,500 cal BP, as suggested by the radiocarbon dates. The latter, upper component fits very well with the Younger Dryas climatic event, and the lower, older component correlates well, for the most part, with a documented resurgence of the pluvial ZX and Winter lakes between ca. 14,500 and 13,600 cal BP (Friedel 2001).

Plotting all acceptable OH dates from the entire site against the radiocarbon dates (Figure 4.7) indicates a large number of hydration dates clustering between 12,000 and 13,140 cal BP (ca. 10,200–11,130 ^{14}C BP), while a smaller number cluster with the 14,280 cal BP (ca. 12,260 ^{14}C BP) dates obtained from a camelid bone and three human coprolites. As noted above, a number of OH readings suggest that older dates of occupations may be represented at the site. However, these OH dates are currently unsupported by radiocarbon dating and cannot be accepted at face value until they are.

CONCLUSIONS

Cultural remains are stratigraphically and chronologically associated with megafaunal remains in the lowest deposits of the Paisley Caves, as indicated by Cressman (1942). However, the method of accumulation in these strata (site formation processes) must be carefully considered beyond stratigraphic associations. Bone distribution and taphonomy suggest that people may well have been the cause of some bone accumulations, but others may have been partially or entirely the product of bone scavenging by carnivores and rodents. Establishing the contemporaneity of cultural and paleontological specimens at the site required more direct proof of their relative ages than stratigraphic associations could provide alone. Radiocarbon dating of cultural and megafaunal remains by two independent AMS radiocarbon laboratories has provided that proof, establishing that humans defecated in the Paisley Caves some 14,280 years ago, at the same time that camelid bones were deposited there. These dates are in turn supported by OH dates that closely predict the distribution of radiocarbon ages throughout the site. As discussed above, OH dating has its own limitations to reliability related to individual artifact depositional histories and unique depositional settings within sites. However, here the OH method, which has the advantage of being impervious to carbon-based contamination, provides verification for the radiocarbon dates. Both dating methods clearly indicate the site was occupied at the time the megafaunal remains were deposited in the Paisley Caves.

It remains puzzling that none of the Western Stemmed or Clovis (Dietz) sites in the region, apparently dating between 13,200 and 10,200 ^{14}C BP (ca. 16,000 and 12,000 cal BP) (Bedwell 1973; Willig 1988, 1989; Willig et al. 1988; Wingard 2001:584), have produced reliable evidence of human association with extinct fauna other than bison (*Bison antiquus*?). Jennings (1986:115) noted some 20 years ago that nowhere in the Great Basin was there irrefutable evidence of the exploitation of Pleistocene megafauna by human populations, despite repeated claims to the contrary (Bryan 1979:244; Cressman 1966:41, 1986:122; Harrington and Simpson 1961; Orr 1956; Shutler 1967; Simpson 1956). Technically, that remains the case today, though considering the potential for DNA and protein residue analyses to disclose individual elements of Paleoindian diets, it seems highly unlikely that it will remain the case for much longer.

There is no practical reason for late Pleistocene hunters to have avoided megafauna—in particular, camelids and horses, which were quite common in the Great Basin—had they been available in the region. Two of the most reasonable past explanations for this situation have been that either these animals were already extinct prior to the arrival of human hunters ca. 11,000 ^{14}C BP (ca. 13,000 cal BP), or their demise occurred shortly after humans arrived. Strong evidence for human/megafauna contemporaneity in the Great Basin spanning a period of about 1,300 years (14,280 to 13,000 cal BP) has now been firmly established. Finding the evidence for the hunting and processing of megafauna is the next challenge to Great Basin Paleoindian studies, a challenge that DNA and protein residue analyses are particularly well suited to meeting.

On a broader scale, the volume of evidence for pre-Clovis occupations like those at the Paisley Caves is gradually increasing. Bettinger and Young (2004:246) have posed a very good question: “If the New World was largely occupied by 14.6 cal kya, why are [pre-Clovis] sites so rare?" As the investigations at Monte Verde (Collins and Dillehay 1986; Dillehay 1997, 2002; Dillehay and Rossen 2002), Meadowcroft (Adovasio et al. 1978), and other early sites (cf. Adovasio and Pedler 2004 for an updated review) have demonstrated, the largest portions of cultural assemblages in the earliest sites are perishable artifacts that generally do not preserve well in open sites. Quite simply put, the majority of tools in the cultural repertoire of the first colonists in the Great Basin have not survived in the earliest open site locations, and assemblages from the earliest components of the cave sites are both exceedingly small and generally indistinguishable from subsequent toolkits (cf. Bedwell 1973:144 for an example). A combination of morphologically and functionally generic stone tool assemblages may make the earliest colonists in the Great Basin archaeologically invisible. This is particularly true for those who expect to find distinctive Clovis progenitor assemblages of some kind in deposits purported to be older than 11,500 ^{14}C BP (ca. 13,500 cal BP), but instead recover tools easily subsumed within early Holocene assemblages. This argument has been made in various forms in the past, perhaps most prominently by Bryan (1980), but it now seems more reasonably supported with the widespread acceptance of a pre-Clovis age for Monte Verde.

High residential mobility, such as that expected of late Pleistocene occupations in the northern Great Basin, results in low archaeological visibility because the quantity of refuse left at briefly occupied sites is generally quite sparse. Tools designed to facilitate high mobility tend to exhibit multifunctional characteristics suited to maintenance of light, portable, and broadly useful toolkits. The result is that small bands of highly mobile hunter-gatherers generally leave thin artifact scatters composed of debitage, a few broken or lost tools with multifunctional capability (scrapers and bifaces), and a few bone scraps that often quickly disappear from open sites. Chronologically diagnostic artifacts tend to be few and often are not found in short-term single occupation sites at all.

The contemporaneity of cultural items found with megafaunal remains in the northern Great Basin (Paisley Caves [Cressman 1942:93], Christmas Valley [Minor and Spencer 1977], Catlow Cave [Cressman 1940:10], Lower Klamath Lake [Cressman 1942:99–100], and Tule Lake [Beaton 1991]) has generally been questioned on the grounds of inconclusive context, tiny assemblage size, and inadequate reporting (Heizer and Baumhoff 1970, for instance). At the Paisley Caves, well-documented Pleistocene lithic assemblages contain few items, most of which are not morphologically distinctive (debitage, bifaces, and test cores). The earliest assemblage could easily have been subsumed within latter Younger Dryas assemblages were it not for the pre-Clovis radiocarbon dates on human feces and the unusually thick hydration rinds on the obsidian artifacts. Had the human coprolites at the Paisley Caves not been analyzed for DNA and subjected to rigorous dating methodology, the pre-Clovis age of the cultural items recovered with the megafaunal remains could not have been conclusively proven. In other words, the pre-Clovis component of this site could very well have been missed or dismissed.

The Paisley Caves example suggests that pre-Clovis assemblages may exist at other sites, but their presence may be masked by mixture with younger assemblages (Madsen 2004:18–19; Dillehay 1997). It is only in very rare instances that cultural deposits of Pleistocene age have remained stratigraphically separated from overlying Holocene deposits in such a way that the two may be reliably proven to be distinct. Even then, as at the Paisley Caves, it will often be necessary to sort them out of Younger Dryas and early Holocene assemblages. Finding pre-Clovis sites in the New World is likely to require the application of new technology and methodology suitable to unmasking these assemblages. In fact, the evidence of pre-Clovis site occupations may be waiting for us to sort out of boxes already on museum shelves.

ACKNOWLEDGMENTS

Field investigations and the preliminary analysis of the Paisley Caves materials were supported by the Lakeview District Office of the Bureau of Land Management. This funding was obtained under the direction of William J. Cannon (district archaeologist), Richard Hanes (state archaeologist), and Tom Rasmussen (Lakeview field office manager). I thank them for all the support, cooperation, and direction they have given to this project. Eske Willerslev (director of the Ancient DNA and Evolution Group, Centre for Ancient Genetics, Niels Bohr Institute and Biological Institute, University of Copenhagen, Denmark) and his international team have donated the exhaustive DNA analysis and seven AMS radiocarbon dates establishing the age and genetic profiles for the coprolites and all personnel (students, supervisors, site visitors, and laboratory personnel) involved. The genetics labs in Germany and Sweden have likewise provided invaluable assistance by freely analyzing key specimens of human coprolites. Tom Higham (deputy director of the Oxford Radiocarbon Accelerator Unit, Research Laboratory for Archaeology and the History of Art, University of Oxford) donated six AMS dates on human coprolites. The Association of Oregon Archaeologists supported the research with grants for AMS dating and blood protein analysis. The Museum of Natural and Cultural History, University of Oregon, paid for part of an AMS date at the direction of museum director Jon Erlandson. Robert Yohe III and Robert Parr, at the Laboratory of Archaeological Sciences, California State University, Bakersfield, donated the blood protein analysis of multiple coprolites. Loren Davis provided the base for the stratigraphic profile and the descriptions of the basic lithostratigraphic units. Thomas Connolly and Brian O'Neill provided site maps and reviewed early drafts of this paper. Tom produced most of the illustrations. L. Suzann Henrikson and Kaylon McAllister produced the map of Cave 5 unit excavations. Jim Martin (South Dakota School of Mines and Paleontology) provided the preliminary identification of the megafauna remains collected in

2002. Students Kelly Shaw, Chrystal Hockett, and Vickey Reed conducted further faunal and paleontological identifications under the direction of William Orr (Department of Geology) and Patrick O'Grady (Museum of Natural and Cultural History) at the University of Oregon. Finally, I would like to thank Gerald Skelton and the Cultural Heritage Committee of the Klamath Tribes for their trust, cooperation, guidance, and field visits.

5

Stratigraphy and Chronology of the Pleistocene to Holocene Transition at Bonneville Estates Rockshelter, Eastern Great Basin

Kelly E. Graf

THIS CHAPTER PRESENTS INITIAL RESULTS OF GEOARCHAEOLOGICAL research at Bonneville Estates Rockshelter (BER), a multicomponent site in the eastern Great Basin. Details on site geomorphology, stratigraphy, and radiocarbon chronology and age of cultural components of the late Pleistocene through the middle Holocene are included. This chapter facilitates further discussion of lithic, faunal, and floral assemblages from BER (see Goebel, this volume; Hockett, this volume; Rhode and Louderback, this volume) to address human adaptations and paleoecology of the east-central Great Basin during the Pleistocene to Holocene transition.

To tackle the question of whether early inhabitants were more "Paleoindian" or more "Archaic" in their adaptations, comparisons of the remains left by early people in a given region or at a given site should be systematically compared with those of later Archaic peoples. For much of the Great Basin this type of comparison is difficult at best. Most sites containing early archaeological remains are surface sites; therefore, their association with later archaeological remains at the site or in the region cannot be known for sure. At BER, however, we have a very detailed record for these periods, so this chapter includes a description of deposits from both the late Pleistocene to early Holocene and the middle Holocene.

SITE LOCATION AND HISTORY OF EXCAVATIONS

BER is located near the Nevada/Utah border in the eastern Great Basin, approximately 50 km south of West Wendover, Nevada, and the famed Danger Cave (Figure 5.1). BER is a rather large shelter, measuring about 25 m wide by 10 m high by 15 m deep near its center. The immensity of the shelter is realized when considering the surface area of deposits measures more than 250 m^2. The shelter's entrance faces to the southeast and overlooks an unnamed wash that

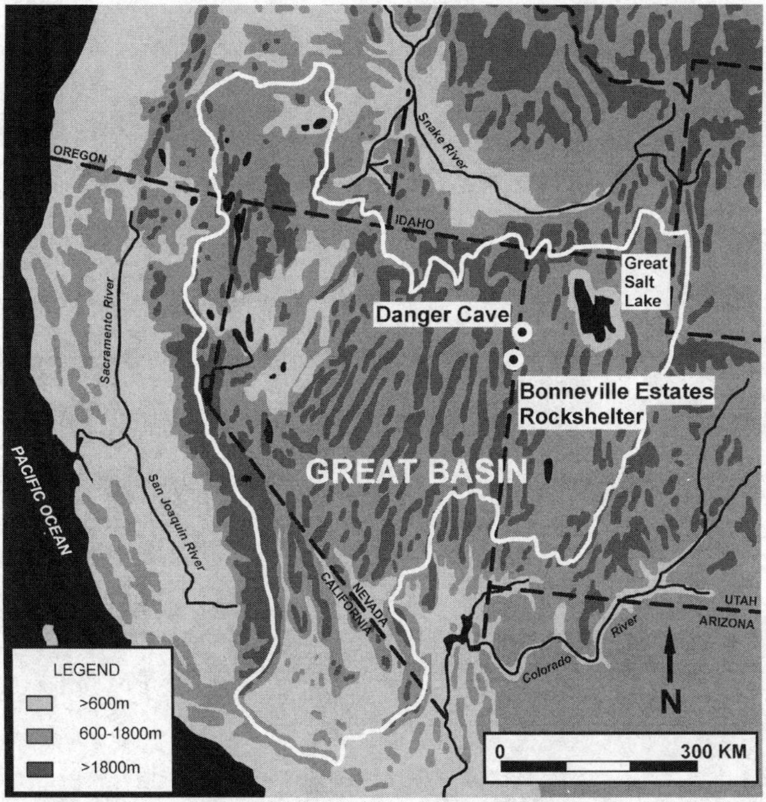

FIGURE 5.1.
Map of the Great Basin showing the locations of Bonneville Estates Rockshelter and Danger Cave.

drains northeast to the Bonneville salt flats. Natural vegetation in close proximity to the site consists of an *Atriplex* and *Purshia* community with several grasses such as *Achnatherum hymenoides* and *Achnatherum speciosum* (Rhode 2002).

BER was discovered and recorded by Steve Dondero and Tim Murphy of the Elko District of the U.S. Department of the Interior Bureau of Land Management (BLM) in 1986. Unfortunately, the rockshelter was actively being looted at this time. In 1988 P-III Associates Inc., directed by Alan Schroedl, conducted test excavations for the BLM in order to establish whether intact cultural deposits still existed at the site. They found a well-preserved sequence of cultural strata spanning the last 6,000 ^{14}C years of prehistory (Schroedl and Coulam 1989), but due to budgetary constraints they were unable to excavate into deeper deposits.

During the summer of 2000, a joint BLM and University of Nevada field crew—led by Bryan Hockett, Ted Goebel, and Kelly Graf—began excavating at BER to determine whether deposits predating 6000 ^{14}C BP existed, and if such deposits possessed the potential for investigating changes in the paleoecology and adaptive strategies of humans in the eastern Great Basin during the late Pleistocene and early Holocene. By the end of the 2001 field season, we had exposed the profile in Figure 5.2, establishing that humans had occupied BER

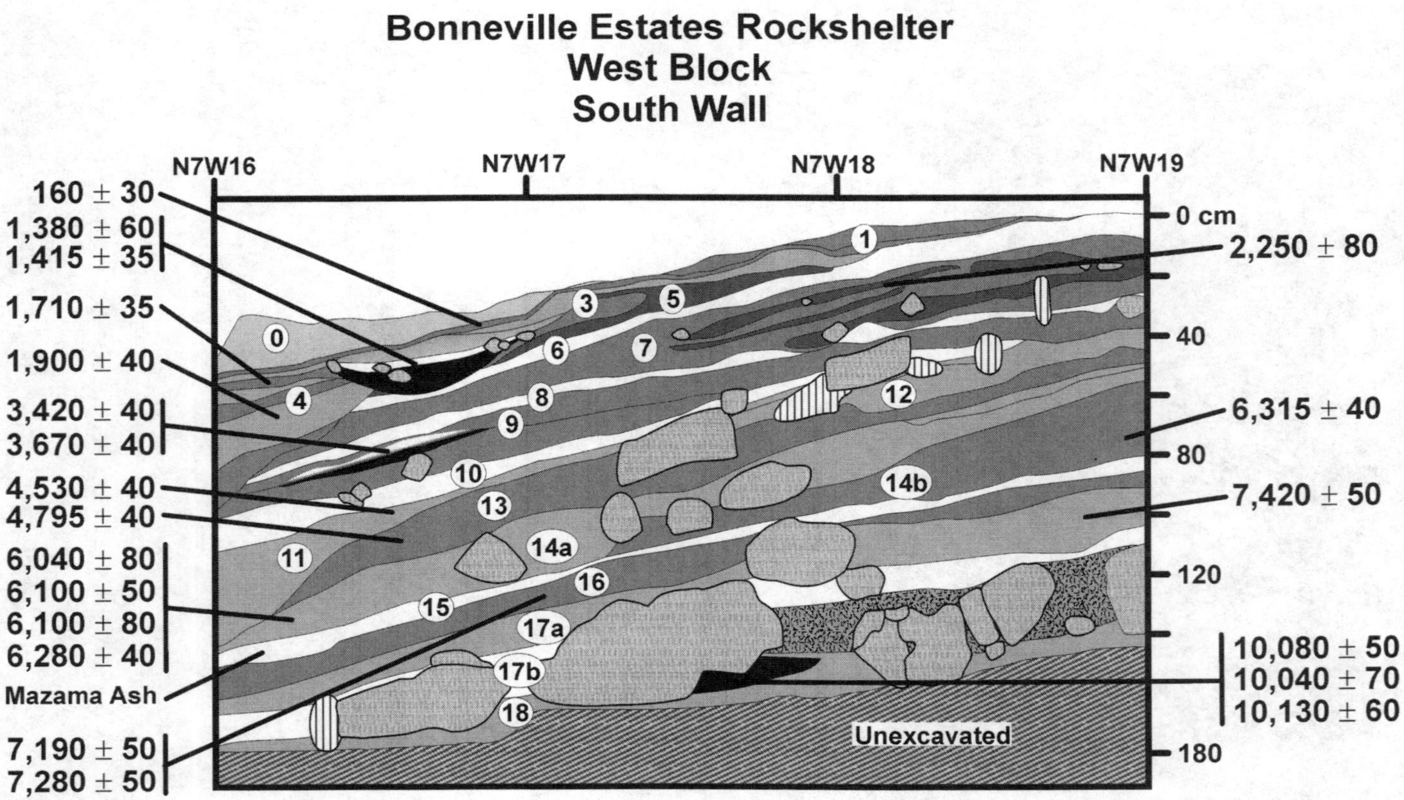

FIGURE 5.2. Stratigraphic profile of the West Block of Bonneville Estates Rockshelter showing the deposits exposed during our 2001 excavations.

from the late Pleistocene throughout the Holocene (Goebel et al. 2003; Graf et al. 2002a, 2002b, 2004). As of 2006, our excavations covered an area of 67 m^2, with 55 m^2 penetrating into or through late Pleistocene to early-middle Holocene deposits (the other 12 m^2 had not been excavated to early-middle Holocene or late Pleistocene depths) (Figure 5.3), exposing several late Pleistocene and early Holocene living floors and activity areas.

Two separate block excavations (West Block and East Block [Figure 5.3]) were initially opened between 2000 and 2002. After three seasons of excavations, it became apparent that the stratigraphy could be generally correlated based mainly on radiocarbon dates and diagnostic artifact types, but the nature of deposition rates and preservation of sediments in these two areas was different. For these reasons, in 2003 we began to excavate a trench connecting the two blocks to expose a profile allowing precise correlations between their stratigraphic profiles. Through 2006, the trench had been excavated into middle Holocene deposits.

All artifacts and ecofacts encountered were three-point provenienced, and all sediments from excavated deposits were screened through ⅛" mesh (screened materials were provenienced to 50 cm^2 horizontal quadrants and 5 cm vertical intervals or the stratigraphic unit, depending on which came first). Four column samples measuring approximately 50 cm^2 were collected, for which all sediment was run through sieves to a size of 1.6 mm; laser ablation analysis was conducted on the fraction smaller than 1.6 mm. Detailed feature and artifact distributions were carefully mapped and photographed.

SITE GEOMORPHOLOGY AND PALEOECOLOGICAL CONTEXT

BER is located in the Lead Mine Hills of the western Bonneville Basin (Figure 5.4). The rockshelter is situated at the late Pleistocene highstand of pluvial Lake Bonneville at approximately 1,580 m in elevation. Before the construction of the Bonneville highstand beach, the lake's waters had eroded the Upper Pennsylvanian limestone and dolomite precipices of the Lead Mine Hills, carving the large opening of BER. This occurred before about 15,000 ^{14}C BP (18,800 cal BP). Also during the lake's last highstand, beach gravels were deposited within and in front of the rockshelter. Soon after 15,000 ^{14}C BP (18,800 cal BP), the lake breached Red Rock Pass near the town of Zenda, Idaho, and the lake's level fell to the Provo shoreline, 1,470 m. The rockshelter, therefore, would have been open for occupation after about 14,500 ^{14}C BP (17,500 cal BP) (Benson et al. 1990, 1997; Currey and Oviatt 1985; Rhode et al. 2005).

During Provo times, as the lake level stabilized, the shoreline would have been within 2.5 km of the rockshelter, and the nearby vegetation community would have been dominated by limber pine and sagebrush (Rhode 2000). Sometime after 14,000 ^{14}C BP (17,000 cal BP), but before about 10,800 ^{14}C BP (12,800 cal BP), the region experienced a long drought in which the lake became severely desiccated to levels lower than the modern-day Great Salt Lake (Currey

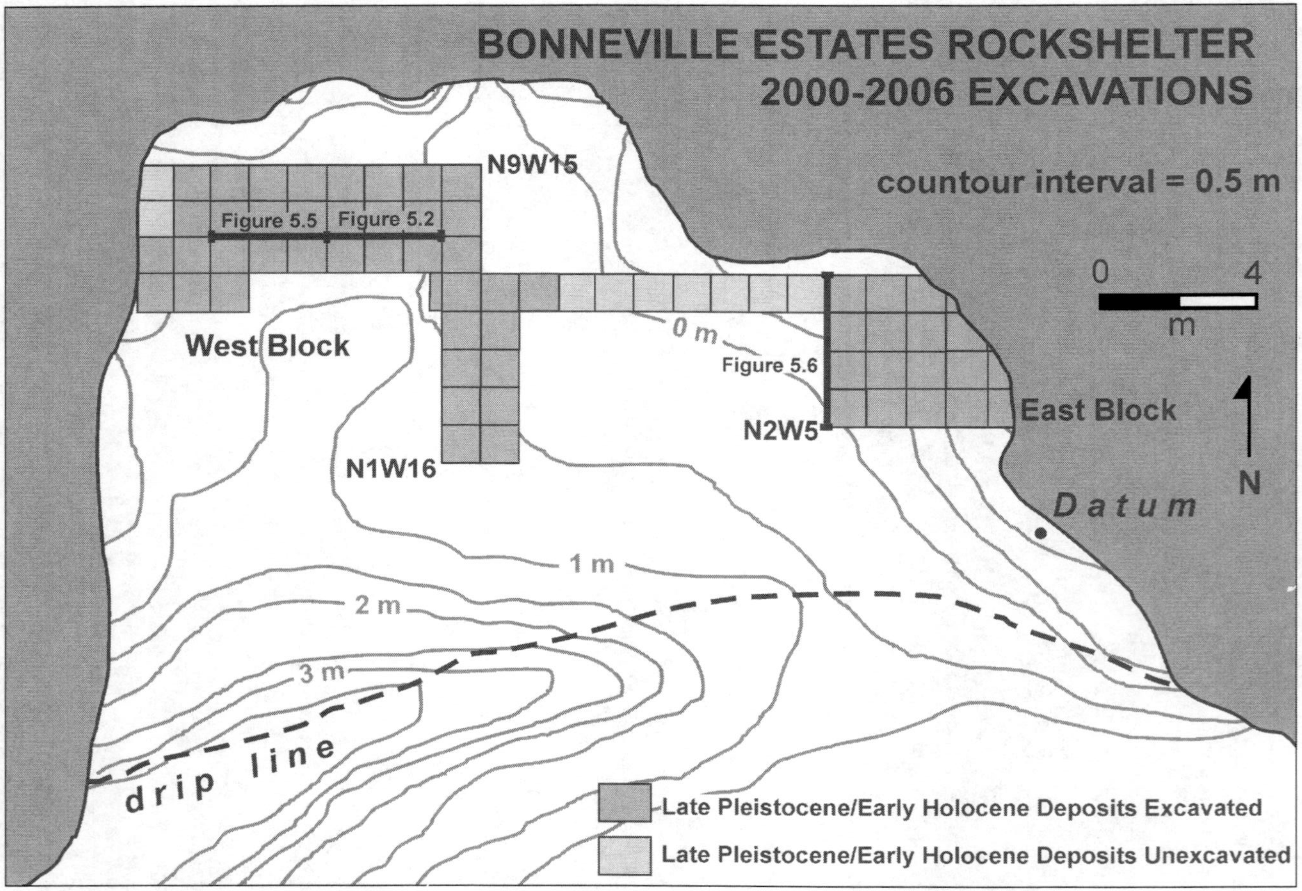

FIGURE 5.3. Excavation map of Bonneville Estates Rockshelter for the field seasons of 2000–2006.

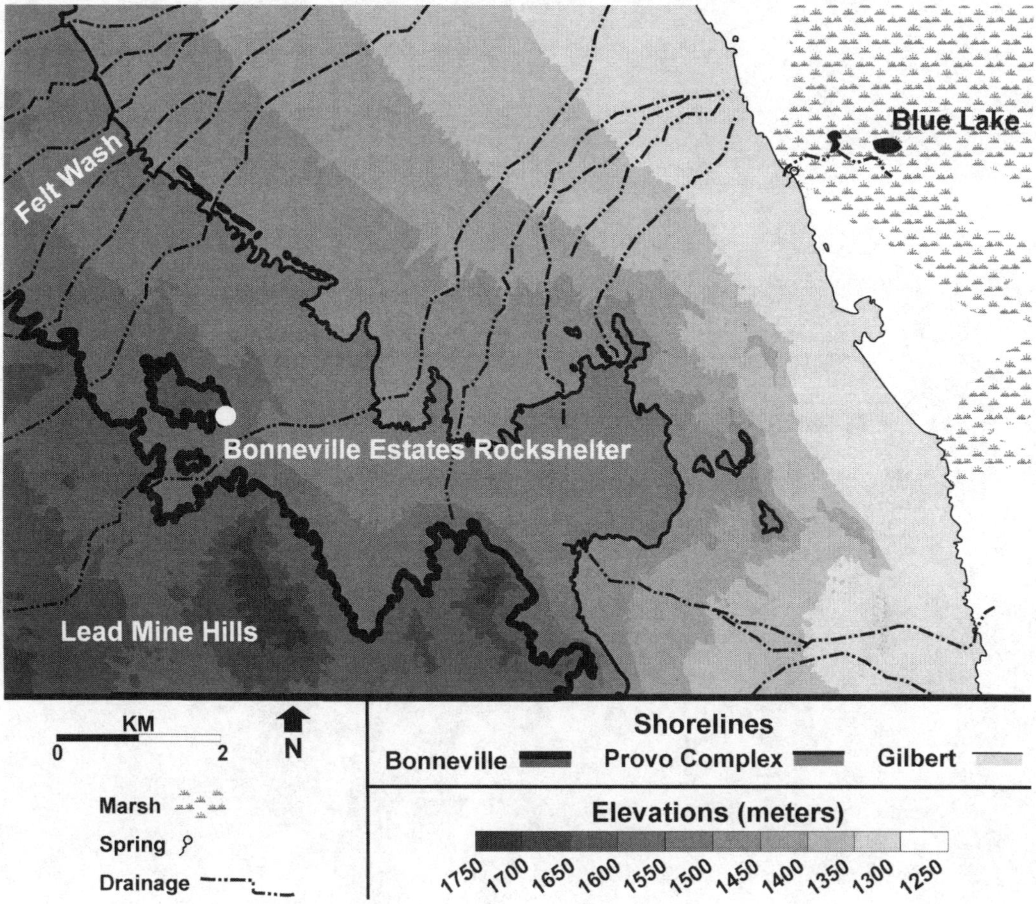

FIGURE 5.4. Map showing the relation of Bonneville Estates Rockshelter to major pluvial shorelines and other geomorphic features.

and Oviatt 1985; Oviatt 1997; Rhode et al. 2005). By 10,800 ^{14}C BP (12,800 cal BP), the lake rose again, forming the Gilbert shoreline at an elevation of about 1,314 m. At this time the lake would have been located nearly 6 km from the rockshelter, and the local vegetation community would have still been dominated by limber pine and sagebrush (Rhode 2000). After about 10,000 ^{14}C BP (11,400 cal BP) the lake receded from the Gilbert shoreline, and the modern-day springs feeding Blue Lake, about 8 km east-northeast of BER, became active. Since these springs are still active, we can assume that during much of the Holocene water would have been locally available to BER occupants.

STRATIGRAPHIC CONTEXT

As mentioned above, excavations of BER have mainly focused on the West and East blocks. Because the stratigraphic profiles of the two blocks cannot yet be directly correlated with confidence, I present each separately. Stratigraphic descriptions are based mainly on field observations and the results of particle size analysis (Farrand 2001).

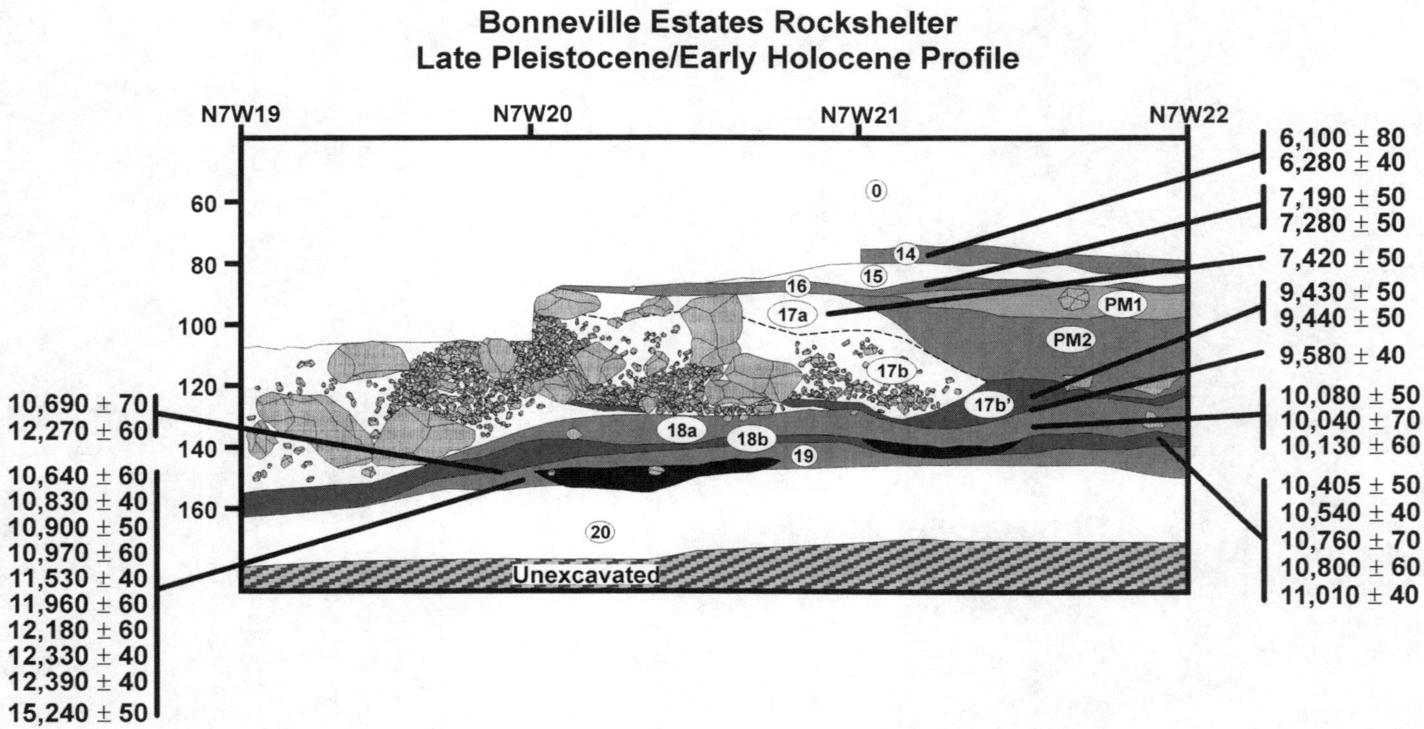

FIGURE 5.5. Lower profile from the West Block showing Stratigraphic Units 20 through 16.

West Block

In the 34 m^2 West Block at BER, seven stratigraphic units represent the late Pleistocene to early-middle Holocene deposits (Figures 5.2 and 5.5). These stratigraphic units, labeled Strata 20–14, make up the lower 200 cm of the West Block deposits, extending to a depth of nearly 300 cm below the surface. Below is a description of each of these stratigraphic units, beginning with the lowest (Stratum 20) and ending with the highest (Stratum 14).

Bedrock in the West Block is mantled by a deposit of well-rounded gravels interpreted as representing the late Pleistocene beach of Lake Bonneville; however, these gravels have been exposed only in 1 m^2 of the West Block and have not undergone detailed description and analysis.

Overlying the gravel is Stratum 20, which has been completely excavated in 1 m^2 and partially exposed in 30 m^2 of the West Block. It is a yellowish-brown, loose sandy loam that contains occasional angular gravel-sized to cobble-sized eboulis clasts and rare boulder-sized clasts. Stratum 20 has been subdivided into two sub-strata, 20a and 20b. The upper of the two (20a) contains particles strongly coated with calcium carbonate; this may represent chemical weathering of the sediment in place or leaching of calcium carbonate from overlying strata. Where excavated, Stratum 20 has been found to be culturally sterile, affected only by geogenic and biogenic processes. Few faunal and floral materials have been found, and when present are poorly preserved.

To date, Stratum 19 has been excavated across an area of about 20 m^2 and is visible in the profiles of the southwestern corner of the West Block. This stratum ranges from 2 to 10 cm thick, and its lower contact with Stratum 20 is diffuse. Stratum 19 is a dark yellowish-brown loam that grades from a loose matrix of loam and eboulis at its base to a compact loam near its upper contact with Stratum 18. The stratum contains many angular gravel-sized to cobble-sized eboulis clasts. It also contains numerous plant macrofossils and animal bones, but these are typically somewhat deteriorated. Formation of Stratum 19 can be attributed mainly to biogenic and geogenic processes; however, as discussed below, anthropogenic processes may have affected the upper portion of Stratum 19.

Stratum 18 extends across the West Block and has been excavated in portions of 31 m^2. This stratum overlies Stratum 19; the contact between the two is gradual in places. Stratum 18 was further subdivided into Sub-strata 18b and 18a in the western half of the West Block excavation. As the stratigraphic profile dipped to the east, 18b and 18a became indistinguishable. During excavation both elevation and slight sediment color changes were used to distinguish between the two units. Sub-stratum 18b was excavated across much of the West Block (portions of 30 m^2). It ranges from 5 to 10 cm thick and consists of a grayish-brown, very compact sandy loam with few angular gravel-sized to cobble-sized eboulis clasts. Sub-stratum 18b's lower contact with Stratum 19 was typically quite clear. Sub-stratum 18a was present in portions of 20 excavation squares of the West

Block. This sub-stratum ranges from 4 to 10 cm in thickness and, where present, has an abrupt contact with the underlying sub-stratum, 18b. Sub-stratum 18a consists of a brown loose-to-compact sandy loam that contains occasional gravel-sized to cobble-sized eboulis clasts. Both 18b and 18a contain numerous well-preserved plant macrofossils, faunal remains, and lithic and perishable artifacts (i.e., cordage, and bone and antler tools). Formation of Stratum 18 appears to have been predominantly anthropogenic, but evidence of biogenic and geogenic processes is also present (see Rhode and Louderback, this volume).

Stratum 17 is prevalent across the West Block of the shelter. It has been excavated in 40 m^2 and is present in all profiles of the West Block. It overlies and has a clear lower contact with Stratum 18. Stratum 17 has been divided into three sub-strata: 17b, 17b', and 17a. Lower-lying Sub-stratum 17b has been excavated across most of the West Block (portions of 40 m^2) and is a 20 to 25 cm thick rubble zone composed predominantly of angular gravel-sized to boulder-sized eboulis clasts, but yellowish-brown silt fills the few voids in this sub-stratum. The formation of 17b can be attributed exclusively to geogenic processes, especially the spalling of large eboulis clasts from the shelter's ceiling, probably during several roof collapse episodes. Sub-stratum 17b' is an organic-rich lens (with numerous very well preserved plant macrofossils and animal bones) near the base (within 10 cm) of Sub-stratum 17b in the western portion of the West Block (10 m^2). Sub-stratum 17b' is generally 2–6 cm thick and consists of a yellowish-brown, compact loam. Few gravel-sized to cobble-sized eboulis clasts are present. Its upper and lower contacts with Sub-stratum 17b are typically very distinct. Sub-stratum 17a has been excavated across much of the West Block (24 m^2) and forms a wavy but clear lower contact with Sub-stratum 17b. It is a yellowish-brown sandy loam containing many angular gravel-sized to cobble-sized eboulis clasts. The formation of Sub-stratum 17a can be attributed mainly to geogenic processes; however, a single hearth feature and a small assemblage of lithic artifacts provide evidence of human activity as well. Vegetal remains are rare.

Stratum 16 directly overlies Stratum 17 and has been found only in portions of 18 m^2 of the West Block. It ranges in thickness from 2 to 10 cm and consists of a brown, loose sandy loam with many angular gravel-sized to cobble-sized eboulis clasts. It is rich in organic remains (plant macrofossils and animal bones), and the formation processes of this stratum appear to have been anthropogenic as well as biogenic and geogenic in nature.

Stratum 15 is a very thin (1–3 cm thick), predominantly silt band. Stratum 15 forms an abrupt boundary with Stratum 16 below and Stratum 14 above and contains few angular gravel-sized eboulis clasts. It also contains several isolated pockets of Mazama ash. This stratum contains no organics and few lithic artifacts and was formed exclusively by geogenic processes.

Stratum 14 has been excavated in 21 excavation squares in the West Block. Unfortunately, two large looters' pits in the northern portion of the West Block erased Stratum 14 from the rockshelter deposits; however, we know that its extent is larger than the excavated area because it is present in the southern and eastern profile walls of the West Block excavation. Stratum 14 has been subdivided into three sub-strata, 14c, 14b, and 14a. While Sub-stratum 14c appeared in only four excavation squares, Sub-strata 14b and 14a are the best known from our excavations (exposed in 21 excavation squares). Sub-stratum 14c is very thick, ranging from about 10 to 30 cm in thickness where excavated. It is a yellowish-brown sandy loam containing rare angular gravel-sized to cobble-sized eboulis clasts. Sub-stratum 14b, ranging from 2 to 20 cm in thickness, consists of a brown sandy loam with abundant to rare angular gravel-sized to cobble-sized eboulis clasts. Sub-stratum 14a has a wavy lower contact with Sub-stratum 14b. It also ranges from 2 to 20 cm in thickness and consists of a yellowish-brown sandy loam with few to many angular gravel-sized to cobble-sized eboulis clasts. All sub-strata of Stratum 14 contain numerous very well preserved vegetal and faunal remains, as well as human coprolites and human-produced textiles (basket and cordage fragments). Stratum 14 was formed through anthropogenic as well as biogenic and, to a lesser extent, geogenic processes.

In the far western portion of the West Block, a prehistoric wood rat midden was found interdigitated with Strata 17 and 16. Present in portions of 6 m^2, the midden appeared to be stratified with several horizontal layers of predominantly unconsolidated vegetal materials and wood rat dung. Two column samples were collected from this area of the shelter. Future paleoecological analyses of these columns will provide important information about the local environment of the shelter during this time.

In sum, seven stratigraphic units have been described from the late Pleistocene to early-middle Holocene deposits in the West Block of BER. These units are well stratified and in some cases can be further divided into sub-strata based on slight textural and/or color changes. The basal unconsolidated deposit is lacustrine in origin and probably represents a relatively stable lake level of Pleistocene Lake Bonneville. Strata 20–14 consist of terrestrial sediments formed from roof fall activities, in-washing of colluvial materials from outside the rockshelter's drip line, deposition of aeolian materials, and possibly in-place weathering of the sediment. In these strata rodent activity is present —sometimes minor, other times predominant (Strata 19, PM2, PM1). Strata 18b–17b' were primarily deposited through human activities, providing a well-preserved record of late Pleistocene occupation of BER, especially in the western half of the West Block. Sub-stratum 17b was formed primarily by geogenic activities related to a major roof collapse episode or series of episodes. Strata 17a, 16, 15, and 14 represent the early to early-middle Holocene. Although 17a contains a few cultural remains, it is primarily geogenically derived silt and rubble. Stratum 15 contains no artifacts and is geogenic. Strata 16 and 14 contain well-preserved cultural remains and represent the major Early Archaic occupations of the West Block.

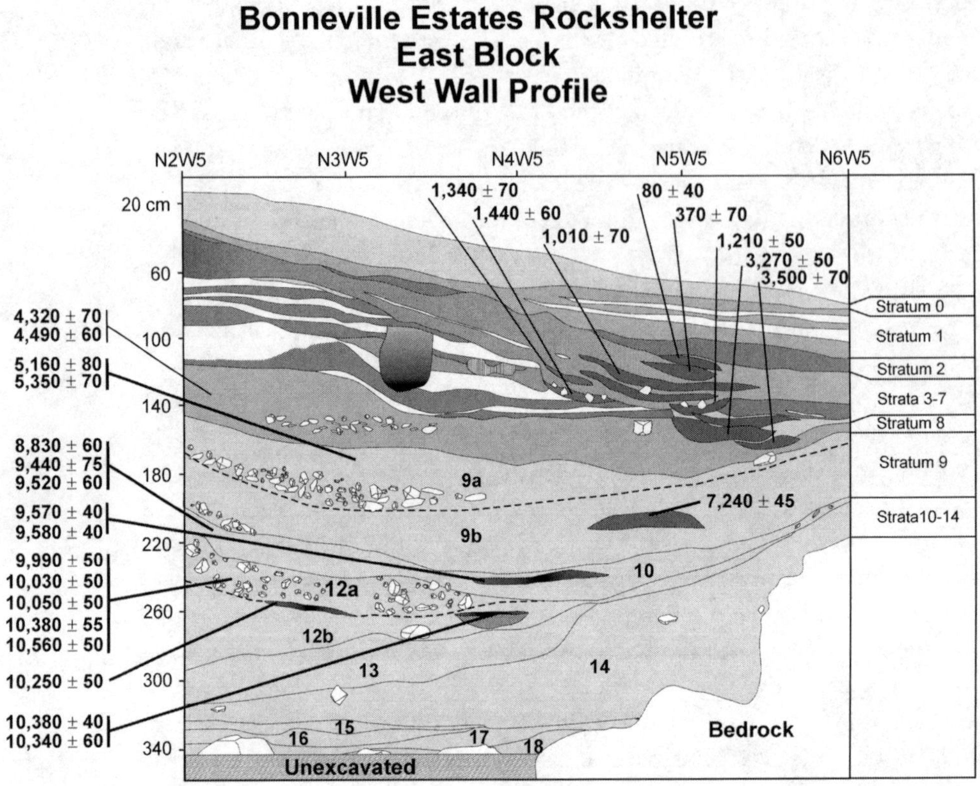

FIGURE 5.6. Stratigraphic profile of the East Block of Bonneville Estates Rockshelter. Substratum 12a was originally designated Stratum 11 but has recently been incorporated into Stratum 12.

East Block

In the East Block excavation at BER, the late Pleistocene to middle Holocene deposits consist of 10 distinguishable stratigraphic units (Figure 5.6). These are represented by Strata 18–9, which extend to a depth of nearly 340 cm below the modern surface of the East Block and have been excavated across an area of 14 m^2. Strata 18–13 are culturally sterile units of late Pleistocene lacustrine gravels and sands, colluvium, and roof fall materials, while Strata 9–12 are culture-bearing stratigraphic units. Because no column samples of sediment from the East Block have been collected yet, descriptions of this set of deposits are less detailed than those for the West Block.

Strata 18–13 are culturally sterile deposits that have been exposed across an area of 6 m^2 in the western portion of the East Block. Stratum 18 ranges from 2 to 10 cm in thickness and contains predominantly sub-rounded to rounded gravel-sized clasts interpreted as Lake Bonneville beach gravels. Along the back wall of the rockshelter, Stratum 18 directly overlies bedrock. Stratum 18 contains a few sub-angular to sub-rounded boulders. Their relatively rounded nature suggests that their presence is probably the result of roof spalling activity prior to the deposition of Stratum 18 gravels. Stratum 17 overlies Stratum 18, forming a gradual contact. This stratum ranges from 2 to 10 cm in thickness and consists of gravel-sized rounded clasts and coarse sands mixed with occasional sub-angular to sub-rounded boulders of roof fall. Strata 18 and 17 were formed through geogenic processes.

Stratum 16 forms a clear boundary with Stratum 17 below it, ranges from 1 to 10 cm in thickness, and consists of a yellowish-brown, loose silt-rubble matrix with many sub-angular to angular gravel-sized eboulis clasts. Stratum 15 has a clear basal contact with Stratum 16, ranges from 1 to 8 cm in thickness, and consists of a yellowish-brown silt-rubble matrix with many sub-angular to angular gravel-sized eboulis clasts with isolated cobble-sized clasts. Stratum 14 overlies Stratum 15; they are separated by a clear contact. Stratum 14 ranges from 10 to 80 cm in thickness and is a yellowish-brown, loose silt-rubble matrix containing predominantly sub-angular to angular gravel-sized eboulis clasts less than 2 cm in diameter. In the southern half of the excavation Stratum 14 contains isolated cobble-sized eboulis clasts. Stratum 13 shares a clear contact with lower-lying Stratum 14, ranges from 1 to 35 cm in thickness, and contains a yellowish-brown silt-rubble matrix with predominantly sub-angular to angular gravel-sized eboulis clasts. Strata 16–13 were mainly formed by geogenic processes.

Stratum 12 is the lowest stratum of the East Block with cultural remains. Its lower contact with Stratum 13 is distinct. Stratum 12 ranges from about 2 to 80 cm in thickness and can be subdivided into two sub-strata based primarily on textural differences observed in the field. Sub-stratum 12b consists mainly of a yellowish-brown, loose silt-rubble matrix with sub-angular to angular gravel-sized clasts. Above this is Sub-stratum 12a, consisting of yellowish-brown, loose rubble, angular cobble-sized eboulis clasts, and isolated gravel-sized eboulis clasts.[1] The few cultural remains from Stratum 12 were located in the lower-lying Sub-stratum 12b. Stratum 10 overlies Stratum 12; they are distinguished by a gradual contact. Stratum 10 ranges from about 2 to 20 cm in thickness and consists of a light yellowish-brown, unconsolidated silt-rubble matrix with many sub-angular to angular gravel-sized eboulis clasts. Stratum 9 overlies Stratum 10, forming a gradual boundary. Stratum 9 ranges from 10 to 60 cm in thickness and can be divided into two sub-strata based primarily on textural differences. Sub-stratum 9b consists of a yellowish-brown silt-rubble matrix containing primarily sub-angular to angular gravel-sized clasts of eboulis with isolated cobble-sized clasts. Sub-stratum 9a also consists of a light yellowish-brown, massive silt-rubble matrix with many sub-angular to angular gravel-sized eboulis clasts and a zone of cobble-sized clasts near its contact with Sub-stratum 9b.

In sum, the lower 100–150 cm of the East Block profile contain culturally sterile deposits that were formed largely through geogenic processes. Strata 18 and 17 are lacustrine beach gravels and sands deposited by Lake Bonneville, presumably as the lake receded from its highstand. Strata 16–9 are terrestrial deposits formed predominantly through two geologic processes: (1) in-washing of colluvial materials from outside the drip line of the rockshelter, and (2) roof spalling and collapse. There is also evidence of aeolian processes and in-place weathering. Rodent activity is present in these strata but does not appear to have significantly impacted the deposits. Even though Strata 12–9 contain evidence of the earliest human occupations in the East Block, human presence and activities were fairly minimal in this area of the rockshelter, leaving behind few lithic artifacts. Faunal remains and features, although present, have been severely affected by in situ weathering of the deposits and thus are not as well preserved as in the West Block.

RADIOCARBON CHRONOLOGY AND AGE AND CONTEXT OF CULTURAL COMPONENTS

Samples of charcoal, bone, and textile fragments from both the West and East Block stratigraphic profiles of BER were dated through accelerator mass spectrometry (AMS) and conventional radiocarbon (^{14}C) procedures. Charcoal was prevalent in the profile and in cultural features encountered during excavations. Because site preservation at BER is very good, and nearly all of the late Pleistocene to early-middle Holocene stratigraphic units were at least partially formed through anthropogenic means, typically only materials clearly associated with human activities were used for radiocarbon dating (e.g., wood charcoal from hearth features and textile fragments). Pretreatment and radiocarbon analysis of samples conducted at the NSF-Arizona AMS Facility (AA) and Beta Analytic Inc. (Beta) followed standard methods described by Jull et al. (1983) and Long et al. (1989) for the AMS radiocarbon dating of charcoal and bone. Samples obtained from the Lawrence Livermore National Laboratories Center for Accelerator Mass Spectrometry (CAMS) and the University of California, Irvine, Accelerator Mass Spectrometry Laboratory (UCIAMS) were first sent to Stafford Research Laboratories Inc. for pretreatment. The radiocarbon chronology for the late Pleistocene to early-middle Holocene of BER is presented below, coupled with a discussion of the cultural components of both the West and East Block excavations.

West Block

As of 2006, we had obtained 31 radiocarbon determinations spanning the late Pleistocene to early-middle Holocene in the West Block (Table 5.1). These determinations have been corrected for ^{13}C fractionation. No radiocarbon estimates have been obtained for Strata 20, 17b, or 15 because these strata contain little to no organic materials for dating.

STRATUM 19

We have radiocarbon dated 10 samples from within Stratum 19 and two from the contact of Stratum 19 with above-lying Sub-stratum 18b. Initially, during

Table 5.1. Bonneville Estates Rockshelter AMS Radiocarbon Age Estimates from the West Block

Stratum	Component	Square	Feature	Lab Number	Material	Age Estimate[a]	Age Calibrated[b]
19	n/a	N8W22	n/a	UCIAMS-22180	Bone	15,240 ± 50	18,950–18,020
19	6	N7W20	04.13b	Beta-195045	Hearth charcoal	12,390 ± 40	15,180–13,960
19	6	N7W20	04.13b	Beta-195046	Hearth charcoal	12,330 ± 40	15,060–13,840
19	6	N7W19	03.17	AA-58587	Hearth charcoal	12,180 ± 60	14,700–13,760
19	6	N8W22	n/a	Beta-209265	Bone	11,960 ± 60	14,320–13,520
19	6	N6W20	04.13b	Beta-210523	Bone	11,530 ± 40	13,650–13,210
19	6	N7W20	04.13b	Beta-200874	Hearth charcoal	10,970 ± 60	13,100–12,700
19	6	N7W19	03.17	UCIAMS-22176	Bone	10,900 ± 50	13,010–12,690
19	6	N7W19	03.17	Beta-210524	Bone	10,830 ± 40	12,940–12,660
19	6	N7W19	03.17	Beta-200875	Hearth charcoal	10,640 ± 60	12,790–12,490
19/18b contact	6	N8W20	03.16/04.13	AA-58595	Hearth charcoal	12,270 ± 60	14,860–13,820
19/18b contact	6	N8W19/20	03.16/04.13	AA-58590	Hearth charcoal	10,690 ± 70	12,810–12,550
18b	6	N6W16	05.06	Beta-207009	Hearth charcoal	11,010 ± 40	13,120–12,720
18b	6	N8W20	03.15a	AA-58594	Hearth charcoal	10,800 ± 60	12,920–12,640
18b	6	N8W20	03.15a	AA-58592	Hearth charcoal	10,760 ± 70	12,880–12,600
18b	6	N7W21	04.14	Beta-195047	Hearth charcoal	10,540 ± 40	12,800–12,220
18b	6	N8W20	03.14	AA-58593	Hearth charcoal	10,405 ± 50	12,685–11,985
18a	6	N7W17	01.01	Beta-170444	Hearth charcoal	10,130 ± 60	12,140–11,340
18a	6	N7W17	01.01	Beta-164229	Hearth charcoal	10,080 ± 50	12,020–11,280
18a	6	N7W17	01.01	Beta-170443	Hearth charcoal	10,040 ± 70	11,950–11,210
18a/17b' contact	6	N6W22	05.02	Beta-207010	Hearth charcoal	9580 ± 40	11,190–10,690
17b'	6	N8W19	03.13	AA-58589	Hearth charcoal	9440 ± 50	10,810–10,550
17b'	6	N8W19	03.13	AA-58588	Hearth charcoal	9430 ± 50	10,780–10,540
17a	5	N8W17	00.01	CAMS-72351	Hearth charcoal	7420 ± 50	8380–8140
16	5	N7W17	n/a	CAMS-72352	Dispersed charcoal	7280 ± 50	8110–7990
16	5	N7W17	n/a	Beta-164230	Textile fragment	7190 ± 50	8120–7920
14	5	N8W20	03.11	AA-58591	Garbage pit charcoal	6315 ± 40	7340–7140
14	5	N7W15	01.04	Beta-164227	Hearth charcoal	6280 ± 40	7290–7150
14	5	N7W15	01.05	Beta-164226	Hearth charcoal	6100 ± 80	7250–6750
14	5	N4W15	n/a	Beta-164224	Cordage fragment	6100 ± 50	7210–6810
14	5	n/a	n/a	Beta-29542	Bulk charcoal	6040 ± 80[c]	7130–6690

[a] Age in ^{14}C BP, reported with 1 σ (standard deviation).

[b] The CalPal online radiocarbon calibration program (CalPal 2006) was used to calibrate radiocarbon age estimates; presented with 2 σ range.

[c] This date was obtained by Schroedl and Coulam (1989) and is a conventional date on bulk charcoal presumably from Stratum 14.

the 2003 and 2004 field seasons, four age estimates were obtained from wood charcoal excavated from a stacked feature, 03.16/04.13/04.13b. These yielded the following ages: 12,390 ± 40 (Beta-195045), 12,330 ± 40 (Beta-195046), 12,270 ± 60 (AA-58595), and 10,690 ± 70 (AA-58590) ^{14}C BP. In 2003 we exposed only a small corner of this feature (then designated 03.16); this produced the age estimates of 12,270 ± 60 and 10,690 ± 70. In 2004 we continued excavating the feature and were able to delineate two stacked features, an upper hearth (03.16/04.13) and a lower burn feature (04.13b). The ages of 12,390 ± 40 and 12,330 ± 40 were obtained on unidentified wood charcoal from the lower part of this feature (04.13b). In fall 2004 we also dated a piece of *Artemisia* wood charcoal from the lower feature, 04.13b, which returned an age estimate of 10,970 ± 60 (Beta-200874).

Another hearth feature, 03.17, was excavated from Stratum 19 during the 2003 season. This feature produced an age estimate of 12,180 ± 60 (AA-58587) on unidentified wood charcoal and a subsequent age of 10,640 ± 60 (Beta-200875) on *Artemisia* wood charcoal.

The results of our 2004 excavations and radiocarbon testing led to some concern and questions regarding the nature of the pre-12,000 ^{14}C BP age estimates and raised the possibility that the unidentified samples were old pine wood. Therefore, five bone samples were submitted (three to Beta Analytic and two to Stafford Research Laboratories) to further explore the discrepancy in ages. Two samples were taken from a portion of Stratum 19 that did not contain cultural materials either horizontally or vertically, while three were taken from either within or adjacent to the hearth features described above. These samples include long bone shaft fragments of medium-bodied to small-bodied ungulates. The first two samples came from excavation square N8W22, which contained no Stratum 19 artifacts or cultural features. N8W22 is two excavation squares west of the nearest hearth feature (04.13b), and vertically the dated bone samples came from just above the contact of Strata 19 and 20. Resulting age estimates are 15,240 ± 50 (UCIAMS-22180) and 11,960 ± 60 (Beta-209265) ^{14}C BP. Of the three bone samples from near the previously dated hearth features, the first came from within hearth feature 04.13b and yielded an estimate of 11,530 ± 40 ^{14}C BP (Beta-210523). The other two bone samples were split from a single bone specimen that was found adjacent to hearth feature 03.17. This bone produced age estimates of 10,900 ± 50 (UCIAMS-22176) and 10,830 ± 40 ^{14}C BP (Beta-210524).

Thus, we have three sets of dates for Stratum 19: (1) 15,240 ^{14}C BP, (2) ~12,300–11,500 ^{14}C BP, and (3) 10,900–10,600 ^{14}C BP. These dates, along with the nature of the deposits, suggest that Stratum 19 is the result of slow, gradual accumulation of natural materials (perhaps through packrat activity). The bone dating to 15,240 ± 50 ^{14}C BP (~18,500 cal BP) predates the last highstand of Lake Bonneville and is clearly discordant. It is either an old bone that was brought into the shelter and incorporated into a nest by a packrat after about 12,300 ^{14}C BP, or it is an aberrant radiocarbon assay. The charcoal and bone producing dates ranging from 12,300 to 11,500 ^{14}C BP are probably the result of natural processes. Most likely, they represent the in situ burning of natural midden organics (both wood and bones) of Stratum 19 by humans making fires on the surface of Stratum 19. If this were the case, then the 10,900–10,600 ^{14}C BP dates more likely reflect initial human activities in this part of the shelter.

As described above, Stratum 19 is relatively thin and forms gradual to diffuse boundaries with the natural strata below and cultural strata above. The stratum's cultural materials include the two ephemeral hearth features discussed above and 34 pieces of nondiagnostic debitage found primarily in the upper 5 cm. Macrobotanical and faunal remains are present throughout Stratum 19; however, many of these are likely the result of natural accumulation processes. These data support the interpretation that this stratum is the result of gradual accumulation of mainly natural deposits between 15,200 and 12,500 cal BP. Based on these interpretations, cultural remains from Stratum 19 are assigned to Cultural Component 6, the Pre-Archaic.

Sub-stratum 18b

Six radiocarbon age estimates have been obtained from Sub-stratum 18b. Sample Beta-207009, a piece of wood charcoal from hearth feature 05.06, yielded an estimate of 11,010 ± 40 ^{14}C BP. Sample AA-58594, a piece of wood charcoal from hearth feature 03.15a, yielded an age of 10,800 ± 60 ^{14}C BP. Sample AA-58592, also wood charcoal from feature 03.15a, yielded an age of 10,760 ± 70 ^{14}C BP. These two dates from the same hearth feature (03.15a) overlap at one standard deviation (or sigma). A wood charcoal sample (Beta-195047) from hearth feature 04.14 provided an age estimate of 10,540 ± 40 ^{14}C BP, and a wood charcoal sample (AA-58593) from hearth feature 03.14 yielded an estimate of 10,405 ± 50 ^{14}C BP. These last two dates overlap at two sigma.

Considering these radiocarbon estimates, three distinct human occupations may be represented, possibly one at about 11,000 ^{14}C BP, a second at about 10,800–10,700 ^{14}C BP, and a third at about 10,500–10,400 ^{14}C BP. After calibration, however, these distinctions are blurred, with each age range overlapping with the next-youngest range. Therefore, these ages cannot be unequivocally separated. The data suggest that Sub-stratum 18b accumulated within an 1,100-year span between 13,100 and 12,000 cal BP. It likely represents a series of short-term human occupations during that time.

No diagnostic artifacts have been found in association with feature 05.06 in the vicinity of square N6W16. However, a single Haskett Stemmed point base was found during the 2006 field season in the western portion of the West Block excavation (N5W21), not far from the radiocarbon-dated hearth features (03.14, 03.15a, and 04.14). On this basis, Sub-stratum 18b is assigned to Cultural Component 6, the Pre-Archaic component of the rockshelter. Both horizontal and vertical placement of the Haskett point, coupled with radiocarbon assays, suggest that it dates to between 10,800 and 10,400 ^{14}C BP (12,900 and 12,000 cal BP). The earliest hearth feature (05.06, dating to about 11,010 ^{14}C BP [~13,000 cal BP]) cannot be attributed to a specific Pre-Archaic complex since no diagnostic points have been found in association with it.

Sub-stratum 18a

Three radiocarbon age estimates have been obtained from Sub-stratum 18a. These three estimates are from a single hearth feature that was excavated during the 2001 and 2002 field seasons. All three samples consisted of wood charcoal from feature 01.01 and resulted in age estimates of 10,130 ± 60 (Beta-170444), 10,080 ± 50 (Beta-164229), and 10,040 ± 70 (Beta-170443) ^{14}C BP. All three of these age estimates overlap at one sigma.

Calibration of these radiocarbon age estimates places the occupation between about 12,000 and 11,200 cal BP, suggesting that Sub-stratum 18a may have accumulated during a 700- to 800-year span of time. Based on radiocarbon determinations and the presence of Parman Stemmed points, Sub-stratum 18a is assigned to the Pre-Archaic, the sixth cultural component.

Sub-strata 18a and 17b' Contact

Sub-strata 18a and 17b' are typically separated by a silt band that forms the base of 17b. In the far western section of our 2005 and 2006 excavations, however, 18a and 17b' were not clearly separated by silt. During excavation of these strata in square N6W22, we encountered hearth feature 05.02. A piece of wood charcoal from this ephemeral hearth feature yielded one radiocarbon assay of 9580 ± 40 (Beta-207010) ^{14}C BP.

Sub-stratum 17b'

Two radiocarbon age determinations have been obtained for Sub-stratum 17b'. Sample AA-58589, wood charcoal from hearth feature 03.13, yielded an age estimate of 9440 ± 50 ^{14}C BP. Sample AA-58588, also wood charcoal from hearth feature 03.13, provided an estimate of 9430 ± 50 ^{14}C BP. The two age estimates overlap at one sigma.

These data, coupled with the age range given for Feature 05.02 from the contact of Sub-strata 18a and 17b', suggest that Sub-stratum 17b' ranges in age from about 11,150 to 10,500 cal BP. Diagnostic cultural materials include stemmed points, which, coupled with the radiocarbon determinations, suggest that Sub-stratum 17b' represents a late Pre-Archaic occupation. It has been assigned to Cultural Component 6.

Sub-stratum 17a

One radiocarbon determination was obtained from Sub-stratum 17a. Sample CAMS-72351, wood charcoal from hearth feature 00.01, provided an age estimate of 7420 ± 50 ^{14}C BP. Calibration of this age estimate shows that Sub-stratum 17a was deposited around 8380–8140 cal BP. These data also suggest that the upper portion of Stratum 17 represents an ephemeral occupation of humans in the West Block area of the shelter after a 2,200-year hiatus following about 10,500 cal BP. The artifact assemblage from Sub-stratum 17a is small and contains no diagnostic artifacts, but based on the radiocarbon determinations, Sub-stratum 17a probably represents the first Early Archaic occupation in BER and is therefore assigned to Cultural Component 5.

Stratum 16

Two radiocarbon age estimates have been obtained from Stratum 16. Sample CAMS-72352, a piece of dispersed wood charcoal found next to a piece of a coiled basket, yielded an age estimate of 7280 ± 50 ^{14}C BP. The other sample, Beta-164230, a fragment of the coiled basket piece, provided an age estimate of 7190 ± 50 ^{14}C BP. Calibration of these age estimates suggests that Stratum 16 accumulated over a 200-year period (8110–7920 cal BP). As discussed above, although well-preserved, this stratum is restricted to a small area of the rock-shelter. No cultural features such as hearths have been found, and cultural materials consist of a small assemblage of nondiagnostic lithic artifacts. Based mainly on the radiocarbon age estimates, we conclude that Stratum 16 represents an Early Archaic occupation of the shelter and assign it to Cultural Component 5.

Stratum 14

Five radiocarbon determinations have been obtained for Stratum 14. Sample AA-58591, wood charcoal from pit feature 03.11, is most likely associated with Stratum 14 and yielded an age of 6315 ± 40 ^{14}C BP. We encountered feature 03.11 just below a large looters' pit. The looters had stopped digging when they hit rocky and sterile Stratum 17, leaving a large portion of Feature 03.11 undisturbed. The top of it was truncated, so we cannot unequivocally say that it represents the Stratum 14 occupation; however, given its radiocarbon age, it likely does. The next sample, Beta-164227, wood charcoal from hearth feature 01.04, yielded an age estimate of 6280 ± 40 ^{14}C BP. These first two estimates overlap at one sigma. A sample of wood charcoal from an adjacent hearth feature, 01.05, provided an estimate of 6100 ± 80 (Beta-164226) ^{14}C BP. Sample Beta-164224, a cordage fragment, yielded an age estimate of 6100 ± 50 ^{14}C BP. Finally, in the late 1980s Schroedl and Coulam (1989) obtained an age estimate of 6040 ± 80 (Beta-29542) ^{14}C BP on bulk charcoal collected from their excavations at the BER. From their descriptions, we conclude this bulk charcoal sample came from Stratum 14. The last three dates overlap at one sigma.

Calibration of the radiocarbon data suggests that Stratum 14 accumulated during a 600-year span of time (7340–6690 cal BP); however, only one sample of the five from Stratum 14 can be clearly assigned to a sub-stratum. This was the cordage sample (Beta-164224) from Sub-stratum 14a. As discussed above, the 6300 ^{14}C BP (7340–7140 cal BP) and 6040 ^{14}C BP (7130–6690 cal BP) estimates cannot be directly tied to a sub-stratum of 14 since they were impacted by looting or were from an earlier phase of excavation, respectively. The two samples (Beta-164227 and 164226) from adjacent hearth features 01.04 and 01.05

TABLE 5.2. Bonneville Estates Rockshelter AMS Radiocarbon Age Estimates from the East Block

Stratum	Component	Square	Feature	Lab Number	Material	Age Estimate[a]	Age Calibrated[b]
12	6	N3W3	D4-10-C8/ C7b	Beta-182931	Hearth charcoal	10,560 ± 50	12,800–12,260
12	6	N4W3	D5-10- C8a/C8d	AA-58600	Hearth charcoal	10,380 ± 60	12,670–11,950
12	6	N3W4 N2W3	E4-12-C5/ D3-10-C10	Beta-195013	Hearth charcoal	10,380 ± 40	12,670–11,970
12	6	N3W4 N2W3	E4-12-C5/ D3-10-C10	Beta-203504	Hearth charcoal	10,340 ± 60	12,680–11,940
12	6	N2W4	E3-13-C4	Beta-206278	Hearth charcoal	10,250 ± 50	12,340–11,680
12	6	N4W3	D5-10- C8a/C8d	Beta-182935	Hearth charcoal	10,050 ± 50	11,920–11,260
12	6	N3W3	D4-12-C9b	Beta-182934	Hearth charcoal	10,030 ± 50	11,870–11,230
12	6	N3W3	D4-10-C8/ C7b	AA-58598	Hearth charcoal	9990 ± 50	11,710–11,230
10	6	N3W4 N4W4	E4-10-C3/ E5-10-C7	Beta-195042	Hearth charcoal	9580 ± 40	11,190–10,690
10	6	N3W4 N4W4	E4-10-C3/ E5-9-C5b	Beta-195044	Hearth charcoal	9570 ± 40	11,170–10,690
10	6	N5W4	E6-10-C10	Beta-161891	Hearth charcoal	9520 ± 60	11,200–10,580
10	6	N5W4	E6-10-C10	AA-58599	Hearth charcoal	9440 ± 80	11,070–10,450
10	6	N3W0	A4-9-C1	Beta-203507	Hearth charcoal	8830 ± 60	10,250–9610
9b	5	N4W4	E5-9-C5a	AA-58596	Hearth charcoal	7240 ± 50	8210–7950
9a	5	N3W2	C4-10-C8g	Beta-182933	Hearth charcoal	5350 ± 70	6330–5950
9a	5	N3W2	C4-10-C8c	Beta-182928	Hearth charcoal	5160 ± 80	6140–5680

[a] Age in ^{14}C BP, reported with 1 σ (standard deviation).
[b] The CalPal online radiocarbon calibration program (CalPal 2006) was used to calibrate radiocarbon age estimates presented with 2 σ range.

were impossible to assign to a specific substratum of 14 because ash smears and burnt materials lying in and around the hearths blurred their precise stratigraphic context. Cultural materials are quite numerous from Stratum 14, and the presence of diagnostic large side-notched points suggests an Early Archaic affiliation. Stratum 14, therefore, is assigned to Cultural Component 5.

East Block

We currently have 16 radiocarbon determinations to help date the late Pleistocene/early-middle Holocene deposits in the East Block. These are presented in Table 5.2. These determinations have been corrected for ^{13}C fractionation. No radiocarbon estimates have been obtained for Strata 18–13 because these strata are noncultural and have yielded no well-preserved organic materials for dating. Presented below are chronology and cultural affiliations of Strata 12, 10, and 9.

Stratum 12

Eight radiocarbon estimates have been obtained for Stratum 12. Sample Beta-195013, wood charcoal from hearth feature E4-12-C5/D3-10-C10, yielded an age of 10,380 ± 40 ^{14}C BP. Sample Beta-203504, a second piece of wood charcoal from hearth E4-12-C5/D3-10-C10, provided an age estimate of 10,340 ± 60 ^{14}C BP. Both ages from this hearth overlap at one sigma. A second hearth feature, E3-13-C4, produced wood charcoal dated to 10,250 ± 50 (Beta-206278) ^{14}C BP. All three dates from these two hearths overlap at two sigma. An age estimate of 10,560 ± 50 (Beta-182931) ^{14}C BP was obtained from wood charcoal excavated from an ephemeral, unprepared hearth feature, D4-10-C8/C7b. A second sample of wood charcoal from this hearth feature produced an age estimate of 9990 ± 50 (AA-58598) ^{14}C BP. Two wood charcoal samples from another ephemeral hearth feature, D5-10-C8a/C8d, produced two age estimates of 10,380 ± 60 (AA-58600) and 10,050 ± 50 (Beta-182935) ^{14}C BP. Just below these hearths, sample Beta-182934, wood charcoal from hearth feature D4-12-C9b, yielded an age estimate of 10,030 ± 50 ^{14}C BP. Most likely, the two older estimates from Features D4-10-C8/C7b (10,560 ± 50) and D5-10-8a/8d (10,380 ± 60) are aberrant ages, given their vertical positions above hearths E4-12-C5/D3-10-C10 and E3-13-C4.

These results suggest the possibility of two human occupations represented in Stratum 12. The first may have occurred sometime between about 10,380 and 10,250 ^{14}C BP, and the second may have occurred at about 10,000 ^{14}C BP. After calibration, however, these distinctions are blurred, with the age ranges overlapping with the next-youngest range. These data suggest that Stratum 12 accumulated over a 1,600-year period from 12,800 to 11,200 cal BP and likely represent a series of short-term human occupations during this time frame. Stratum 12 contains a very small lithic assemblage, including one Parman Stemmed point fragment. The artifact assemblage and the radiocarbon ages presented above suggest a Pre-Archaic cultural affiliation. Therefore, Stratum 12 is assigned to Cultural Component 6.

Stratum 10

Five radiocarbon age estimates have been obtained for Stratum 10. Sample Beta-195042, wood charcoal from hearth feature E4-10-C3/E5-9-C5b, yielded an age estimate of 9580 ± 40 ^{14}C BP. Another wood charcoal sample (Beta-195044) from hearth feature E4-10-C3/E5-9-C5b produced an age of 9570 ± 40 ^{14}C BP. These two estimates overlap at one sigma. Wood charcoal from hearth feature E6-10-C10 yielded two age estimates of 9520 ± 60 ^{14}C BP (Beta-161891) and 9440 ± 80 ^{14}C BP (AA-58599). Both estimates overlap at one sigma, and all four ages presented above overlap at two sigma. A final radiocarbon age estimate for Stratum 10 was obtained on wood charcoal from hearth feature A4-9-C1, 8830 ± 60 (Beta-203507) ^{14}C BP. This estimate does not overlap with the other dates from Stratum 10.

These results suggest that Stratum 10 may contain two separate human occupations. The earlier occupation may have taken place between about 9580 and 9440 ^{14}C BP (11,190–10,450 cal BP), and the later occupation may date to approximately 8800 ^{14}C BP (10,250–9610 cal BP). The small lithic assemblage from this stratum includes a Parman-like stemmed point. The artifact assemblage and radiocarbon ages suggest a late Pre-Archaic cultural affiliation, and therefore Stratum 10 has been assigned to Cultural Component 6.

Stratum 9

A single radiocarbon age estimate was obtained from Sub-stratum 9b. The sample consisted of wood charcoal from hearth feature E5-9-C5a and yielded an age of 7240 ± 50 (AA-58596) ^{14}C BP. Two radiocarbon ages were obtained from a stacked set of hearth features in Sub-stratum 9a. Sample Beta-182933, wood charcoal from lower hearth feature C4-10-C8g, provided an age of 5350 ± 70 ^{14}C BP. The second sample, wood charcoal from upper hearth feature C4-10-C8c, yielded an age estimate of 5160 ± 80 (Beta-182928) ^{14}C BP. These two ages overlap at two sigma.

Calibration of these radiocarbon estimates suggests that Stratum 9 accumulated over a 2,300-year period, with Sub-stratum 9b dating to 8210–7950 cal BP and Sub-stratum 9a dating to 6330–5680 cal BP. Unfortunately, the artifact assemblage from Sub-stratum 9b contains no diagnostic lithic artifacts; however, based solely on the radiocarbon date presented above, an Early Archaic cultural affiliation is suggested. Sub-stratum 9b contains a rich lithic assemblage, including large side-notched projectile points clearly representing an Early Archaic cultural occupation. Therefore, Stratum 9 is assigned to Cultural Component 5.

DISCUSSION AND CONCLUSIONS

BER became available for human occupation shortly after about 15,000 ^{14}C BP (18,800 cal BP) as Lake Bonneville receded from its high shoreline. The wash located directly in front of the rockshelter may have provided an intermittent source of water during relatively cool, wet episodes in the past, and the springs at Blue Lake would have provided a relatively close source of permanent water. Due to its large size, south-facing entrance, and close proximity to water throughout much of the late Pleistocene and Holocene, BER would have made an excellent occupation site for humans (as well as other animals).

Both the West and East Block excavations at BER contain a series of late Pleistocene to middle Holocene deposits in sealed primary contexts. In the West Block these deposits include 11 stratigraphic units. Because preservation is very good in the West Block, seven of these stratigraphic units contain cultural materials—not just lithic artifacts but also perishable artifacts, well-preserved faunal remains, floral remains, and features. In the East Block, the late Pleistocene to middle Holocene deposits include 10 stratigraphic units. Of these, three contain cultural materials. Although lithic artifacts and features occur, faunal and floral remains are not well preserved. Clearly, geogenic processes have greatly affected the preservation level in the East Block; consequently, separation of cultural occupations is largely based on concentrations of lithic and faunal materials, and associated radiocarbon dates from ephemeral hearth features.

Evidence for the earliest unequivocal human occupations of the rockshelter comes from Stratum 18b in the West Block excavation and dates to between about 13,100 and 12,000 cal BP. The earliest of these dates, about 13,120–12,720 cal BP (11,010 ^{14}C BP), falls into the established time of Clovis (Fiedel 1999; Haynes 2002); however, to date no diagnostic projectile points have been found associated with the feature that produced this date. Interestingly, a Clovis point base was found during the summer 2004 field season in Stratum 1, associated with the historic occupation of the site. This point was obviously removed from its original context, which was either in the rockshelter or somewhere nearby. It must be stated that the equivocal nature of the dating of Stratum 19 does not exclude the possibility of an earlier occupation at BER. Further excavations should help to clarify this issue.

Pre-Archaic hunter-gatherers wielding Great Basin stemmed points and an associated toolkit appear to have repeatedly used the rockshelter on a frequent basis, utilizing both the West and East Block areas from about 12,900–12,000 cal BP (10,800–10,300 ^{14}C BP), 12,000–11,200 cal BP (10,200–9990 ^{14}C BP), and again 11,000–10,500 cal BP (9580–9430 ^{14}C BP).

After about 10,600 cal BP, there was a hiatus in human occupation of the West Block until after 8400 cal BP; however, a single hearth in the East Block (but with no associated artifacts) may suggest that Pre-Archaic people used the rockshelter again about 9700 cal BP. This is not surprising because soon after 9500 cal BP a shift to drier and possibly warmer climatic conditions in the Bonneville Basin had begun (Rhode et al. 2005; Schmitt and Madsen 2005).

Immediately following a 1,400-year hiatus of human activity, Archaic people began to make use of the rockshelter, discarding their artifacts in both the West and East blocks. Two very small lithic assemblages, each associated with a single hearth, mark human occupations between about 8300 and 8000 cal BP. Paleoecological records for the region suggest that between about 8100 and 6800 cal BP the climatic transition to modern conditions had occurred, with pinyon pine and Utah juniper migrating into the Bonneville Basin (Rhode et al. 2005). After a period of relatively dry conditions, during which Great Salt Lake desiccated, the climate became slightly moister about 6800 cal BP (Schmitt and Madsen 2005). These changes in vegetation and climate likely correlated with the rich Early Archaic occupations of Strata 16 and 14 in the West Block of the site.

BER contains a wealth of evidence of human occupations during the transition from the terminal Pleistocene to early Holocene. It presents some of the first detailed evidence of Pre-Archaic hunter-gatherer adaptations in the Great Basin and offers a precise record of human technological and subsistence activities as climate and environment changed during the early Holocene. Chapters in this volume by Goebel, Hockett, and Rhode and Louderback provide further details of some of this record.

NOTE

1. During excavation, Sub-stratum 12a was originally designated Stratum 11; however, due to its ephemeral nature, it has been incorporated into Stratum 12 as Sub-stratum 12a.

ACKNOWLEDGMENTS

I would like to thank several people and organizations who have made this work possible. The Jonathan O. Davis Scholarship Program at the Desert Research Institute (DRI), the National Science Foundation (grants BCS-0313813, BCS-0514504), the Bureau of Land Management, and Sundance Archaeological Research Fund all provided funding for this research. I would also like to thank David Rhode for identifying charcoal samples, Lisbeth Louderback and John Varnon for sieving column sample sediment, the Department of Atmospheric Science at DRI for laser ablation analysis, and Ted Goebel and Bryan Hockett for helpful comments on earlier versions of this chapter.

6

Late Pleistocene/Early Holocene Geomorphology and Human Occupation of the Old River Bed Delta, Western Utah

Dave N. Schmitt, David B. Madsen, Charles G. Oviatt, and Rachel Quist

During the regressive phase of the last major highstand of Lake Bonneville, a river connected the two major sub-basins of the Bonneville Basin. Beginning sometime around 12,000 ^{14}C BP (13,900 cal BP) (Godsey et al. 2005; Oviatt et al. 1992), the river ran north, draining the southern Sevier Basin and emptying into the Great Salt Lake along its southwestern margin, in what is now the Great Salt Lake Desert of western Utah. The end of this ancient river, known simply as the Old River Bed (ORB), formed a vast, digitated delta (Figure 6.1) across an area that now includes the northern U.S. Army Dugway Proving Ground (DPG) and the southern reaches of the U.S. Air Force Utah Test and Training Range (UTTR). Sometime after about 8700 ^{14}C BP (9600 cal BP), water ceased to flow in the ORB (Oviatt et al. 2003), and environmental conditions along the channel began to approach those found at present. During the ~3000 years of its existence, however, the water in the ORB fed a large deltaic marsh/wetland system covering about 700 km^2 and supported a riverine environment along its length. This ~3000-year interval corresponds almost exactly to the earliest phase of human occupation in the Bonneville Basin (e.g., Grayson 1993; Schmitt and Madsen 2005 and references therein). Foragers have been drawn to these rich marsh/wetland ecosystems throughout the human history of the Great Basin (e.g., Fawcett and Simms 1993; Janetski and Madsen 1990; Kelly 2001; Madsen 2002; Raven and Elston 1988), but sites of the Paleoarchaic period are particularly associated with Great Basin wetlands (Beck and Jones 1997; Jones and Beck 1999; Willig et al. 1988).

Over the past five years we have conducted archaeological inventories of approximately 60 km of the more than 250 linear km of exposed deltaic channels within Dugway Proving Ground and have located more than 70 Paleoarchaic sites directly associated with deltaic channel remnants or adjacent wetlands (Madsen et al. 2000; Madsen et al. 2005; Schmitt et al. 2002a; Schmitt et al. 2003). This density, together with archaeological investigations at the

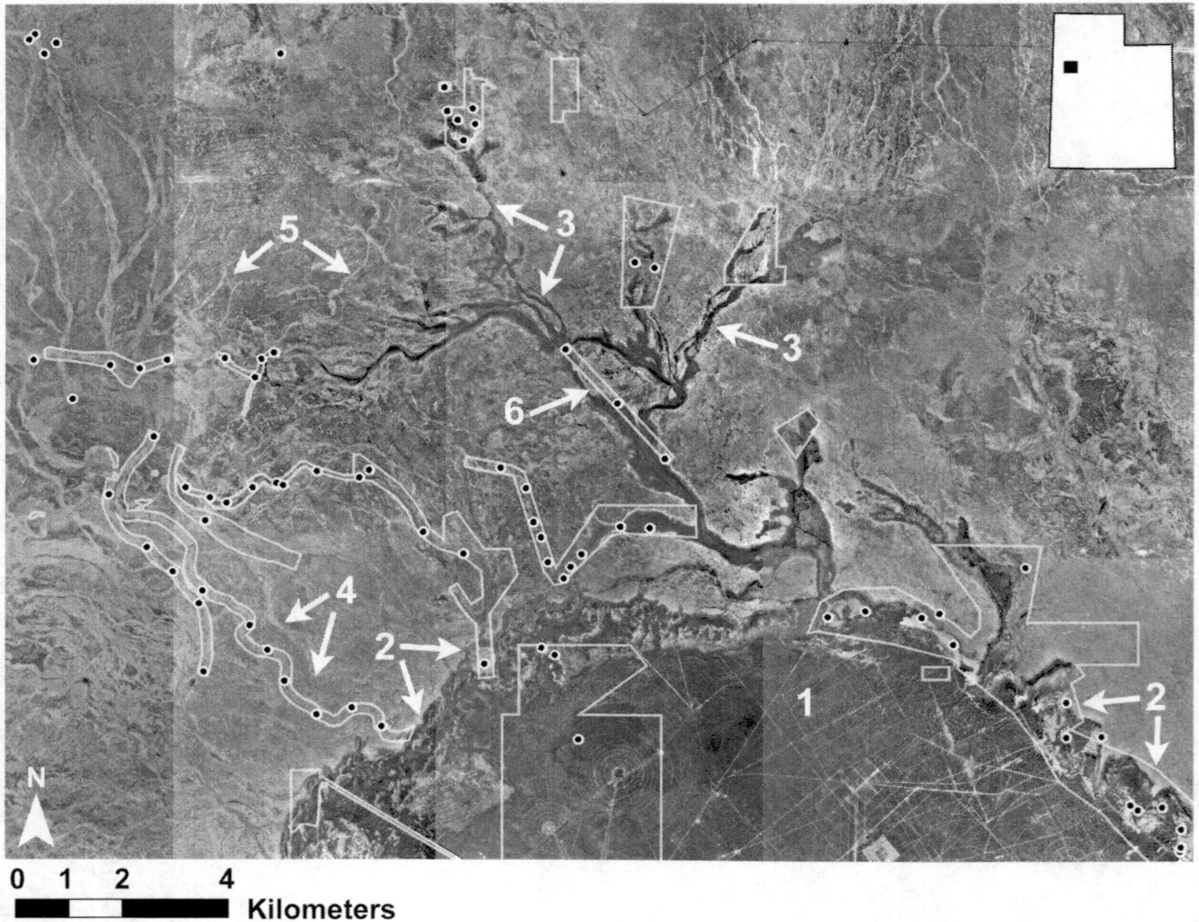

FIGURE 6.1. Satellite image of the Old River Bed delta showing survey areas (white outlines) and identified Paleoarchaic sites (dots) in the northern Dugway Proving Ground. 1: fine-grained underflow fan deposits of the flat plain; 2: sand silt dunes; 3: gravel channels; 4: intermediate channels; 5: sand channels; 6: the Lake Öferneet playa basin.

extreme northern toe of the delta (e.g., Arkush and Pitblado 2000; Carter and Young 2001), suggests that over 500 Paleoarchaic sites may be present within the delta area. A paleoenvironmental project running concurrently with this archaeological evaluation is geared toward understanding and dating changes in the hydrological system that governed flow in the ORB (e.g., Madsen et al. 2000; Oviatt 1999; Oviatt et al. 2003). While these studies are ongoing, some aspects relevant to human occupation of the delta are already evident.

DELTA GEOMORPHOLOGY AND CHRONOLOGY

The delta is composed, at least in part, of silts and clays in an underflow fan deposited on the bottom of Lake Bonneville after the lake regressed below the ORB threshold and the ORB river began to dump fine sediment into the lake. As the lake regressed across this underflow fan, it prograded to the north, wave action cut a relatively flat surface across its top, and the river channel was incised increasingly deeper into the fan as the lake level dropped. During this initial regressive phase, the water in the river was probably derived from overflow from the Sevier Basin (and, by extension, from the Sevier and Beaver rivers), and currents in the river were sufficient to support a high sediment load, including coarse sands and gravels in the deeper and swifter currents. Sometime prior to 10,500 ^{14}C BP (12,500 cal BP), stream flow in the river was substantially reduced, and as the lake retreated to the north, the river ceased to empty into it. By the time the gravel channels had stopped forming, the lake had dropped to an elevation of 1,300 m. After 10,500 ^{14}C BP (12,500 cal BP) stream flow in the lower ORB increased. Whether this increased flow was due to overflow from the Sevier Basin or from groundwater sources along the ORB channel is not yet clear, but it was substantial enough to carry coarse sands in channels 4–5 m wide and 2–3 m deep.

The primary geomorphic features of the ORB delta consist of the delta itself, mudflats to the north, and a series of dunes that mark the transition between the two. An abrupt boundary in the delta separates the groundwater-discharge mudflats from the well-drained, fine-grained sheetflow and aeolian deposits. We informally use the terms *gravel channel*, *sand channel*, and *intermediate channel* for fluvial landforms and deposits on the mudflats. Gravel channels are deposits of coarse sand and gravel that in plan view are straight to curved and digitate, and have abrupt bulbous ends. In transverse cross section, gravel channels are topographically inverted, with the crests of the gravel deposits standing 1–4 m higher than the surrounding mudflats (Figure 6.2). They are identified as fluvial in origin by their plan view form, the composition of the gravel, and their longitudinal profiles (Oviatt et al. 2003). The coarse sand and gravel of gravel channels is trough cross-bedded and overlies lacustrine mud and marl of Lake Bonneville.

Reduced flow in the ORB produced what we have identified as sand channels, which are found in the same general area of the mudflats as the gravel

FIGURE 6.2. View to the north of a topographically inverted gravel channel at Site 42To1668. Wildcat Mountain is in the center background. The top of the inverted channel in this location is approximately 3.5 m above the surrounding mudflats.

channels. They are less topographically inverted and are truncated by the mudflat surface. Where they have been protected, they may stand as much as 1.2 m above the surrounding mudflats, and in other areas they typically stand about 0.5 m above the mudflats. Sand channels are not easy to identify on the ground due to mudflat deflation, but from the air the preserved sand channel roots exhibit well-developed meander-scroll patterns (Figure 6.1). Sediments in sand channels consist of fine to coarse cross-bedded sand, and locally include mud containing organic mats, mollusks, and bones of Utah chub (*Gila atraria*), a fish adapted to warm and slightly saline waters.

Intermediate channel forms, which are straighter and smaller in width than sand channels and locally contain some gravel, can be traced north through DPG and the UTTR to altitudes as low as 1,285 m in the west-central Great Salt Lake Desert. Most of these intermediate channel forms originate on the far western edge of the uneroded delta margin. In some satellite images these intermediate channels can be traced into the uneroded delta where they, and any associated Paleoarchaic sites, remain buried.

Gravel channel characteristics indicate that the ORB river emptied into a shallow Lake Bonneville. They can be traced to an altitude of 1,299 m, where they end abruptly. By about 11,300 ^{14}C BP (13,200 cal BP) the lake had retreated to elevations well below this point (Broughton et al. 2000; Madsen et al. 2001), and the gravel channels were likely partially exhumed at this time. Most sand channels end at an altitude between 1,301 and 1,303 m. Sand channels are younger than gravel channels; the scoured bottoms of sand channels are topographically lower than those of gravel channels, and sand channel deposits are inset into gravel channel deposits. Sand channels were produced by perennial rivers that were active during the period from at least 11,000 to 8700 ^{14}C

BP (12,900 to 9600 cal BP). Gravel channel deposits are not directly dated, but their ages are bracketed by other radiocarbon ages. Lake Bonneville had dropped to approximately the level of the ORB threshold (~1,390 m) by about 12,500 ^{14}C BP (14,600 cal BP) (Oviatt 1988), and possibly not until after 12,000 ^{14}C BP (13,900 cal BP) (Godsey et al. 2005). The gravel channels were produced by river discharge from the Sevier Basin across the threshold after this time. The age of the Gilbert shoreline is approximately 10,300 ^{14}C BP (12,100 cal BP) (Currey 1982; Oviatt et al. 2005), and the oldest sand channels and overbank deposits in the DPG area (~11,000 ^{14}C BP [~12,900 cal BP]) predate it. Therefore, the gravel channels predate the Gilbert shoreline and were formed after 12,500 and prior to 11,000 ^{14}C BP (14,600 and 12,900 cal BP). During the later gravel channel phase, a period of higher river flow produced wider and deeper gravel channels that crosscut earlier channels in a number of places, forming shallow ponds. Archaeological sites around at least two of these ponds, one of which we informally are calling Lake Öferneet (see Figure 6.1), may be associated with the very end of gravel channel flow and may thus predate 11,000 ^{14}C BP (12,900 cal BP).

The age of the sand channels is relatively well controlled by seven radiocarbon dates obtained on materials in the channels and eight dates on marsh/wetland deposits associated with the channels (Oviatt et al. 2003). These 15 dates place the formation of the sand channels between about 9800 and 8700 ^{14}C BP (11,200 and 9600 cal BP). We are less certain of the age of the intermediate channel forms, as they were probably produced by moderate water flow at the end of the gravel channel phase and again during the transgressions and regressions of the Gilbert level. It is likely that most of the evident intermediate channels date to the period between about 10,300 and 9500 ^{14}C BP (12,100 and 10,800 cal BP) since wave action during the Gilbert transgression would likely have removed the relatively fine-grained materials of these intermediate channel forms. Eight additional dates from marsh/wetlands not directly associated with either gravel or sand channels fall between about 11,400 and 10,000 years ^{14}C BP (13,300 and 11,500 cal BP). Three of these are associated with material representing shallow/stagnant water in the fill of an abandoned intermediate channel in the flat underflow fan upstream from the exhumed channels. The earliest intermediate channel, dated to ~10,600 ^{14}C BP (12,700 BP), exits the uneroded delta margin to the north and appears to be associated with increased stream flow during the rise of the lake to the Gilbert level. Marsh deposits collected from an excavation in the western part of the uneroded delta (GP-5A; Oviatt et al. 2003) date to 9,520 ± 40 ^{14}C BP (11,000–10,700 cal BP) and suggest the intermediate channels to the west are associated with the regression of the lake from the Gilbert level.

After ~8700 ^{14}C BP (9,600 cal BP), deflation of the mudflats accelerated. This deflation continued to exhume the gravel channels and partially exhumed the sand channels, and may be related to the proximity of the water table to the surface. The mudflats are a landscape that is actively undergoing denudation through several processes. The mud is moistened by groundwater discharge, which probably occurs continuously, but the surface dries out during the hot summer when evaporation exceeds the groundwater discharge rate. This wetting and drying causes the mud to swell and shrink and to pelletize, thus creating loose particles that can easily be entrained by the wind and removed. The larger particles accumulate in dunes at the mudflat margin, and the smaller particles are blown much farther. Salt precipitation in the mud may contribute to the breakup of the mud. Shallow films of water also accumulate on the mudflat surface after heavy rains and are agitated by the wind, thus loosening particles that are easily deflated when the surface dries.

The dunes along the margin between the delta plain and the mudflats (see Figure 6.1) thus likely formed sometime during the early-middle Holocene, and have probably been in place since then. Exposures within the dunes suggest they have been stabilized a number of times during this period. While these periods of stabilization are not yet dated, they may correspond to periods of increased precipitation ~6,000, ~3,400, and ~1,000 years ago (Madsen 2000).

SITE DISTRIBUTIONS

Generally speaking, the delta area is composed of four major geomorphic zones: the flat delta plain, channels, dunes, and mudflats. The distribution of archaeological sites is quite distinctive among these four zones, and in most cases site types and locations parallel the results of our initial surveys. With the exception of a very small fringe area related to the location of deflated sand channels, there are no sites in the mudflats. Scatters of artifacts do occur on mudflats in some areas, but in most cases these materials have clearly eroded out of nearby sites associated with channel margins and/or more recent dunes. However, an informal survey conducted during one of our projects (Schmitt et al. 2003) suggests that some mudflat artifact scatters may represent a series of excursions by Paleoarchaic foragers who ventured far out into the shallow wetlands, most likely to procure resources, and either lost or discarded stone tools.

Very few sites are evident on the flat delta plain, although the possibility of intact sites remains higher than on the mudflats. We have found only three sites in this zone during survey of the area (Madsen et al. 2000; Schmitt et al. 2002a), and large surveys north of Granite Peak (Page et al. 2003) identified three sites that contain Pinto points associated with ORB channels evident in enhanced satellite images. This does not mean that early prehistoric sites are rare in the flat delta plain. Indeed, as discussed below, the uneroded portions of the plain may contain some of the most pristine Paleoarchaic sites on all the lands under DPG control.

Archaeological sites in the ORB delta are largely concentrated in the two remaining zones, although the sites in the two areas have very different chronological signals and were used in very different ways by the foragers who occupied them. Exactly when the dunes that mark the transition between the delta plain and the mudflats began to develop is unclear. Most likely these began to build once lake waters regressed from the Great Salt Lake Desert to the north and west of the ORB delta. Madsen (2000) and colleagues (2001) suggest the early Holocene (~10,000–8000 ^{14}C BP; ~11,500–8900 cal BP) in the Bonneville Basin was substantially cooler and wetter than at present (see also Schmitt et al. 2002b). This period largely corresponds to the time water was flowing through the later ORB sand channels and intermediate channel forms. While we are not yet sure if these waters reached Great Salt Lake throughout the post-Gilbert interval, at the very least they likely fed a shallow lake and/or a series of ponds covering most of what is now the Great Salt Lake Desert. Lake Öferneet (Schmitt et al. 2003) appears to manifest a good example of such features. Thus, it is likely that dune formations began to build sometime after about 8000 ^{14}C BP (8900 cal BP) but probably built relatively quickly as the mudflats deflated (Madsen and Schmitt 2005). This means that dunes, and the biotic communities they support, have been available to human foragers on DPG from at least the early-middle Holocene to the present. What we do not know is how the location of these dunes may have changed, and how much, or how little, the position of the interface between the mudflats and the flat delta plain may have migrated.

With few exceptions, the chronologically diagnostic cultural materials seen in the dunes so far are limited to those associated with the Late Archaic through late prehistoric periods, and there is little direct evidence of Early Archaic occupation in the dunes themselves. These materials may simply be buried in early dune horizons, but there is some limited evidence that the dunes are migrating, albeit slowly. As this happened, the earliest Archaic materials appear to have been deposited on deflation zones along the ORB delta margin. All of the sites that contain such Early Archaic diagnostics (e.g., Pinto points) are in these deflation zones next to dunes that appear to be migrating to the south and east. The clearest examples of such sites are 42To1689 (Schmitt et al. 2003), 42To1358, 42To1359 (Schmitt et al. 2002a), 42To1000, and 42To1182 (Madsen et al. 2000). At each of these sites, the sand channels that are well removed from the dunes tend to be more heavily deflated than those at the dune face, suggesting that the dunes are gradually exposing the channels. Cultural materials are clearly being deposited on the deflated surfaces as the dunes move. This may help explain the absence of Early Archaic diagnostic artifacts in the dunes themselves, and their presence on sites associated with sand channels containing even earlier Paleoarchaic artifacts.

The most complex sites in the ORB delta are associated with the sand and gravel channels most evident along the eastern and northern margins of the flat delta plain (Madsen et al. 2000). More precisely, they are associated with the sand channels and intermediate channel forms; given the high-energy environment that created them, sites on or immediately adjacent to the gravel channels must necessarily postdate the channels. In most cases the topographically inverted gravel channels formed a barrier along which the stream that created the sand channels flowed, and they were apparently used by human foragers as high ground overlooking the sand channels and their associated resources. As noted above, most of the sand channels were created sometime between ~9800–8700 BP (~11,200–9600 cal BP), although wetlands in the low areas of the entrenched channels may have persisted until after ~8300 ^{14}C BP (~9200 cal BP). Most of the archaeological materials associated with the sand channels likely date to this period, but since some marsh/wetland habitats in the ORB delta were present prior to this time, there may be older materials as well. These sand channel rivers appear to have meandered extensively through the delta, creating a vast wetland system with relatively few areas suitable for habitation or activities not directly associated with foraging in the marsh itself. The only dry areas were probably the gravel channels and natural levees formed along the sand channel margins, and it is these areas where human activities appear to have been concentrated. This occupational pattern is clearest at 42To1358 (see Appendix; Schmitt et al. 2002a), where there is a rather dense linear distribution of flakes and tools along the northern gravel channel and along the southern banks of the last sand channels created by the ORB river.

A number of other sites are apparently associated with sand channels, but a definitive relationship to the natural levees along the channel margins is less clear. Most are very large sites that have sand channels either running through them or immediately adjacent to them, but there is no clear association of lithic debris with the channels themselves. Rather, tools and detritus are widely scattered over the surface of these sites, with no apparent concentrations in any one location. There are a couple of possible explanations for these distributional patterns. These sites may be products of the erosional forces that are exhuming the sand channels, with wind and wave action in ephemeral ponds scattering debris across a wide area. However, it is unclear why these sites should be disturbed in this way, while the linear sites appear to be merely deflated, with little lateral movement of artifacts. More likely, the sites are the result of actual foraging activities in the marsh/wetlands along the sand channels. The relatively low density of debitage and its distribution over very broad areas suggest this may be the case.

All but one of the sites we identified along the western intermediate channels are of the linear variety and appear to have been associated originally with the partially exhumed channels or natural levees. The exception is 42To2559, a site consisting of multiple artifact loci on the multiple margins of a migrating channel (Madsen et al. 2005). Lithic material extends well out onto adjacent mudflats, and the site appears to be a composite form resulting from occupation of the channel or channel margins and foraging in the adjoining wetland.

All of the sites we identified along the later channel forms have been exhumed by erosion of the flat delta plan and deflation to the elevation of the mudflats adjacent to the delta; they otherwise would not have been evident from the surface. Given the density (as well as the size) of sites along the exposed channels, it is highly probably that others occur along channels that have not yet been exhumed. We think the surface of the flat delta plain is a product of sheetwash and aeolian deposition, and that these Holocene materials have buried a number of ORB distributaries that flowed through the delta. As noted above, these buried channels are visible in some satellite imagery, and a few sites containing Pinto points have been discovered in direct association. At paleoecological location GP-5, near one of these buried channels in an area well away from the eroded delta margin, the marsh deposits are buried below about a meter of redeposited aeolian and colluvial materials (see Oviatt et al. 2003). It seems highly probable that other marsh deposits along the sand channels running through the delta are buried to a similar depth, and that sites associated with the channels and wetlands can be found there as well. Clearly, if any of these buried contexts contain cultural features, datable materials and/or subsistence residues, the archaeological significance of such associations for Great Basin prehistory would be profound (e.g., Beck and Jones 1997; Elston and Zeanah 2002).

Sites in the ORB delta area take one of two principal forms. The majority are linear features along the margins of topographically inverted gravel channels or adjacent to sand channels. These linear sites have a high ratio of finished tools to flaking debris (e.g., Oviatt et al. 2003: Figure 6.8), and most of the tools consist of stemmed bifaces and simple, edge-modified flakes. Debitage and some tools feather out onto adjacent mudflats that were probably wetlands when the sites were occupied. The other, less common, site type consists of broad, diffuse scatters of lithic debitage and tools that largely occur on sand channel and mudflat surfaces. These are often hundreds of meters in diameter and, in places, merge into a background scatter of cultural materials covering much of the mudflats within the delta area. Tool-to-debitage ratios in these sites are reduced (see, for example, 42To1671 and 42To1689 in the appendix), and they appear to represent resource procurement activities within the wetlands themselves.

SITE CONTENTS

Archaeological sites along the ORB channels are characterized by the presence of Paleoarchaic and Early Archaic diagnostic artifacts. The Paleoarchaic materials include numerous Great Basin Stemmed points/point variants (Figure 6.3) dating to ~10,000 to 7500 BP. (e.g., Beck and Jones 1997), and a recent survey discovered a possible fluted point fragment (Figure 6.4) associated with stemmed points on an intermediate channel. Pinto points are also common, and since they occur in the eastern Great Basin prior to 8000 BP (8900 cal BP), they may have been deposited at the same (or nearly the same) time the stemmed points were. As noted above, it is also possible that these Pinto points and other Early Archaic point types (such as Elko-eared and Humboldt points) were deposited in dunes overlying the sand channels and associated marsh deposits, and were deflated down to the same surface where the Paleoarchaic materials were found. We have no evidence that the sand channel marshes

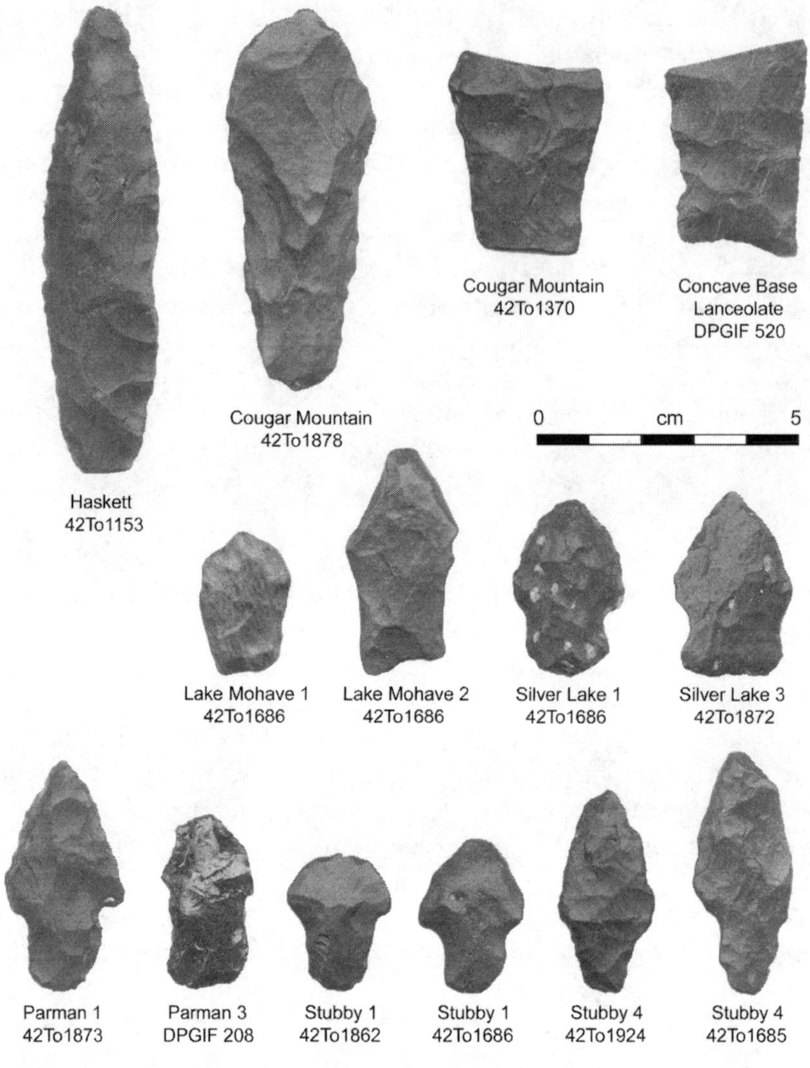

FIGURE 6.3. Selected Great Basin Stemmed bifaces collected from Old River Bed delta sites. Note the generally irregular forms, extensive reworking, and weathering of all of these tools.

persisted much beyond ~8700 BP (9600 cal BP), so the Early Archaic use of the ORB delta area likely differed from the way Paleoarchaic foragers used the area (see discussions in Elston and Zeanah 2002; Rhode et al. 2006). Whether or not this use included biotic communities common to dunes or was the result of foraging for other resources is not yet clear.

Additional artifact classes (see Appendix) include gravers, scrapers, chisels, crescents, an array of simple flake tools, and two Cody knives. Although we have no direct dates for any of these tool forms, the sample of points is sufficiently large at 23 sites to determine a relative chronology based on seriation (Figure 6.5). Many of the later forms appear to have been scavenged from earlier deposits and repeatedly resharpened to such an extent that the amount of cutting edge above the hafting element is minimal. This is particularly true for a unique form of Great Basin Stemmed biface we term *Dugway Stubbies* (Figure

FIGURE 6.4. Possible fluted projectile point fragment from Site 42To2553.

6.3). Toolstone sources are limited in comparison to other Great Basin Paleoarchaic sites, and mobility patterns also appear to have been much different. Specifically, toolstone consists primarily of local basalts, including identified sources immediately west of the delta in the Deep Creek Range and south of the Cedar Mountains in eastern DPG, and Topaz obsidian from a source less than 50 km southwest of the delta (Hughes 2002). Chert debitage is extremely rare in ORB sites, although some finished chert tools, notably crescents (Figure 6.6), have been identified. Browns Bench obsidian from the junction of the Utah/Nevada/Idaho borders is also present but rare, and single pieces of obsidian have been sourced to the Wild Horse Canyon (southwestern Utah) and Malad (southeastern Idaho) areas (Hughes 2002).

One of the more common artifact types in the delta consists of large, thinned basalt bifaces from which large flakes were detached. These large flakes, by far the most common artifacts in ORB sites, often show edge modification apparently for (or from) their use in processing marsh resources. Although fragments and manufacture detritus are common, complete thinned bifaces are extremely rare, as these portable bifacial cores were continually reduced to the point where they eventually broke and were abandoned, or were modified into another tool form. Figure 6.7 presents one of the few complete basalt bifaces recorded. It is characterized by the removal of large flakes, most ranging from 2 to 4 cm in width, but larger flakes, including one overshot flake covering nearly half of one face, also occur. Smaller flakes have been removed to shape the biface, but there is no evidence of use-wear. Altogether, this specimen is reminiscent of large Clovis bifaces (M. Collins, personal communication 2005), although basalt toolstone is rare in Clovis assemblages.

Some additional Clovis-like lithic artifacts in the delta include a few cores and prismatic flakes reminiscent of core-and-blade technology. A few basalt and obsidian blades found in ORB sites further reflect this technology. These specimens are elongated, struck from flat platforms, prismatic in cross-section,

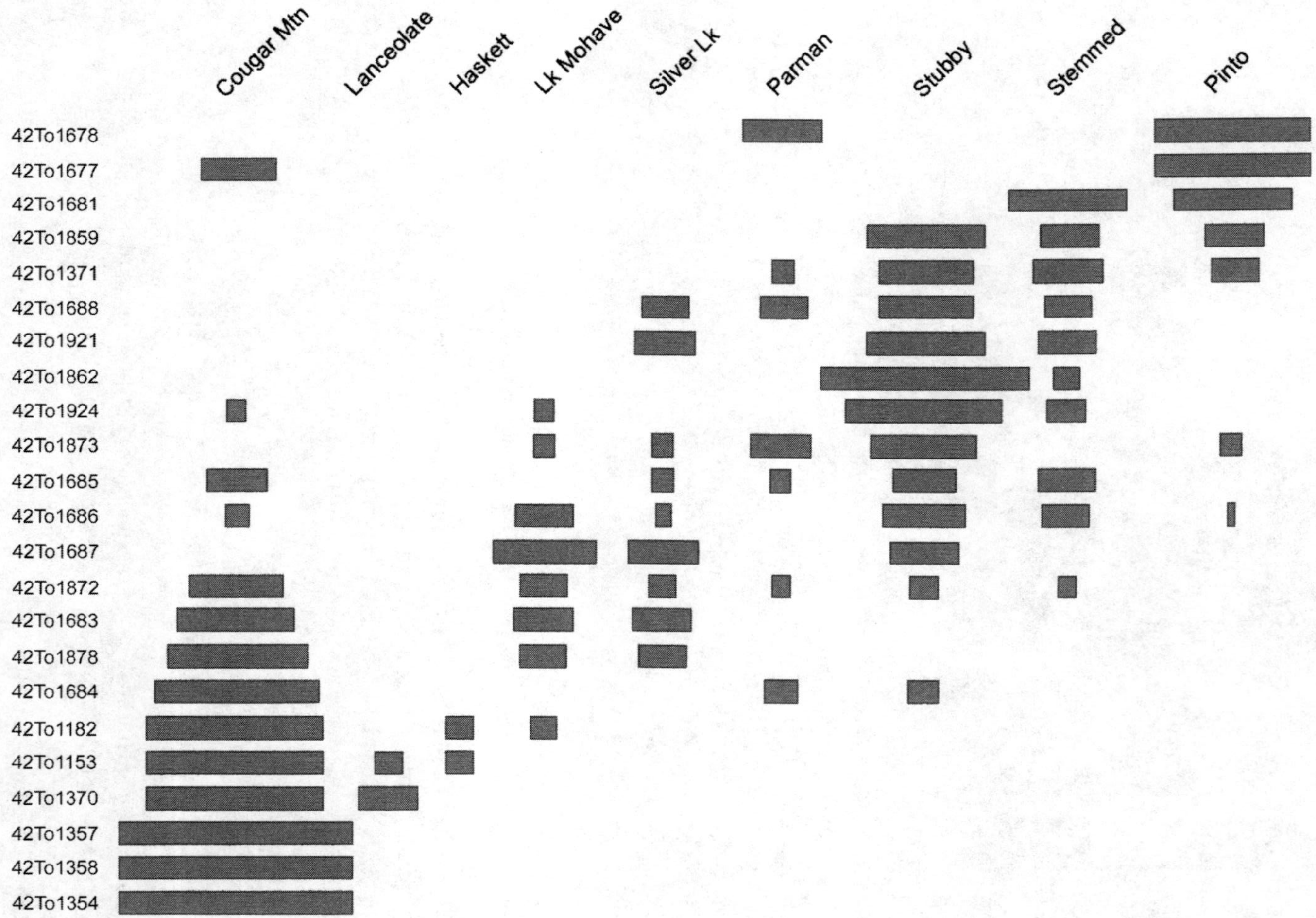

FIGURE 6.5. Seriation of Great Basin Stemmed bifaces from 23 Old River Bed delta sites (after Jones et al. 2002: Figure 31). Pinto points are generally thought to date after ~9000 ^{14}C BP. The ages of the other types are unknown, but the time span represented by this seriation is ~11,000–8800 ^{14}C BP (~12,900–9800 cal BP).

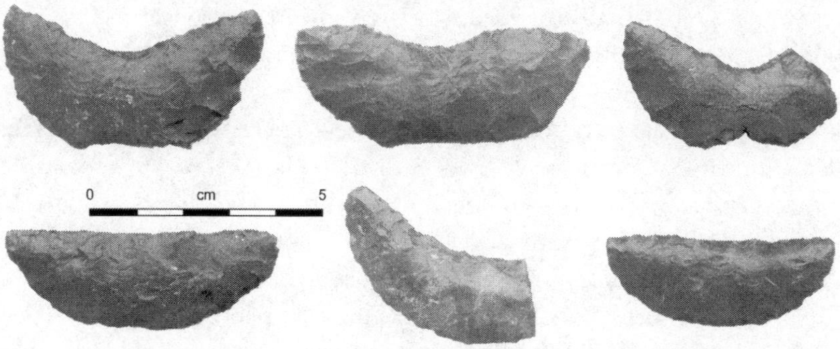

FIGURE 6.6. Chert crescents collected from the Old River Bed delta.

FIGURE 6.7. Example of a large basalt biface from the Old River Bed delta. Thinning flakes from large bifaces, such as the one shown here, were clearly being used as tool blanks by Paleoarchaic foragers in the delta. Overshot flakes, similar to the large flake removed from the center of this artifact, are characteristic of Clovis biface forms.

and display evidence of use-wear along one or both linear margins. We again note that these artifacts are few in number and reminiscent of Clovis lithic technology, but the presence of one or more Clovis sites in this late Pleistocene wetland would not be surprising.

DISCUSSION AND SUMMARY

Most major wetlands in the Great Basin lie at the end of major river systems—such as those of the Humboldt, Bear, and Carson rivers—and have been in existence for at least the period of human occupation in the region. As a result of both continuous use of these marshes by foragers (e.g., Madsen 2002) and erosional/depositional cycles associated with Holocene climatic changes, intact Paleoarchaic sites are relatively rare (e.g., Raven 1990; Raven and Elston 1988). The ORB delta differs in that it was forming for only a limited time during the Paleoarchaic, and although erosion has taken place since that time, there has been no movement of the marsh ecosystem back and forth across the landscape that would result in the disturbance of early sites. Moreover, as warmer and drier middle Holocene climates took hold just prior to 8000 BP (8900 cal BP) (Grayson 2000b; Huckleberry et al. 2001; Madsen et al. 2001; Schmitt et al. 2002b), the area became relatively unattractive to hunter-gatherers, and the impact that human activities generally have had on early sites in most major wetland systems was probably much more limited. In addition, DPG has been closed to the general public for more than 50 years, and disturbance by relic hunters has likely been less than that found outside the installation boundaries.

We have little direct evidence of the foraging activities of the Paleoarchaic peoples who occupied the sand channel sites, but it is reasonable to assume that these activities largely focused on the marsh/wetland resources that dominated the ORB delta landscape at that time. A key aspect of the artifact complex we identified is that it lacks ground stone. There is no evidence for Paleoarchaic ground stone at more than 100 such sites in the western Bonneville Basin, and metates and manos did not appear at Hogup and Danger caves until after about 8600 ^{14}C BP (~9500 cal BP) (Aikens 1970; Jennings 1957; Rhode et al. 2005; Rhode et al. 2006). Based on this and other evidence, Rhode et al. (2006) conclude that seed collecting and processing were not part of the regional subsistence focus until after the desertification that marked the end of the early Holocene wet phase eliminated the ORB wetlands ~8700 ^{14}C BP (9600 cal BP). Whether foraging was limited to large and small game animals or included other plant resources such as marsh rhizomes is not yet known. However, it is reasonable to assume that a variety of wetland resources were pursued, including waterfowl, artiodactyls, muskrat (*Ondatra zibethicus*), beaver (*Castor canadensis*), voles (*Microtus* sp.), roots, tubers, eggs, and insects (see discussions in Byers and Ugan 2005). In the western Paleoarchaic Great Basin, foragers around the Stillwater Marsh area were eating small fish at the time the ORB sand channel sites were occupied (Eiselt 1997b; Napton 1997), and it is possible that the fish in the ORB river were also being exploited.

Sites in the ORB delta fit easily within the Paleoarchaic or Western Stemmed Tradition characterized by an adaptation to wetland ecosystems around the many shallow lakes on valley floors in the Great Basin during the Pleistocene/Holocene transition (Beck and Jones 1997; Willig et al. 1988). The kind of mobility pattern characteristic of foragers in the ORB delta, while similar to that found among regional Paleoarchaic foragers in a general sense, appears to have differed due to the size of the marsh ecosystem. In most of the isolated Great Basin valleys, wetland ecosystems were small relative to the ORB delta, and both theoretical models and limited empirical data suggest foragers employed a high mobility pattern characterized by the use of a variety of widely spaced toolstone sources, large flake and bifacial tools, and a narrow diet breadth (e.g., Elston and Zeanah 2002; Graf 2001; Jones et al 2003). In the ORB delta area, on the other hand, bifacial tools are extremely small and often extensively reworked, and toolstone sources are predominantly local and limited in number, suggesting a more restricted pattern of movement between sites. As Madsen suggests in Chapter 1, much of this difference may be due to patch size. Elsewhere in the Great Basin, wetland ecosystem patches are small, and resources were quickly depleted, necessitating frequent moves of relatively long distances between foraging patches (Elston and Zeanah 2002; Jones et al. 2003; Madsen 2002). The wetland ecosystem on the ORB delta was enormous relative to these other Paleoarchaic wetland foraging localities, and while movement within the wetland may have been almost as frequent, the distances involved were much shorter. As a result, movement outside the ORB delta area to other foraging locations occurred only after extended periods of stay within the marsh. In turn, the ability to refurbish toolkits frequently at a variety of widely separated toolstone sources was limited, as reflected in the extensive reworking of tools and the few toolstone types found at ORB delta sites.

ACKNOWLEDGMENTS

Funding for ORB fieldwork and analyses was provided by the Directorate of Environmental Programs, U.S. Army Dugway Proving Ground. A number of individuals provided valuable insights and support; we thank Charlotte Beck, Mike Collins, Jason Fancher, Don Grayson, Bo Hadley, Chris Hall, Jeff Hunt, Kristen Jensen, Tom Jones, Rebecca Kessler, Tom Larkin, Karen Lupo, Dave Page, Evie Seelinger, Monson Shaver, and especially Kathy Callister and Dave Rhode. The Desert Research Institute and Utah Geological Survey offered crucial administrative assistance, and we thank Mark Teters, Jaynie Hirschi, and all of the folks at the UTTR 388th Squadron for their ongoing support.

PART III

Technological Organization and Settlement Studies

7

Episodic Permanence in Paleoarchaic Basin Selection and Settlement

Daron G. Duke and D. Craig Young

SETTLEMENT AND SUBSISTENCE ARE CRITICAL ISSUES TO ARCHAEOLOGISTS studying early Great Basin occupations. This is no clearer than in the term *Paleoarchaic*, usually applied to an essentially Paleoindian time frame. A subsistence focus on basin wetland plants and small animals is the main link to Archaic adaptations, contrasting with the large migratory game orientation typically used to define North American Paleoindians (Beck and Jones 1997; Jones et al. 2003; Willig and Aikens 1988). Settlement patterns have been difficult to clarify with data that seem more contradictory than complimentary: obsidian sourcing suggests large tracts of land were covered by highly mobile people (e.g., Jones et al. 2003), while the emphasis on basin wetlands suggests some degree of settlement permanence (e.g., Willig and Aikens 1988). Building from extensive assemblages in western Utah's Bonneville Basin, we pull technological organization into existing notions of Paleoarchaic mobility to generate a set of expectations for the archaeological record.

This chapter is concerned with residential mobility as evidenced through two Paleoarchaic toolstone favorites: obsidian and basalt. Obsidian needs little introduction, and its tendency to have been transported great distances lends it to the study of peoples' geographic movements (Jones et al. 2003). Basalt and related grainy volcanics (e.g., dacite and andesite) were exploited more locally (Beck et al. 2002; Jones and Beck 1999; Elston 1994) and are more indicative of the immediate needs people had in specific basins. Using patterns in obsidian and basalt reduction, we argue that while Paleoarchaic groups ranged farther than later peoples, they could also spend considerable time in individual basins as conditions warranted. These stays amount to repeat episodes of occupation that leave signatures of their duration in the lithic record.

PALEOARCHAIC MOBILITY AND LITHIC TECHNOLOGY

Jones et al. (2003; also see Beck and Jones 1997) provide the most up-to-date overview of Paleoarchaic mobility. Their obsidian sourcing data from hundreds of Western Stemmed Tradition (WST) bifaces across the Great Basin indicate at least four geographically distinct territories, or what they call lithic "conveyance zones," within which obsidian types appear to be restricted. An eastern Great Basin conveyance zone encompasses western Utah and eastern Nevada on a generally north-south axis. Regarding chronological change, Jones et al. (2003:32) state: "The actual geographic extent of these territories appears to have changed as drying conditions during the early Holocene altered the distribution and quality of resource zones, and particularly as wetlands containing rich concentrations of animal and plant foods disappeared from many valleys."

What is not present in this and other portrayals of Paleoarchaic mobility is a good sense of the structure of peoples' movements. Jones et al. (2003) argue that early Great Basin inhabitants were "travelers" (sensu Bettinger and Baumhoff 1982) focusing on high-ranking plant and animal resources as they moved quickly through resource patches. More critical to characterizing Paleoarchaic foragers, however, is how they managed select basins once they were in them, and what we should expect of the archaeology surrounding and between them. While the use of other landscape settings is noteworthy, especially along the margins of the Great Basin, it is appropriate to orient this discussion toward the lake margin/basin drainage wetland settings that broadly characterized the region at the time. Whether people were exclusively tied to these areas (e.g., Bedwell 1973) or tethered to them in a looser sense (e.g., Willig and Aikens 1988) remains a crucial question for archaeologists. The nature of how activities were organized from residences within logistical strategies bears directly on this issue. For example, one basin may show relatively intensive, long-term occupation, while the next seems a virtual way station with only temporary use. The differences may result from the duration of residential stays or from task-oriented use by individuals living in neighboring basins.

There is no shortage of archaeological models and ad hoc interpretations about the Paleoarchaic that draw from lithic data, but little in the way of new theory has been advanced (but see Amick 1999; Beck and Jones 1990a, 1990b; Beck et al. 2002; Elston 1994). Enough data have been collected throughout the Great Basin to identify several underlying themes, perhaps the most fundamental being that obsidian and basalt, alongside chert to varying degrees, formed the foundation of technological organization but served different, although overlapping, purposes. Obsidian was often transported great distances, can be sourced, and can be relatively dated using hydration analysis. In this respect it deserves much of the focus placed on it; however, it is basalt that dominates many Paleoarchaic assemblages, especially outside of the northwestern Great Basin. Research emphasis was different prior to the 1970s, when basalt tools formed the basis for cultural-historical descriptions of the earliest Great Basin cultures, such as San Dieguito (e.g., Rogers 1958; Warren 1967) and the Western Lithic Co-tradition (Davis et al. 1969); while the crudish technologies detailed in these studies suited researcher interest in typological diversity, they prevailed at the expense of less variable components. Likewise, archaeologists since that time have emphasized sleeker, curated (and usually obsidian) tools because they relate to processual questions of advance planning and stages of manufacture, a perspective that tends to assimilate expedient artifacts as low-grade variations on these themes. Defining the technological roles of raw materials on their own terms helps us avoid these mischaracterizations.

We do not doubt the chronological trend described by Jones et al. (2003) and others, but believe that much remains to be sorted out. The extent to which Paleoarchaic groups moved through large annual ranges, moved more in some years than others, moved long distances at one time, or moved in any annually routine manner whatsoever all remain open questions. Much like later Great Basin hunter-gatherers, it is likely that early peoples saw great variability in basin productivity year to year and season to season (Mehringer 1977). It is also likely that they dealt with this variability in a similar manner—by being flexible in their adaptation to the environment (Jennings 1957; Steward 1938). In this chapter we consider individual basins as units of residence, movements within them notwithstanding, and refer to peoples' stays as occupational episodes. The duration of these episodes is sometimes referred to as "permanence" to set it apart from any usual notions of extended occupations that might be more sedentary in nature. We are working under the assumption that basins represent units of association with technological patterning. The following analysis relates obsidian and basalt tools and their provisioning to the short- and long-term settlement plans of Paleoarchaic peoples. Theoretical relationships between stone tools and inferred behaviors will be detailed and cited as necessary, but the reader is referred to the extensive technological organization literature for additional background.

PALEOARCHAIC ARCHAEOLOGY IN THE BONNEVILLE BASIN

Our study area is on Air Force lands in western Utah, in and around the Wild Isle dune field, 10 km west of Wildcat Mountain (Figure 7.1). Covering approximately 13,000 acres, the Holocene dunes of Wild Isle are anchored by remnant levees and distal channels formed by interplay between the extinct Old River delta and regressing Lake Bonneville (Figure 7.2). This area is now a barren desert, but prior to about 8500 ^{14}C BP (9700 cal BP) the distal delta of the Old River supported an expansive wetland environment (Young 2000; also see Oviatt et al. 2003); the Holocene dunes of Wild Isle are part of the later landscape and overlay the archaeology. It should be noted that this area represents a northern reach of the Old River deltaic system, and research reported in this volume by David Madsen, Dave Schmitt, and colleagues is being conducted to the south, or upstream, on the same system. While there are great similarities between these two areas, conversations and exchanged field visits suggest that we should not consider our study area representative of all the archaeology associated with the Old River system.

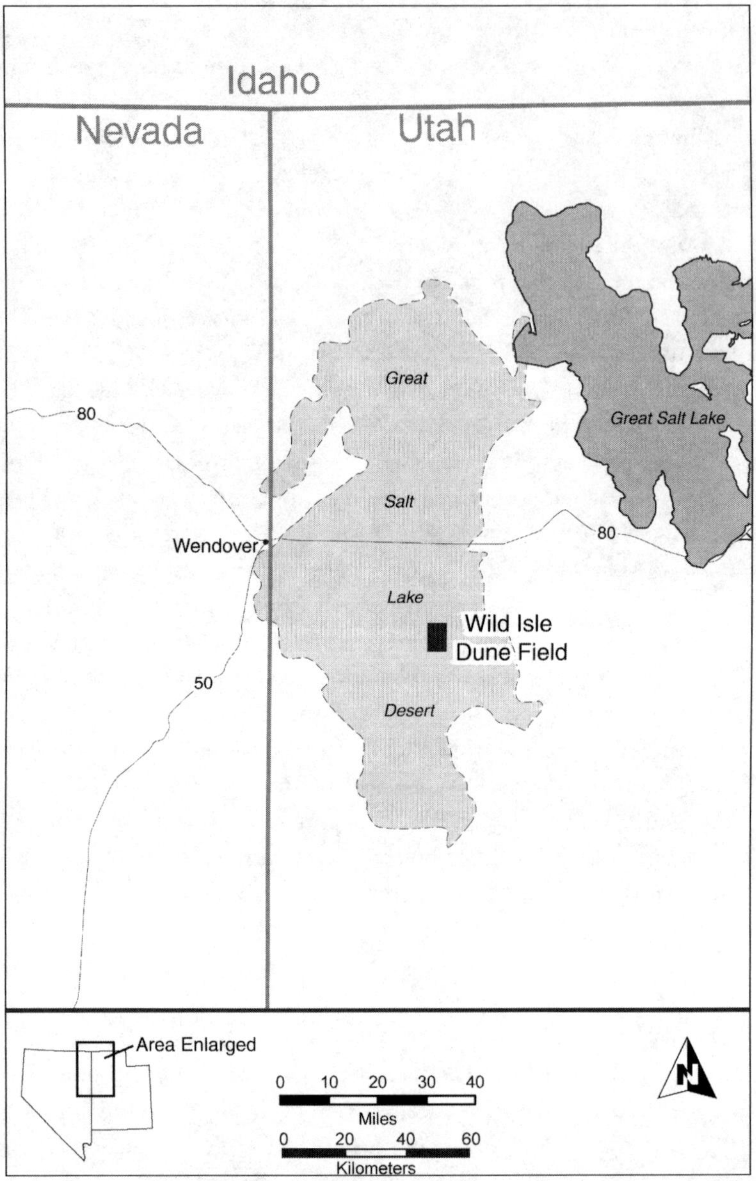

FIGURE 7.1.
Wild Isle study area.

This vast wetland was surrounded by the southern Great Salt Lake Desert, with the primarily sedimentary rocks of Wildcat Mountain being the only solid rock nearby. Topaz Mountain, 80 km to the south, and Browns Bench, roughly 200 km to the northwest, are the closest obsidian sources. There is no other bedrock geology within 30 km of Wild Isle, making this a minimum distance for as yet unknown sources of basalt, but new discoveries of primary basalt sources near Dugway, Utah, 65 km to the east-southeast, and Deep Creek near the Utah-Nevada border, 60 km to the southwest, now provide important anchor points.

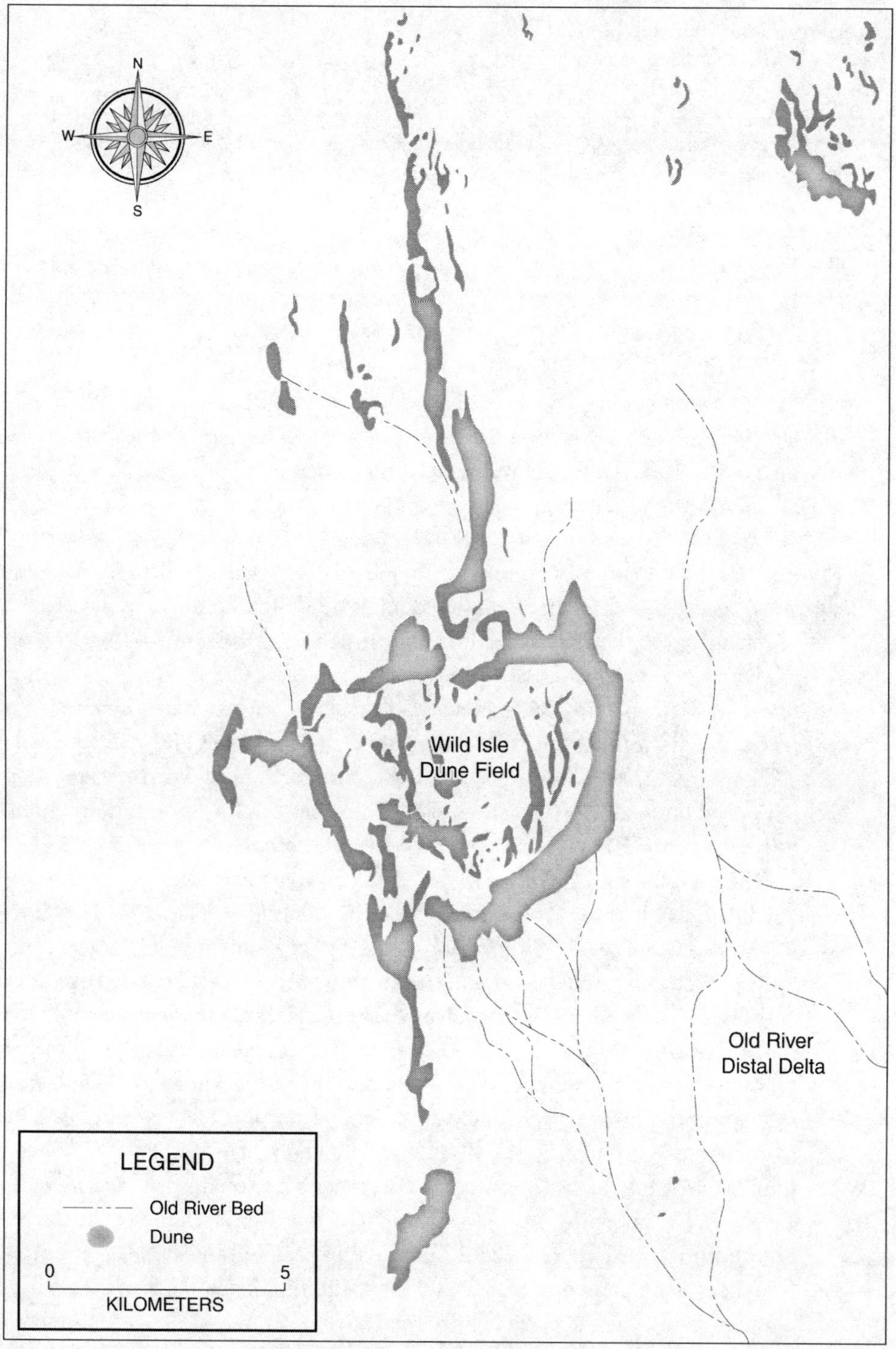

FIGURE 7.2. Wild Isle dune field and relationship to prominent remnant channel features of the Old River Bed.

TABLE 7.1. Radiocarbon Dates Collected at Wild Isle

SAMPLE NO.	YEAR	LAB NO.	MATERIAL	RADIOCARBON AGE	δ^{13}C ‰	CALIBRATED INTERCEPT
44.0.1 MARSH1	2003	Beta-179083	organic sediment	8890 ± 60	-14.7	10,045
L12-367-1	2001	Beta-154760	shell	9480 ± 60	-10.9	10,710
42TO1009 5.0	1998	Beta-120199	shell	9640 ± 60	-9.5	10,905
19.0.1 MARSH2	2003	Beta-179083	organic sediment	10,000 ± 80	-19.4	11,340
L30-368-1	2001	Beta-154760	shell	11,080 ± 70	-9.9	13,080

Description of Technology

The archaeology of Wild Isle is all about Paleoarchaic basalt and obsidian artifacts; tools made from chert are rare. Results from five years of research at or near Wild Isle document more than 100 Paleoarchaic sites, 350 WST bifaces, and 10,000 other artifacts (Carter 1999; Carter and Young 2001; Carter et al. 2004; Duke 2003; also see Arkush and Pitblado 2000). Basalt dominates all but one class, stemmed bifaces, roughly 60 percent of which are made from obsidian. Otherwise, abundant basalt tools and debitage litter dune blowouts and the adjacent mudflats. Our best estimate for time of occupation is 11,000 to 8500 ^{14}C BP (13,000 to 9700 cal BP), but the last 1,500 years of this range seems most likely for intensive use. Radiometric dates collected from marsh deposits and terrestrial gastropod shell associated with ancient channels and levees provide some indirect chronological insight (Table 7.1). Portions of our study area may have been inundated by the Gilbert Lake transgression for some period centering on 10,300 ^{14}C BP (11,700 cal BP) (Oviatt et al. 2003).

The technology of Wild Isle indicates different priorities depending on whether one is looking at obsidian or basalt. Reduction of obsidian WST bifaces was in most cases exactly as might be expected of highly mobile people: carefully worked and resharpened artifacts are indicative of long intended use-lives and distant transport, but extensive reworking indicates a definite strain on toolstone reserves (Figure 7.3). Geochemical sourcing supports this, showing that obsidian points were made from sources 85 to 500 km away (Figure 7.4). Other points, however, were made expediently from small flake blanks, usually from the closest source, Topaz Mountain. Few larger bifacial artifacts have been found at Wild Isle, but we know of their presence based on the debitage. In many cases, WST bifaces probably served as knives (Jones and Beck 1999), with larger items ultimately being reduced to stemmed bifaces as a toolstone economization measure. That people were concerned with their obsidian usage is clear in another class of artifacts we have named "jots" (Figure 7.5). Jots are the small remnant pieces that result from biface smashing, a reduction technique that was sometimes used prehistorically to maximize otherwise expended stone. These may or may not have been tools themselves, and any smaller, sharper fragments have long since blown across the playa. Jots are defined as possessing more than one break and being non-orientable at a minimum, with some

FIGURE 7.3. Examples of Western Stemmed Tradition bifaces of obsidian and basalt collected at Wild Isle.

possessing direct evidence of radial fracture. They are virtually always pressure flaked and made from obsidian, and we treat them as a diagnostic artifact class to keep them separate from nondistinctive biface fragments.

The important thing to take from obsidian artifacts at Wild Isle is that most can be produced one from another in a single reduction sequence. To a lesser degree, the presence of pebble cores, core reduction debitage, and expedient stemmed bifaces and flake tools indicates that additional stone was procured at times. These dichotomous reduction strategies are also noted by Arkush and Pitblado (2000) from the nearby Wildcat Mountain vicinity.

Basalt artifacts, by contrast, exhibit all the traits of a local industry even though this material is not immediately available. Bifaces play a central role

FIGURE 7.4. Sources of obsidian for Western Stemmed Tradition bifaces collected at Wild Isle. Triangles represent source localities.

FIGURE 7.5. Examples of biface jots collected at Wild Isle.

FIGURE 7.6. Examples of basalt bifaces collected at Wild Isle.

in basalt technology and are usually crude, at times grading into more irregular semi-bifaces that are generally ovate in form (Figure 7.6). Some are simply shaped differently than others, implying distinctive functional roles. Adding to this are various flake tools with more definitive modification or use-damage on select margins. These include large scrapers, various irregular tools, and smaller edge-damaged flakes. Crescents—or "crescentic artifacts," so-called for their lack of refinement compared to classic Great Basin crescents—are a subcategory of both bifaces and flake tools (Figure 7.7).

Finally, there are the basalt WST bifaces, which take on a different character than most of those made from obsidian (Figure 7.4). There is little in the way of the raw material conservation seen with obsidian as virtually all basalt stemmed bifaces are made from large flakes with no more reduction than necessary to put

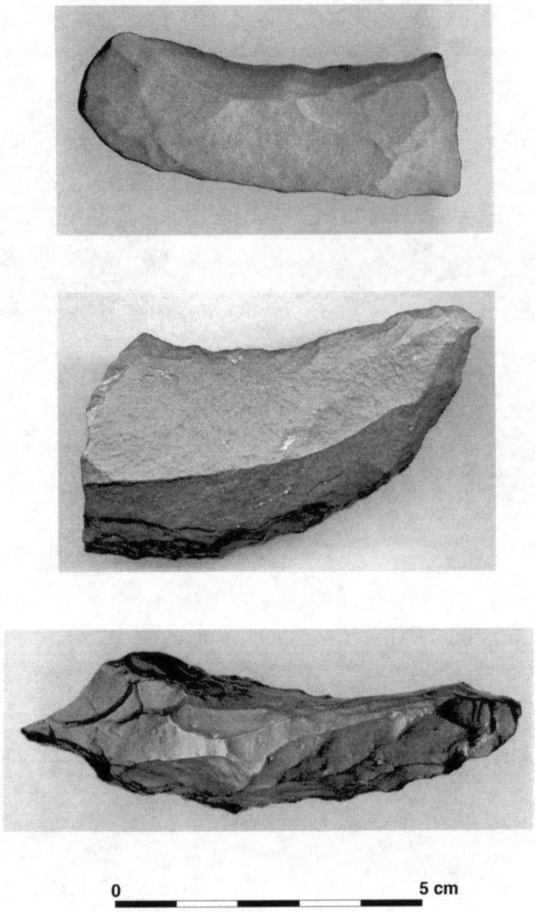

FIGURE 7.7. Examples of crescentic artifacts collected at Wild Isle.

them into working order. It is apparent that some have been resharpened and reworked, but this is by no means a rejuvenation tactic used on the same order as with obsidian. (The term *Dugway Stubbies* has been used by others working on the Old River system [see Oviatt et al. 2003] to distinguish the "stubby" look of the WST bifaces of this area resulting from extensive reworking. From the Wild Isle perspective this applies mostly to obsidian specimens.) The common theme for basalt tools is that they are crudely made. They are not, as a rule, related in an extended reduction trajectory. Different sizes of bifaces and flake tools are made from different sizes of flake blanks, and core reduction and early biface reduction debitage indicate that many of these artifacts were made onsite from generalized blanks either brought from the quarry or generated from large tabular pieces and bifacial cores.

Interpretation

Obsidian and basalt appear to have played very different roles in the lithic technology at Wild Isle, the only significant overlap being among WST bifaces, where basalt items tend to replace expended obsidian pieces. Obsidian artifacts often follow a reduction pattern easily attributable to mobile peoples: thinned, finely flaked bifacial tools were designed for long-term, multipurpose use and transport, although it appears that Wild Isle was the end of this journey in many cases. Basalt use epitomizes an opposing set of motives centering on the short-term, expedient, and local aspects of stone tool needs. From this perspective, people saw nothing unpredictable about their plans for residential movement or encountering resources. These distinctions are considered applicable to many places in the Great Basin.

While obsidian is easy to relate to current reduction models, basalt deserves additional attention here for unique aspects of its reduction that affect how we perceive its relationship to mobility. Basalt is coarse and brittle compared to obsidian and requires unique flaking strategies to make its use in biface technology worthwhile. Biface thinning models often break down when applied to basalt assemblages, as bifaces rarely get proportionately thinner as they get smaller (Carter et al. 2004; Duke 1998). This is because the risk of breakage, particularly by percussive shock, is too great for thinning to be strictly relied upon in most cases. (Extraordinary levels of high mobility and toolstone conservation might have warranted the use and fostering of exceptional flint knapping skill to integrate thinning.) Obsidian and basalt are both brittle, but the fact that basalt is so dense and crystalline compounds this risk because of how hard it must be struck.

Fortunately, basalt is usually abundant and comes in large pieces, which allows for an alternative method of reduction. People systematically procured various sizes and thinnesses of blanks to generate crude bifacial tools, which they simply resharpened until the tool was broken or expended (Carter et al. 2004; Duke 1998). This is obvious by the remnant flake blank evidence present on most artifacts and at all sizes. Thus, rather than a single trajectory in which a biface becomes progressively thinner and incorporates several tool types (e.g., knife to projectile point), different sizes of bifacial tools were made for different tasks. We call this *thinness grading*, and it has the effect of generating expedient-looking assemblages even at great distances from toolstone sources. The importance of recognizing this strategy is that it emphasizes an investment of energy in transport as opposed to reduction.

With this perspective, we can begin to understand the settlement rationale of people in the Wild Isle area. Basalt tools are generally fashioned from flake blanks in a manner consistent with thinness grading. People surely knew what they could get away with in terms of thinning, but it is this rough-and-ready method that provides the reduction paradigm in which they typically worked. Ready-made flake blanks were the focus of this reduction. It also appears that giant bifacial cores, estimated to be as long as 20 to 25 cm, served as the source of some flake blanks. These blanks also functioned as the starting point for many non-bifacial flake tools. A lack of emphasis on refinement diminishes the need to go beyond basic shaping of bifaces and quick production of flake blanks at quarries, and shifts that labor investment to the transport of material for later use. Basalt must be procured more often because it is used expediently, but this expediency offsets transport costs to some extent. This is only viable when stone tool needs can be predicted.

We think Kuhn's (1995) concept of lithic provisioning provides the most robust means of relating reduction patterns to the broader organizational motives of Paleoarchaic groups at Wild Isle. Kuhn (1995) makes a distinction between the provisioning of a place and the provisioning of an individual. Places used for prolonged periods become central decision-making points where it becomes cost-effective to stockpile lithic resources, making tools as necessary to replenish working toolkits. It is the versatile capacity of the stockpile to meet tool user needs that is important rather than the versatility of tools themselves. Individuals become the focus of provisioning when people are more mobile, because the burden of reliable tool use and stone transport is at the individual level; there are more unpredictabilities, and therefore tools must be loaded with greater functional capacity. Wild Isle lithic technology generally indicates the provisioning of a place. Obsidian is the most individually provisioned material type at Wild Isle, but it is typified by its expenditure and expedient replacement with local material. These tactics certainly do not suggest gearing up for distant travel.

It is clear through the lithic technology and the lack of distinctive residential features that people were mobile, but the Old River delta wetlands were apparently extensive enough that moving within and around them could have gone on longer than in other basins. The result is a technology that reflects episodes of great occupational permanence, even if it is not associated with long-term residence in a single spot.

RECOGNIZING THE PERMANENCE OF OCCUPATION EPISODES

Paleoarchaic technology contrasts significantly with that of most North American Paleoindians, who wielded refined, highly curated, and toolstone-efficient technologies to follow migratory large game (Goodyear 1979; Kelly and Todd 1988). The Great Basin pattern of basalt and obsidian use so well represented at Wild Isle is indicative of residential stability punctuated with the potential for distant travel. We expect that the productivity of wetland systems was the primary consideration in the length of basin occupations and for the level of importance placed on the curated aspects of lithic technology.

This characterization is consistent with the interpretations of others (e.g., Beck and Jones 1997; Jones et al. 2003; Oviatt et al. 2003; Willig and Aikens 1988), but further predictive value is needed. We have chosen to consider Paleoarchaic basin occupations as episodes of occupation through which to find predictable patterning in the archaeological record. This perspective places importance on basin occupations and thereby minimizes the spaces between these places as influences on lithic technology; when travel is more common, shorter stays would be implied for basins, and the more-curated aspects of lithic technology would be expected. The assumption used here is that, at their most mobile, Paleoarchaic people were practicing residential mobility in which they acquired resources, including toolstone, on an encounter basis. Logistical mobility (i.e., task-oriented overnight trips away from residences by certain individuals) would have been emphasized when there was some degree of permanence to basin occupations, and this is the perspective favored here for interpreting the more-curated aspects of lithic technology. Elston and Zeanah (2002) envision a scenario for the Paleoarchaic in terms of a sexual division of labor that places young men away from residences on hunting trips. These trips would extend into upland areas and neighboring basins, resulting in low-visibility assemblages comprised of curated tools. By implication, the copious expedient lithic remains we see in lake margin settings across the Great Basin during the Paleoarchaic are a record of predominantly women's activities.

Obsidian and basalt represent two cornerstone materials of Paleoarchaic technology and, combined with chert, provide a tripartite foundation for the structure of lithic economy (Beck and Jones 1990a). Settlement mobility is one important factor that influences this structure. We have discussed a situation at Wild Isle that we believe represents the dynamics of Paleoarchaic lithic resource use when episodes of basin occupation have a great degree of permanence. These patterns contrast with a more commonly envisioned pattern of technological organization for Paleoindians where toolstone selection is oriented toward the spatial unpredictabilities that come with high residential mobility. Assuming these characterizations are near the extremes for Paleoarchaic technological organization, we offer some predictions relative to the duration of basin occupation episodes in Table 7.2. It should be noted that these predictions are directed at basin occupations; people on long-distance logistical forays and in remote upland settings would be expected to carry toolkits akin to those associated with short duration occupations.

Two important influencing factors that affect these predictions are toolstone proximity and chronological change. The first factor is an obvious one. For example, if people occupying a basin for an extended period live near a primary source of high-quality toolstone, this material will be used; it is expected, however, that other technological factors, as indicated in Table 7.2, will reflect their settlement intentions. In terms of chronology, the labels "early" and "late" Paleoarchaic could almost be considered synonymous with the "short" and "long" designations given in Table 7.2. This is an appropriate overlay in the general sense, and a major thrust of our ongoing work, but it is the primary aim of this study to provide a set of independent and archaeologically identifiable behaviors that reflect Paleoarchaic mobility. This allows for more discerning analysis when chronological data are available. For example, we would expect that an increasingly local settlement pattern through time would encourage tool expedience and reliance on local materials. Overarching technological

TABLE 7.2. Expected Implications of Paleoarchaic Basin Settlement Behavior on Lithic Technology

Technological Aspect	Duration of Occupation Episodes: Short	Duration of Occupation Episodes: Long	Data Sources
Toolstone selection	Emphasis on fine-grained materials such as obsidian and chert	Emphasis on coarser-grained materials such as basalt and related grainy volcanic rocks (e.g., dacite, andesite)	Stone identification
Toolstone proximity	Exotic	Local	Geologic source identification; XRF chemical sourcing
Toolstone acquisition	Residential moves to the vicinity of material source	Logistical forays to distant high-quality stone sources; day trips to local sources	Toolkit structure and those aspects that follow from it (below)
Lithic procurement	Gearing up at primary sources, otherwise supplementary	Embedded and ongoing routine activities to maintain via stockpile	Toolkit structure and those aspects that follow from it (below)
Quarry reduction	Extended: carried out to provide working tool	Minimal: sufficient to make suitable blanks or cores for multiple tool types	Debitage and discarded or rejected tools at quarries
Toolkit structure	Curated: multipurpose tools made well in advance of use and designed to perform reliably through moves	Expedient: made for immediate use without special attention to design beyond immediate functional requirements	Toolkit diversity, multiple residential artifact refinement, artifact reworking, artifact replacement, (below); debitage
Toolkit diversity	Low: multipurpose and flexible tools part of a standardized type set	High: functionally discrete tool types made as needed	Morphological and functional typological analysis; use-wear
Artifact refinement	High: reflecting manufacture investment in curated tool	Low: reflecting procurement and transport investment to supply stone	Flaking pattern and intensity; remnant blank evidence
Artifact reworking	High	Extreme on high-quality exotic materials that will not be available soon Low on local materials stockpiled for use	Stem to length ratio; beveling; differential flaking patterns and intensity; inter-material comparisons
Artifact replacement	Exotic material tools replaced with similar material type when source encountered	Exotic material tools replaced with lesser material type locally available	Toolkit structure and artifact reworking (above); material proportions

adjustments to this mobility system and changes in regional population density are potential influencing factors for this expectation that are not discussed here. As always, additional chronological data are needed to help us understand how these factors affect the relationships detailed in this study.

CONCLUSIONS

Residential mobility among Paleoarchaic hunter-gatherers has proven to be difficult to define in the Great Basin. Large geographic territories and frequent moves are widely accepted based on lithic sourcing studies and the notion that people sought high-ranking wetland resources, but strong correlations with the material record are lacking. In this chapter, we examined Paleoarchaic settlement by way of the duration of basin occupation and its implications for lithic technology. The nature of Paleoarchaic technological reliance on basalt and obsidian suggests that people moved about select basins with varying degrees of permanence before undertaking other distant, or not so distant, moves. It is argued that viewing Paleoarchaic mobility as an always flexible series of basin occupations has the potential to yield insight into settlement behavior via predictable technological patterning.

The Paleoarchaic record at Wild Isle represents an extended episode of occupation in the Bonneville Basin and corresponds neatly with the characterization of this place as the preeminent wetland in the eastern Great Basin. Exotic obsidian WST bifaces were expended, and basalt from 30 to 65 km away was brought in to replace these tools and generate new ones. By contrasting these patterns with traits typically equated with high mobility, we provided a set of expectations for lithic technology that should help measure the permanence of occupations. The remaining questions that haunt our research are chronological. Relative to human needs, Paleoarchaic people experienced possibly the most marked environmental deterioration in the Holocene as the wetland resource patches they relied on dried up and disappeared. There can be no doubt that land-use strategies were modified within a broad level of consistency that endured until about 8500 ^{14}C BP (9700 cal BP), and later in some areas. It is expected that the interplay detailed here between basin occupation and lithic technology relates to changing trends in residential mobility through the Paleoarchaic.

Lithic technology is capable of showing us the push and the pull of basic economic considerations through predictable variability in raw material selection and reduction strategies. Obsidian and basalt represent the two sides of Paleoarchaic mobility that seem contradictory at first glance: obsidian for the distant trekking lifestyle that is often envisioned, and basalt as a reflection of the stay-at-home underpinning and routine that probably best captures day-to-day life for these people. Understanding technological details allows us to go beyond material proportions and geographic range to put technological relationships in a more dynamic and archaeologically visible context.

ACKNOWLEDGMENTS

We would like to thank Jaynie Hirschi, Hill Air Force Base archaeologist, for her continued support of our research in the Wild Isle–Wildcat Mountain area. Jerome King, Tammara Norton, and Heather Baron of the Far Western Anthropological Research Group Inc. provided GIS, graphical, editing, and formatting support that is greatly appreciated. Craig Skinner of Northwest Research Obsidian Studies Laboratory provided the figure of obsidian source locales. Finally, we thank James Carter for his contributions to this work and his support of our involvement.

8

Pre-Archaic Mobility and Technological Activities at the Parman Localities, Humboldt County, Nevada

Geoffrey M. Smith

RECENT HUNTER-GATHERER STUDIES HAVE FOCUSED ON THE TECHnological organization of lithic assemblages, or the way groups procured, used, and discarded lithic materials (Kelly 1992). From this work, it appears that at least two factors, mobility and toolstone availability, significantly influenced the way prehistoric populations organized their lithic technologies (Andrefsky 1994; Bamforth 1986, 1991; Kelly 1988, 1992; Kuhn 1994; Odell 1996). While the relationship between these variables remains poorly defined, many researchers argue that such studies have the potential to contribute to our understanding of past lifeways.

These topics are of particular interest to researchers studying the archaeological record of the terminal Pleistocene/early Holocene and are especially relevant in the Desert West, where a substantial lithic record has helped offset a paucity of early-period subsistence remains (Beck and Jones 1997; Jones and Beck 1999). This lack of subsistence data has hindered our ability to infer the adaptive strategies of the Great Basin's first inhabitants and currently prevents us from determining if early-period populations in the region were more like highly mobile hunter-gatherers (Paleoindian) or limnosedentary foragers who moved between highly productive wetland "patches" (Paleoarchaic) (Beck and Jones 1997). This chapter considers results from recent analyses of two major categories of lithic artifacts, bifaces and unifaces, from the four extensive Parman Lake archaeological sites in northwest Nevada occupied during the terminal Pleistocene/early Holocene (Layton 1970, 1979; Smith 2006). These results indicate that the occupants were far-ranging hunter-gatherers who repeatedly stopped at these sites. A major technological activity carried out there was the reprovisioning of individuals with stone implements manufactured on locally available raw material. Such reprovisioning behavior has been associated with residentially mobile hunter-gatherers (Goebel 2005; Graf 2001; Kuhn 1995; McGuire 2002) and suggests that early-period visitors to the Parman Localities were more similar to highly mobile Paleoindians than limnosedentary Paleoarchaic foragers.

THE PARMAN SITES AND THEIR ASSEMBLAGES

Four archaeological sites in northwest Nevada have been incorporated into this study. Referred to as the Parman Localities, these sites are located in Five Mile Flat, a small upland basin in Humboldt County that once contained pluvial Lake Parman (Mifflin and Wheat 1979). The basin's archaeological resources were originally brought to the attention of Thomas Layton in the late 1960s by geologist Martin Mifflin. Upon visiting Five Mile Flat, Layton identified four discrete clusters of early-period artifacts along the basin's western and southwestern margins (Layton 1970, 1979). He referred to these clusters as Parman Localities 1 through 4 and observed that the sites were located at and slightly above 1,786 m (5,857 ft) in elevation (Figure 8.1). Using data collected by Mifflin and Wheat (1979), Layton hypothesized that the sites represented Pre-Archaic occupations along the shores of pluvial Lake Parman. Based on the location of the sites and the presence of weakly developed beach terraces at roughly the same elevation, Layton suggested that the lake would have covered 16.8 km^2 (6.5 mi^2) and reached up to 4 m (13 ft) in depth, providing abundant lake and marsh resources for early-period hunter-gatherers to exploit.

As part of his reconnaissance of the Parman Localities, Layton collected approximately 70 lithic artifacts, including Great Basin Stemmed projectile points (Tuohy and Layton 1977), crescents, unhafted bifaces, and other tools (Layton 1979). In June 2004, the Parman Localities were revisited by a crew supported by the Sundance Archaeological Research Fund of the University of Nevada, Reno, as part of an ongoing effort to assess the potential for Pre-Archaic resources in northwest Nevada (Smith 2006; Smith et al. 2004). Close-interval pedestrian survey of the sites revealed the presence of almost 900 additional lithic tools, including Great Basin Stemmed and Black Rock Concave Base (Clewlow 1968) projectile points, crescents, bifaces, steep-edged side and end scrapers, and other artifacts characteristic of early-period assemblages in the Great Basin. These were combined with the artifacts originally collected by Layton, which together represent the data set used to construct the arguments put forth in this chapter. The bulk of these tools (n = 859; 92.07 percent) are obsidian, with lesser numbers manufactured on cryptocrystalline silicate (CCS) (n = 53; 5.68 percent) and fine-grained volcanic rock (FGVR) (n = 21, 2.25 percent) (Table 8.1). In addition to lithic tools, the Sundance crew collected approximately 2,600 pieces of debitage from the four localities through a total collection within one arbitrary 5 m^2 surface area at each site (Table 8.2). As with the lithic tools in the Parman assemblages, debitage from the sites is predominantly obsidian (n = 2,420; 92.68 percent), with lesser amounts of CCS (n = 130; 4.98 percent) and FGVR (n = 61; 2.34 percent).

Although the precise timing of the occupations at Five Mile Flat has yet to

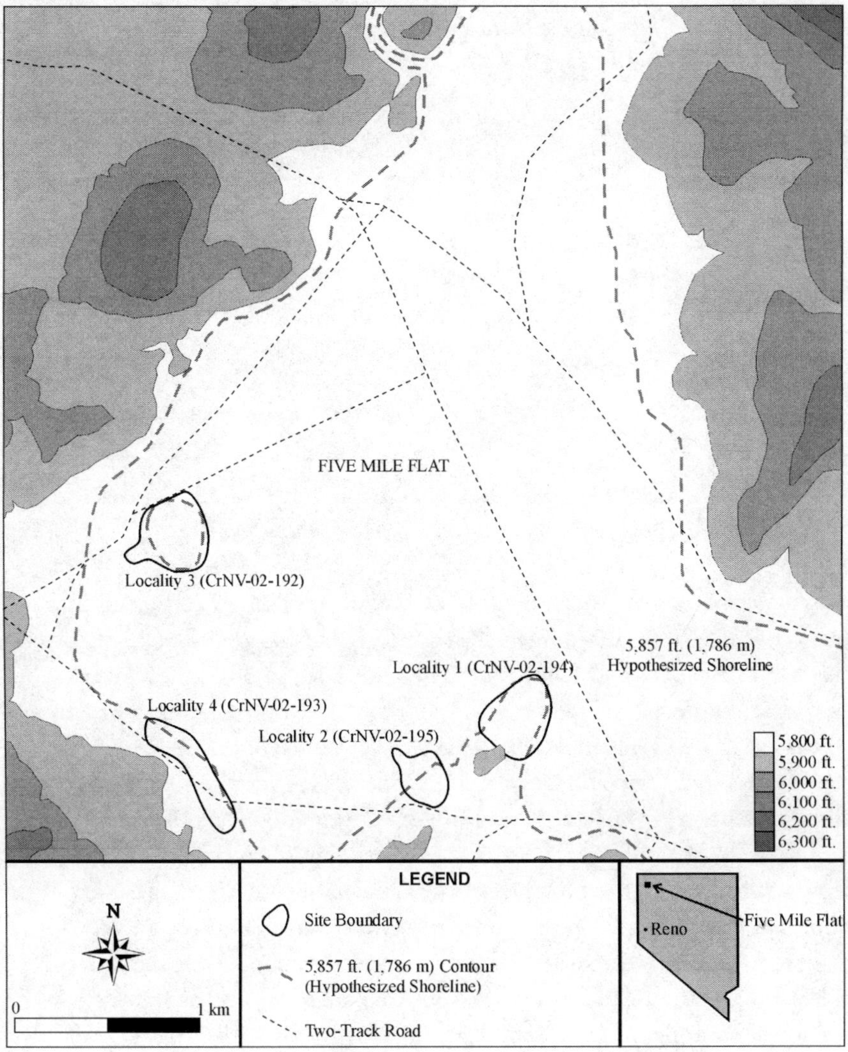

FIGURE 8.1. Map of Five Mile Flat showing locations of the Parman Localities, Humboldt County, Nevada.

TABLE 8.1. Lake Parman Lithic Artifacts

	Raw Material Type			
Tool Type	Obsidian	CCS	FGVR	Total
Projectile points	379	5	6	390
Bifaces	288	18	7	313
Scrapers	55	11	5	71
Flake tools	29	8	1	38
Retouched flakes	78	3	—	81
Combination tools	27	8	—	35
Cores	3	—	2	5
Total	859	53	21	933

Note: CCS = cryptocrystalline silicates; FGVR = fine-grained volcanic rock.

TABLE 8.2. Debitage Sample from the Parman Localities.

	Raw Material Type					
	Obsidian		CCS		FGVR	
Debitage Type	N	%	N	%	N	%
Cortical spalls	290	12.0	1	0.8	5	8.2
Simple platform flakes	155	6.4	21	16.2	9	14.8
Biface thinning flakes	342	14.1	13	10.0	3	4.9
Flake fragments	1,098	45.4	67	51.5	23	37.7
Retouch chips	511	21.1	27	20.7	21	34.4
Angular shatter	14	0.6	1	0.8	—	—
Overpass flakes	9	0.4	—	—	—	—
Split cobble	1	< 0.1	—	—	—	—
Total	2,420	100.0	130	100.0	61	100.0

Note: CCS = cryptocrystalline silicates; FGVR = fine-grained volcanic rock.

be established, similar stemmed point assemblages have been dated in other areas of the Great Basin to between ca. 11,500 and 7500 ^{14}C BP (ca. 13,400 and 8400 cal BP) (Beck and Jones 1997; Jones and Beck 1999). At nearby Last Supper Cave, located about 32 km (20 mi) north of Five Mile Flat, stemmed projectile points have been dated to between ca. 9150 and 8170 ^{14}C BP (ca. 10,800 and 9000 cal BP) (Layton and Davis n.d.). While the Parman Localities do contain projectile points from later periods, these represent only about 6 percent (n = 56) of the total number of lithic tools in the assemblages and are outnumbered by diagnostic early-period bifaces (e.g., Great Basin Stemmed and Black Rock Concave Base projectile points, and crescents) (n = 272) by a ratio of 4.9 to 1.0. Therefore, the bulk of the assemblages can be confidently assigned to terminal Pleistocene/early Holocene occupations that are likely coeval with sites in the region containing similar assemblages.

MATERIALS AND METHODS

Two groups of artifacts from the Parman Localities were analyzed in this study in order to consider the mobility and technological activities of Pre-Archaic hunter-gatherers in the region. The first group, collectively referred to as bifaces, includes Great Basin Stemmed and Black Rock Concave Base projectile points, crescents, and unhafted bifaces (Figure 8.2). Stemmed projectile points from the Parman Localities are classified as Parman (Layton 1970, 1979), Cougar Mountain (Layton 1970, 1979), Windust (Leonardy and Rice 1970), and/or Haskett (Butler 1965) varieties. Unhafted bifaces are categorized by a series of four reduction stages: (1) early stage; (2) mid-stage; (3) late stage; and (4) finished (but with no evidence of hafting elements). These stages are similar to those in reduction sequences developed for Great Basin Stemmed point assemblages by Beck and Jones (1990a), Beck et al. (2002), and Pendleton (1979).

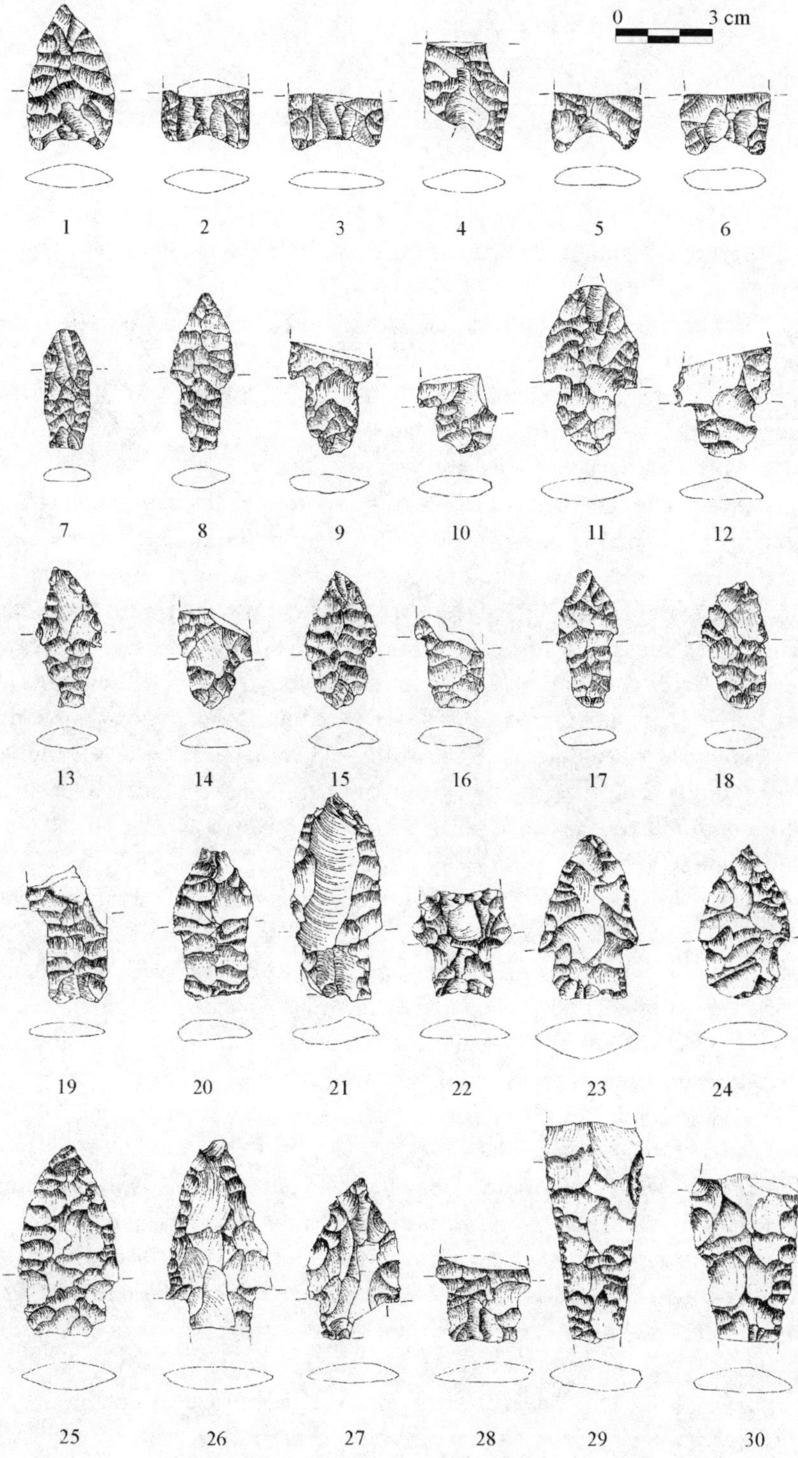

FIGURE 8.2. Selected Pre-Archaic projectile points from Parman Localities 1–4.

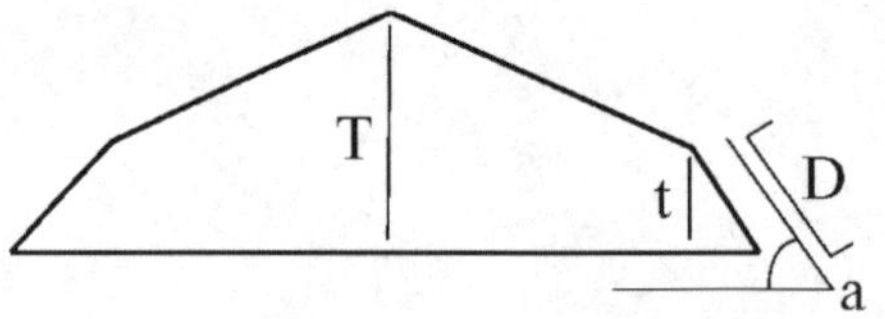

$$URI = \frac{\sin a(D)}{T}$$

FIGURE 8.3. Uniface reduction index (URI) used for measuring use-intensity (after Kuhn 1990, 1995).

The second group of artifacts included in this study is collectively referred to as unifaces. This group is subdivided into two categories, formal and informal. Formal unifaces (manufactured with more effort expended in their production) include end scrapers, side scrapers, and combination tools (artifacts possessing elements designed for two or more uses). Informal unifaces (manufactured with less effort expended in their production) include flake tools (e.g., gravers, notches, backed knives) and retouched flakes.

Unifaces were classified according to a standard typology developed by the Sundance program for early-period assemblages in the Great Basin and were categorized primarily according to their morphology and number/shape of retouched margins. Although a suite of metric and nonmetric attributes were recorded for unifaces, the measures on which this study primarily focuses are the uniface reduction index (URI) and the number of retouched margins per tool. The URI (Figure 8.3) was developed by Kuhn (1990, 1995) as a measure of use-intensity for unifaces in Old World Mousterian assemblages; however, it has recently been successfully used to measure uniface retouch intensity in Paleoindian/Paleoarchaic assemblages in western North America (Graf 2001; Smith 2006).

To calculate the use-intensities of unifaces using the URI, each retouched margin on a tool was measured separately; that is, a unilateral side scraper would have one margin scored, a bilateral side scraper would have two, and so forth. For unifaces possessing more than one retouched margin, the highest URI value was used. As an alternative approach to determining the use-intensity of unifaces (Kuhn 1995), the number of retouched edges on each artifact was also scored. Both measures provide a sense of how Pre-Archaic groups used and discarded unifaces.

A sample of obsidian artifacts from the Parman Localities consisting of projectile points, unhafted bifaces, formal and informal unifaces, and other types were submitted to Craig Skinner at the Northwest Research Obsidian Studies Laboratory in Corvallis, Oregon, for nondestructive X-ray fluorescence (XRF) spectrometer analysis. In the past several decades, similar provenience studies of obsidian artifacts have played a major role in Great Basin research, allowing researchers to examine patterns of raw material procurement and mobility (Hughes 1984, 1986; Jones et al. 2003; Young 2002). The results of the XRF analysis were used to supplement the data provided by technological analyses of obsidian artifacts from the Parman Localities.

TABLE 8.3. Results of XRF Analysis of Obsidian Artifacts from the Parman Localities

	Artifact Class							
Obsidian Source	Hafted Bifaces	Unhafted Bifaces	Formal Unifaces	Informal Unifaces	Cores	Debitage	Total (%)	Distance from Sites
ML/GV	58	72	46	18	3	2	199 (65.89%)	3 km
Coyote Spring	3	12	1	3	—	—	19 (6.30%)	28 km
Badger Creek	5	7	2	—	—	—	14 (4.64%)	37 km
Hawks Valley	7	—	1	—	—	—	8 (2.65%)	40 km
Bog Hot Springs	13	2	3	—	—	—	18 (5.97%)	45 km
BS/PP/FM	8	—	1	—	—	—	9 (2.98%)	50 km
Long Valley	3	—	—	—	—	—	3 (0.99%)	63 km
DH/WH	1	2	1	—	—	—	4 (1.32%)	86 km
Cowhead Lake	4	—	—	—	—	—	4 (1.32%)	87 km
Surveyor Spring	1	—	—	—	—	—	1 (0.33%)	89 km
Beattys Butte	5	—	—	—	—	—	5 (1.66%)	94 km
Buck Mountain	2	—	—	—	—	—	2 (0.66%)	98 km
Whitehorse	7	2	—	—	—	—	9 (2.98%)	113 km
Mount Majuba	—	—	1	—	—	—	1 (0.33%)	113 km
ICB	1	—	—	—	—	—	1 (0.33%)	191 km
Venator	1	—	—	—	—	—	1 (0.33%)	229 km
Unknown 1	1	—	—	—	—	—	1 (0.33%)	Unknown
Unknown 2	1	—	—	—	—	—	1 (0.33%)	Unknown
Unknown 3	1	—	—	—	—	—	1 (0.33%)	Unknown
Unknown 4	1	—	—	—	—	—	1 (0.33%)	Unknown
Total	123	97	56	21	3	2	302 (100.0%)	—

Note: ML/GV = Massacre Lake/Guano Valley; BS/FM/PP = Bordwell Spring/Fox Mountain/Pinto Peak; DH/WH = Double H/ Whitehorse; ICB = Indian Creek Butte.

RESULTS

The combined results of the XRF and technological analyses indicate that far-ranging Pre-Archaic foragers repeatedly stopped at the Parman Localities and reprovisioned their lithic toolkits using locally available raw material.

X-Ray Fluorescence (XRF) Analysis

As shown in Table 8.1, 859 (92 percent) of the artifacts in the Parman assemblages are obsidian, a ubiquitous material in the "tri-state" area of northwest Nevada, northeast California, and southern Oregon, and one that Pre-Archaic groups in the region relied on heavily for tool production (Amick 1995, 1997; Beck and Jones 1997; Hildebrant and King 2002; Jones and Beck 1999; Smith 2005). As part of the Sundance program's ongoing effort to refine our understanding of early-period lifeways in the Great Basin, 302 obsidian artifacts were submitted for nondestructive XRF analysis (Table 8.3).

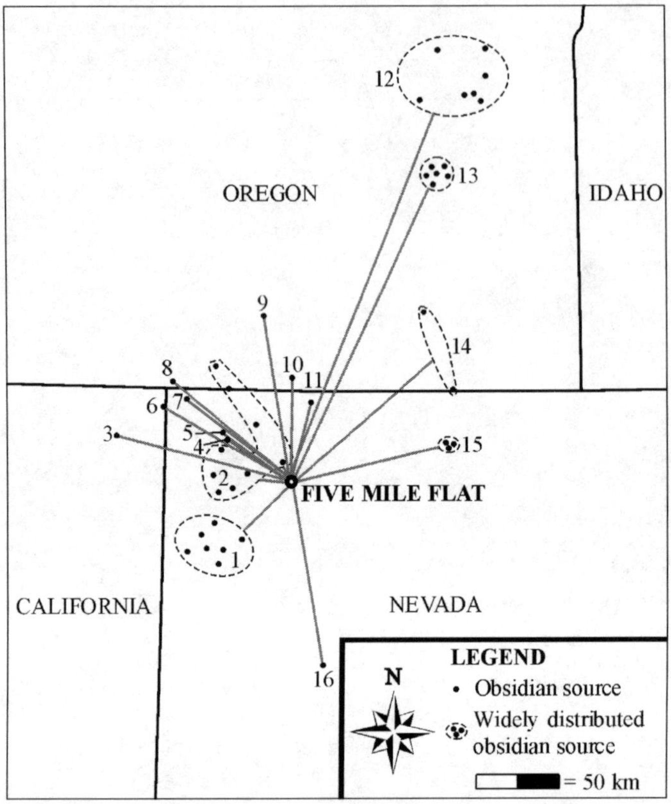

FIGURE 8.4. Locations of obsidian represented in Parman Localities sample. 1 = Bordwell Spring/Pinto Peak/Fox Mountain; 2 = Massacre Lake/Guano Valley; 3 = Buck Mountain; 4 = Coyote Spring; 5 = Badger Creek; 6 = Cowhead Lake; 7 = Long Valley; 8 = Surveyor Spring; 9 = Beattys Butte; 10 = Hawks Valley; 11 = Bog Hot Springs Unknown 1; 12 = Venator; 13 = Indian Creek Butte; 14 = Whitehorse; 15 = Double H; 16 = Mount Majuba.

A total of sixteen known and four unknown geochemical types were identified in the sample. The locations of the known sources of these materials range between 3 and 229 km from the Parman Localities (Figure 8.4). The northernmost obsidian type represented in the sample is from Venator, Oregon, located 229 km from the Parman Localities. The southernmost obsidian type is from Mount Majuba, Nevada, located 113 km from the sites. The westernmost obsidian type represented in the sample is from Buck Mountain, California, located 98 km from the sites, and the easternmost obsidian type is from Double H, Nevada, located 86 km from the Parman Localities.

As Table 8.3 indicates, 65.9 percent (n = 199) of the obsidian artifacts submitted for XRF analysis were manufactured using the widespread Massacre Lake/Guano Valley geochemical type. These consist primarily of projectile points and bifaces, with fewer numbers of scrapers, cores, combination tools, flake tools, retouched flakes, and cores. This preponderance of Massacre Lake/Guano Valley obsidian is not surprising because outcrops of the material are located within 3 km of the Parman Localities. Given this type's proximity to the Parman Localities, the material is considered "local." Fifteen other known and four unknown obsidian types are also represented in the sample. None of these types are currently known to occur near (or closer than 28 km from) the Parman Localities and are therefore considered "nonlocal."

The results of the XRF analysis provide a good sense of the ranges of prehistoric groups in the region and permit examinations of how people procured, used, and discarded lithic material. These results constitute a primary source of data for the analyses put forth in the next section of this chapter.

Integrative Analysis: Effects of Raw Material Procurement on Bifaces and Unifaces

The combination of geologic source provenience data and information obtained via analyses of bifaces and unifaces from the Parman Localities presents a compelling picture of Pre-Archaic technological organization and mobility. These data provide evidence that hunter-gatherer groups in northwest Nevada during the terminal Pleistocene/early Holocene were far-ranging, and also that such groups stopped at the Parman Localities and reprovisioned depleted lithic toolkits with artifacts manufactured on locally available material. Support for these interpretations is presented below.

Bifaces

A total of 15 known and four unknown obsidian types are represented in the sample of hafted and unhafted bifaces (n = 220) (see Table 8.3). The known sources of these materials are located in northwest Nevada, northeastern California, and southern Oregon and are situated between 3 and 229 km from Five Mile Flat. These data suggest that visitors to the basin utilized a large lithic procurement range encompassing much of the aforementioned territory.

When the lithic procurement range represented by the sample of diagnostic Pre-Archaic projectile points (n = 87) is compared with that of diagnostic later Archaic projectile points (n = 36)—including Northern Side-notched, Humboldt, Gatecliff, Elko, and Rose Spring types—a contrasting pattern emerges. The sample of Pre-Archaic projectile points at the Parman Localities is manufactured on 17 different obsidian types (15 known and two unknown) (Figure 8.5). Conversely, the sample of later Archaic projectile points at the sites includes only seven obsidian types (five known and two unknown) (Figure 8.6). This difference in the number of obsidian types represented in each sample may indicate that Pre-Archaic populations possessed more diverse lithic procurement ranges; however, it may also simply be a function of sample size. A basic comparison of the diversity (richness:sample size) of the Pre-Archaic and later Archaic samples (0.195 vs. 0.194, respectively) suggests that the latter possibility may be the case.

While a comparison of the number of obsidian types in each sample may not provide conclusive evidence for a possible reduction in lithic procurement ranges over time, a comparison of obsidian source locations for the two samples is much more informative. Pre-Archaic projectile points were manufactured on obsidian located between 3 and 229 km from the Parman Localities, whereas later Archaic points were manufactured on obsidian located between 3 and 50 km from the sites. These data point to a substantial reduction in extralocal

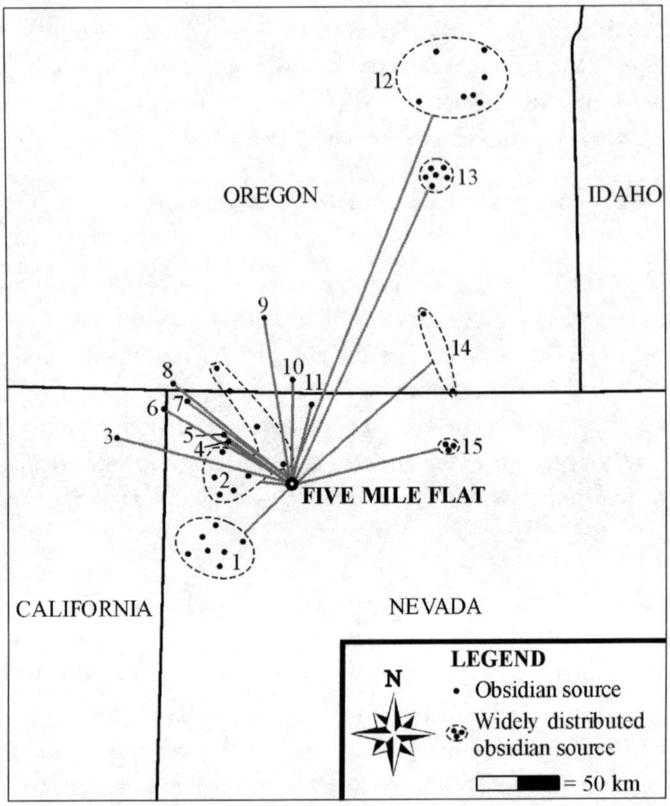

FIGURE 8.5. Locations of obsidian represented in sample of Pre-Archaic artifacts from the Parman Localities. 1 = Bordwell Spring/ Pinto Peak/Fox Mountain; 2 = Massacre Lake/ Guano Valley; 3 = Buck Mountain; 4 = Coyote Spring; 5 = Badger Creek; 6 = Cowhead Lake; 7 = Long Valley; 8 = Surveyor Spring; 9 = Beattys Butte; 10 = Hawks Valley; 11 = Bog Hot Springs Unknown 1; 12 = Venator; 13 = Indian Creek Butte; 14 = Whitehorse; 15 = Double H.

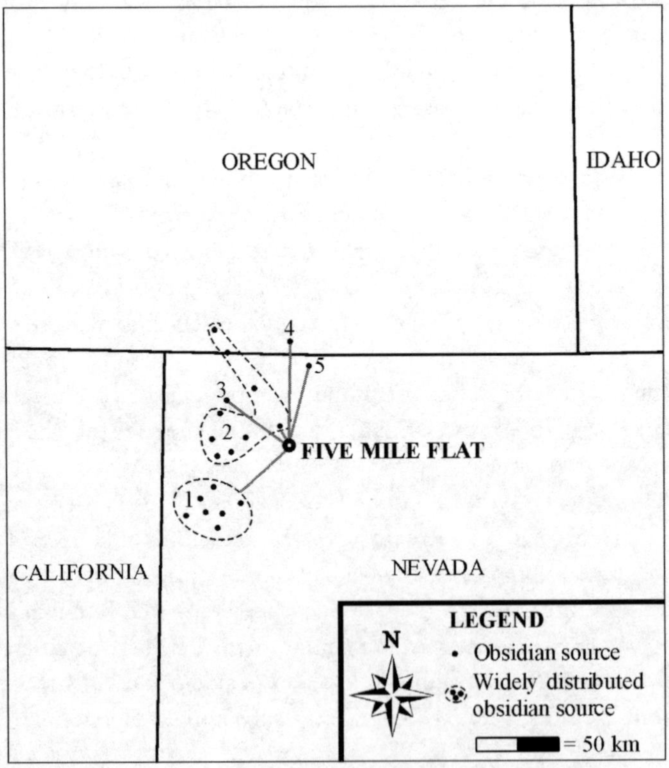

FIGURE 8.6. Locations of obsidian represented in sample of later Archaic artifacts from the Parman Localities. 1 = Bordwell Spring/ Pinto Peak/Fox Mountain; 2 = Massacre Lake/ Guano Valley; 3 = Coyote Spring; 4 = Hawks Valley; 5 = Bog Hot Springs.

TABLE 8.4. Pre-Archaic and Archaic Projectile Points in XRF Sample

	Projectile Points				
	Pre-Archaic		Archaic		
Raw Material	N	%	N	%	Total
Local[a]	36	42.4	22	64.7	58
Nonlocal[b]	49	57.6	12	35.3	61
Total	85	100.0	34	100.0	119

[a] Source located < 5 km from the Parman Localities.
[b] Sources located > 5 km from the Parman Localities.

source diversity across time in the Lake Parman assemblages. Similar reductions in source diversity and/or lithic procurement ranges between Pre-Archaic and later Archaic assemblages have been documented at other sites in the Great Basin (Beck and Jones 1992; Connolly 1999; Goebel 2005; Jones et al. 2003; McGuire 2002); these changes may be related to shifting adaptive strategies, regional demographics, and/or other factors.

The hypothesis that the Pre-Archaic occupants of the Parman Localities were more far-ranging than later Archaic visitors to the sites is further supported by differences in the source provenience of materials used to manufacture different-aged projectile points from the Parman Localities (Table 8.4). Excluding specimens manufactured on obsidian types that currently possess unknown geographic distributions (n = 2), the majority of Pre-Archaic projectile points in the sample (49 of 85; 57.6 percent) were manufactured on nonlocal material. In contrast, only 35.3 percent (12 of 34) of later Archaic projectile points (excluding two specimens of an unknown obsidian type) are manufactured on nonlocal obsidian. These differences are statistically significant ($X^2 = 4.857$, df = 1, p = 0.028) and, like the variation in source diversity between the Pre-Archaic and later Archaic projectile point assemblages, indicate that Pre-Archaic groups traversed larger and more diverse lithic procurement ranges than later populations.

While examinations of lithic procurement ranges provide a general sense of where and how far early-period hunter-gatherers traveled, data from the XRF analysis of unhafted bifaces and Pre-Archaic projectile points from the Parman Localities provide a more fine-grained picture of some of the technological activities such groups carried out during their travels. As part of this study, a total of 181 bifaces—including 97 unhafted bifaces and 84 diagnostic Pre-Archaic hafted bifaces (i.e., projectile points and crescents)—were identified as coming from known sources (Table 8.5).

As Table 8.5 shows, most of the 73 bifaces manufactured on nonlocal obsidian (84.9 percent; n = 62) are projectile points and/or finished bifaces. However, all stages of reduction are evident on bifaces manufactured on locally available Massacre Lake/Guano Valley obsidian (n = 108). A chi-square test indicates that this difference is highly significant ($X^2 = 26.947$, df = 4, p = 0.000). Closer

TABLE 8.5. Lake Parman Bifaces in XRF Sample

	Biface Reduction Stage											
	Early		Mid		Late		Finished		Projectile Point		Total	
Raw Material	N	%	N	%	N	%	N	%	N	%	N	%
Local[a]	20	18.5	18	16.8	17	15.7	17	15.7	36	33.3	108	100.0
Nonlocal[b]	2	2.7	3	4.1	6	8.2	14	19.2	48	65.8	73	100.0
Total	22	—	21	—	23	—	31	—	84	—	181	—

[a] Source located < 5 km from the Parman Localities.
[b] Sources located > 5 km from the Parman Localities.

examination of the bifaces across reduction stages provides additional detail. Of the 73 bifaces manufactured on nonlocal obsidian, two (2.7 percent) are early-stage, three (4.1 percent) are mid-stage, six (8.2 percent) are late-stage, 14 (19.2 percent) are finished unhafted, and 48 (65.8 percent) are projectile points. Clearly, bifaces manufactured on nonlocal obsidian were predominantly at the end of the reduction sequence when discarded at the Parman Localities. Alternatively, bifaces manufactured on locally available Massacre Lake/Guano Valley obsidian were discarded in a range of reduction stages. Of the 108 bifaces manufactured on local obsidian, 20 (18.5 percent) are early-stage, 18 (16.8 percent) are mid-stage, 17 (15.7 percent) are late-stage, 17 (15.7 percent) are finished, and 36 (33.3 percent) are hafted projectile points. These data indicate that using local material to manufacture bifaces for use and/or transport was a major technological activity at the Parman Localities. This hypothesis is supported by the character of the obsidian debitage recovered from the sites (Table 8.2). Biface thinning flakes (n = 342) outnumber flakes possessing simple platforms (n = 155) by more than two to one, suggesting that biface production was the dominant technological activity carried out at the Parman Localities.

Unifaces

An analysis of unifaces from the Parman Localities also permits inferences regarding Pre-Archaic mobility and technological activities in northwest Nevada. As with the sample of bifaces from the Parman Localities, data obtained from analysis of these artifacts suggest that groups arrived at the sites with depleted lithic toolkits and reprovisioned themselves with new tools manufactured on locally available Massacre Lake/Guano Valley obsidian.

Formal Unifaces

Formal unifaces included in the XRF sample (n = 55) were considered to determine if there were differences in the ways artifacts manufactured on local and nonlocal obsidian were used and discarded. First, URI values were calculated for each specimen. Values for specimens manufactured on locally available Massacre Lake/Guano Valley obsidian were then averaged, as were values for

TABLE 8.6. URI Values and Number of Worked Margins on Formal Unifaces in XRF Sample

SOURCE PROVENIENCE	MEAN URI	MEAN NUMBER OF WORKED MARGINS
Local[a]	0.705	1.80
Nonocal[b]	0.879	2.10

[a] Source located < 5 km from the Parman Localities.
[b] Sources located > 5 km from the Parman Localities.

those manufactured on nonlocal obsidian. Specimens (excluding two retouched flake/notch combination tools) manufactured on local Massacre Lake/Guano Valley obsidian (n = 43) possess a mean URI value of 0.705, whereas specimens manufactured on nonlocal obsidian (n = 10) possess a mean URI value of 0.879 (Table 8.6). A one-way ANOVA test was used to determine that this difference is statistically significant ($F = 6.048$, $df = 1$, $p = 0.013$). The results indicate that formal unifaces manufactured on nonlocal obsidian were more extensively retouched than those manufactured on locally available obsidian.

The mean number of retouched margins on formal unifaces in the XRF sample was also considered. Again, specimens manufactured on locally available Massacre Lake/Guano Valley obsidian (n = 45) were compared with specimens manufactured on nonlocal obsidian (n = 10). Formal unifaces manufactured on local obsidian possess a mean number of retouched margins of 1.80, whereas formal unifaces manufactured on nonlocal obsidian (n = 10) possess a mean number of retouched margins of 2.10 (Table 8.6). These data indicate that formal unifaces manufactured on nonlocal obsidian do possess higher numbers of retouched margins than those manufactured on local obsidian; however, according to a Kruskal-Wallis H test, this difference is not significant ($H = 1.102$, $df = 1$, $p = 0.294$).

Informal Unifaces

Informal unifaces included in the sample submitted for XRF analysis (n = 16) were also considered to determine if there were differences in the ways artifacts manufactured on local and nonlocal raw materials were used and discarded. URI values were calculated and averaged for specimens (excluding gravers and notches) manufactured on local (n = 13) and nonlocal (n = 3) obsidian. Informal unifaces manufactured on local Massacre Lake/Guano Valley obsidian possess a mean URI value of 0.423, whereas informal unifaces manufactured on nonlocal obsidian possess a mean URI value of 0.525 (Table 8.7). These samples are too small to permit statistical comparisons, but it does appear that informal unifaces manufactured on nonlocal obsidian were reworked and/or used to a greater extent than those specimens manufactured on locally available Massacre Lake/Guano Valley obsidian.

Finally, the mean numbers of retouched margins on informal unifaces manufactured on local and nonlocal obsidian were considered. Informal unifaces

TABLE 8.7. URI Values and Number of Worked Margins on Informal Unifaces in XRF Sample

SOURCE PROVENIENCE	MEAN URI	MEAN NUMBER OF WORKED MARGINS
Local[a]	0.423	1.29
Nonlocal[b]	0.525	2.00

[a] Source located < 5 km from the Parman Localities.
[b] Sources located > 5 km from the Parman Localities.

manufactured on Massacre Lake/Guano Valley obsidian (n = 17) possess a mean number of 1.29 retouched margins, whereas informal unifaces manufactured on nonlocal obsidian (n = 3) possess a mean number of 2.00 retouched margins (Table 8.7). Like the comparison of URI values outlined above, these data suggest that informal unifaces manufactured on nonlocal obsidian were utilized to a greater degree than those specimens manufactured on local Massacre Lake/ Guano Valley obsidian; however, the sample of informal unifaces is also too small to permit statistical analysis of this difference.

Like the data provided by the analysis of bifaces, analysis of formal and informal unifaces suggests that groups arrived at the Parman Localities with depleted lithic toolkits. Specimens manufactured on nonlocal obsidian, procured between 28 and 113 km from the Parman Localities, typically possess greater amounts of retouch, as indicated by higher mean URI values and numbers of retouched margins, than specimens manufactured on local Massacre Lake/Guano Valley obsidian. This suggests that unifaces manufactured on nonlocal obsidian were transported to the Parman Localities and discarded there. Subsequently, new implements were manufactured on locally available obsidian and used at the sites and/or transported to other locations.

DISCUSSION AND CONCLUSION

Although a relatively simple study, this technological analysis of two common components of early-period toolkits in the Great Basin, bifaces and unifaces, provides a good sense of hunter-gatherer technological organization and mobility during the terminal Pleistocene/early Holocene. Coupled with the XRF analysis of these artifacts, it permits several inferences regarding the lifeways of Pre-Archaic populations in northwest Nevada.

Temporal and Spatial Variability in Lithic Conveyance Zones

Based on data obtained from analysis of the Lake Parman assemblages, it appears that Pre-Archaic hunter-gatherers in northwest Nevada utilized larger and more diverse lithic procurement ranges than later Archaic groups. This hypothesis is supported by the fact that: (1) the sample of Pre-Archaic projectile points contains a more numerous and geographically widespread collection of nonlocal obsidian than the sample of later Archaic projectile points; and (2) the sample of Pre-Archaic projectile points are predominantly manufactured on nonlocal obsidian, whereas those in the sample of later Archaic projectile points are predominantly manufactured on locally available Massacre Lake/ Guano Valley obsidian. Whether such materials were obtained via direct procurement or exchange, it appears that Pre-Archaic populations were more far-ranging than later groups.

Jones et al. (2003) posit that lithic conveyance zones such as those inferred from the Lake Parman assemblages may represent Pre-Archaic subsistence and/or territorial ranges. If this is the case, then it appears that such ranges were markedly different in northwest Nevada and the northern Great Basin than in the central and western Great Basin. The truncated "radial" lithic procurement pattern (extending approximately 340 km north-south and 180 km east-west) suggested by the Lake Parman data is more consistent with those observed at other early-period sites in northwest Nevada/southern Oregon (Amick 1997, 2004; Connolly 1999; Oetting 1993; Smith 2005) and contrasts greatly with the linear north-south patterns observed at Pre-Archaic sites in the western and central Great Basin (Goebel 2005; Graf 2001, 2002; Jones et al. 2003). There, lithic conveyance zones generally possess extended north-south axes, restricted east-west axes, and can span over 450 km in length (Jones et al. 2003).

While it remains unclear exactly what influenced this variation in Pre-Archaic lithic conveyance zones, it appears that at least two factors—topography and the distribution of lithic resources—affected where early-period hunter-gatherers moved about. For example, the Parman Localities are in an area characterized by rugged volcanic tablelands. This region lacks the prominent basin-and-range system and north-south valleys present in much of Nevada, which Jones et al. (2003) and Graf (2002) suggest may have influenced the lithic procurement patterns at Pre-Archaic sites in western/central Nevada. Similarly, there are marked differences between the distribution of lithic raw material in northwest Nevada/southern Oregon and western/central Nevada. The Parman Localities are in a region noted for its abundance of high-quality obsidian. Conversely, the majority of Pre-Archaic sites in western/central Nevada on which Jones et al. (2003) base their conclusions are located in a relatively obsidian-poor region. There, obsidian may have been transported over greater distances by Pre-Archaic groups because the material was not as readily available as in other parts of the Great Basin.

Reprovisioning Activities at the Parman Localities

Based on data obtained from the analysis of biface reduction and XRF sourcing, it appears that groups arrived at Five Mile Flat with worn-out and depleted implements. This hypothesis is supported by the fact that bifaces manufactured on nonlocal obsidian are predominantly late stage and/or finished implements and appear to be at the end of their use-lives. These data also indicate that the production of new bifaces and projectile points from locally available obsidian was a major technological activity carried out at the Parman Localities. All stages of reduction occur on bifaces manufactured on locally available Massacre Lake/Guano Valley obsidian. Analysis of the debitage from the Parman Localities appears to confirm this hypothesis, with biface thinning flakes outnumbering flakes possessing simple platforms by a ratio of greater than 2 to 1.

Similarly, considerations of the extent to which unifaces were retouched coupled with XRF sourcing suggest that Pre-Archaic hunter-gatherers occupying the Parman Localities replaced exhausted implements manufactured on nonlocal obsidian with new ones manufactured on locally available Massacre Lake/Guano Valley obsidian. Evidence for this hypothesis is found in the differences between the URI values and number of retouched margins of specimens manufactured on nonlocal and local obsidian. Such reprovisioning behavior by early-period groups is not unique to the Parman Localities and has also been observed at the Sadmat and Coleman sites in the western Great Basin (Graf 2001, 2002), sites in Butte Valley in the central Great Basin (Beck and Jones 1990b), Bonneville Estates Rockshelter in the eastern Great Basin (Goebel 2005), and archaeological sites in other regions of North America (Andrefsky 1994).

CONCLUSION

Data from the study of bifaces and unifaces in the Parman assemblages point to a strategy of provisioning individuals rather than places. Such a strategy has been associated with high degrees of residential mobility, requiring individuals to transport critical items as they move across the landscape (Goebel 2005; Graf 2001; Kuhn 1995; McGuire 2002). This model of early-period hunter-gatherer mobility complements other current lines of evidence supporting the notion that Pre-Archaic populations practiced high residential mobility, including low site density, minimal inter-assemblage variability, lithic toolkit design, limited evidence of residential structures and midden accumulation, and substantial nonlocal toolstone source diversity (Beck and Jones 1990a; Elston and Zeanah 2002; Graf 2001; Jones and Beck 1999).

In regard to their behavior of reprovisioning themselves with stone tools at the Parman Localities, Pre-Archaic populations in northwest Nevada appear to have behaved more like highly mobile Paleoindian hunters than limnosedentary Paleoarchaic foragers. However, it is somewhat problematic to rely exclusively on a single facet of the archaeological record to reconstruct prehistoric lifeways. While this study may provide data regarding one aspect of Pre-Archaic behavior, it almost certainly ignores others. Therefore, it is critical that we continue to approach the study of Pre-Archaic lifeways from a variety of perspectives in order to develop a more complete model of early-period adaptation. The works presented in this volume represent an important step toward this goal and will undoubtedly contribute to the ongoing debate over whether Pre-Archaic populations were more "Paleoindian" or "Paleoarchaic."

ACKNOWLEDGMENTS

This project was supported by the Sundance Archaeological Research Fund; the University of Nevada, Reno, Anthropology Department; the Desert Research Institute; the UNR Graduate Student Association; the Bureau of Land Management's Winnemucca field office; and Am-Arcs of Nevada. Special thanks to Craig Skinner of the Northwest Research Obsidian Studies Laboratory, Corvallis, Oregon, for conducting XRF analysis of obsidian artifacts from the Parman Localities; Tom Layton for reporting the Parman Localities to the archaeological community; Ted Goebel for providing comments on this manuscript; and Linsie Lafayette for her support.

9

Pre-Archaic and Early Archaic Technological Activities at Bonneville Estates Rockshelter

A First Look at the Lithic Artifact Record

Ted Goebel

LATE PLEISTOCENE/EARLY HOLOCENE ARCHAEOLOGICAL STUDIES IN the Great Basin have been hampered by a paucity of Pre-Archaic sites from sealed, datable contexts. Most sites in the region are from shallow, unsealed settings where organic materials such as bone and charcoal are not preserved. The few known early-period sites that do occur in well-stratified and well-preserved contexts typically have artifact assemblages that are small and unexpressive (Beck and Jones 1997) or faunal assemblages that are not clearly the product of human activity (Haynes 1988, 2002; Hockett 1991; Jennings 1986). As a result, much of what we know about late Pleistocene/early Holocene human behavior comes from surface lithic assemblages that cannot be precisely dated and are often mixed with later, Archaic artifacts (Jones and Beck 1999; Jones et al. 2003; Smith, this volume).

To address this problem, Kelly Graf, Bryan Hockett, David Rhode, and I initiated excavations in 2000 at Bonneville Estates Rockshelter (located near the town of West Wendover, northeast Nevada) in the hope that a terminal Pleistocene cultural occupation could be unearthed. In 2001 we found just that: a 10,100 ^{14}C BP (11,600 cal BP) hearth feature associated with stemmed bifacial points, lithic debitage, and well-preserved faunal and floral remains (Goebel et al. 2003). Since then, our excavations in the rockshelter have exposed an area of about 55 m^2 in which lithic artifacts, perishable artifacts, ecofacts, and features spanning from 12,000 to 6000 ^{14}C BP (15,000 to 7000 cal BP) have been found in a series of stratified cultural layers (Graf, this volume; Hockett, this volume; Rhode and Louderback, this volume; Rhode et al. 2005). With these materials, we are beginning to reconstruct human adaptive change in the eastern Great Basin during the terminal Pleistocene and early Holocene.

This chapter focuses on the lithic artifact assemblages recovered from the Pre-Archaic layers (through 2006) and Early Archaic layers (through 2005) in Bonneville Estates Rockshelter. I have three goals: (1) to present details of the

Pre-Archaic lithic assemblages from Strata 19, 18b, 18a, and 17b', and the Early Archaic lithic assemblages from Strata 17b, 17a, 16, 15, 14, PM2, and PM1; (2) to interpret Pre-Archaic and Early Archaic lithic procurement patterns and technological activities carried out in the vicinity of the rockshelter; and (3) to address whether Pre-Archaic foragers were operating in a "Paleoindian" or "Archaic" manner (Haynes, this volume).

Geochronological and contextual specifics of the rockshelter's early cultural deposits can be found in Graf (this volume), while detailed descriptions of associated paleofaunal and paleobotanical assemblages can be found in Hockett (this volume) and Rhode and Louderback (this volume), respectively. Analysis of perishable artifact assemblages from these layers is being undertaken by Catherine Fowler, Eugene Hattori, and Edward Jolie; their results will be presented in a future publication. Briefly, perishable materials recovered from the Pre-Archaic layers include nine pieces of cordage, a small textile fragment, four bone awl fragments, an antler pressure flaker, a piece of worked wood, several knotted feather quills, and two bone awl fragments. From the Early Archaic layers there are numerous cordage pieces, coiled basket fragments, and a bone promontory peg.

BONNEVILLE ESTATES ROCKSHELTER'S PRE-ARCHAIC LITHIC ASSEMBLAGES

Through the 2006 excavations, the Pre-Archaic assemblages from Bonneville Estates Rockshelter consisted of an isolated Clovis fluted point base, a small assemblage of lithic artifacts from the rockshelter's East Block, and a more extensive assemblage of lithic artifacts from the West Block (Figure 9.1). These are described below.

Fluted Point

In 2004, we unearthed the base of a fluted point in Bonneville Estates Rockshelter (Figure 9.2a). This artifact was recovered from excavation square N5W5, the easternmost square of the stratigraphic trench currently under excavation, near the base of Stratum 1, the shelter's historic dung layer. The fluted point is made on green chert, bears a single flute on each face (as well as additional basal trimming scars around each flute), and is heavily edge-ground along its lateral and basal margins. Morphologically and metrically it falls within the range of Clovis points found elsewhere in western North America.

Given the fluted point's association with historic artifacts and sheep dung, we conclude that it occurred in a secondary context. The point may have originated from the rockshelter's terminal Pleistocene deposits, or it may have been carried into the rockshelter during late prehistoric or historic times. The point fragment, though, does not bear signs of having been reworked or recycled, suggesting that there is a good chance that its original context was within the rockshelter's early deposits.

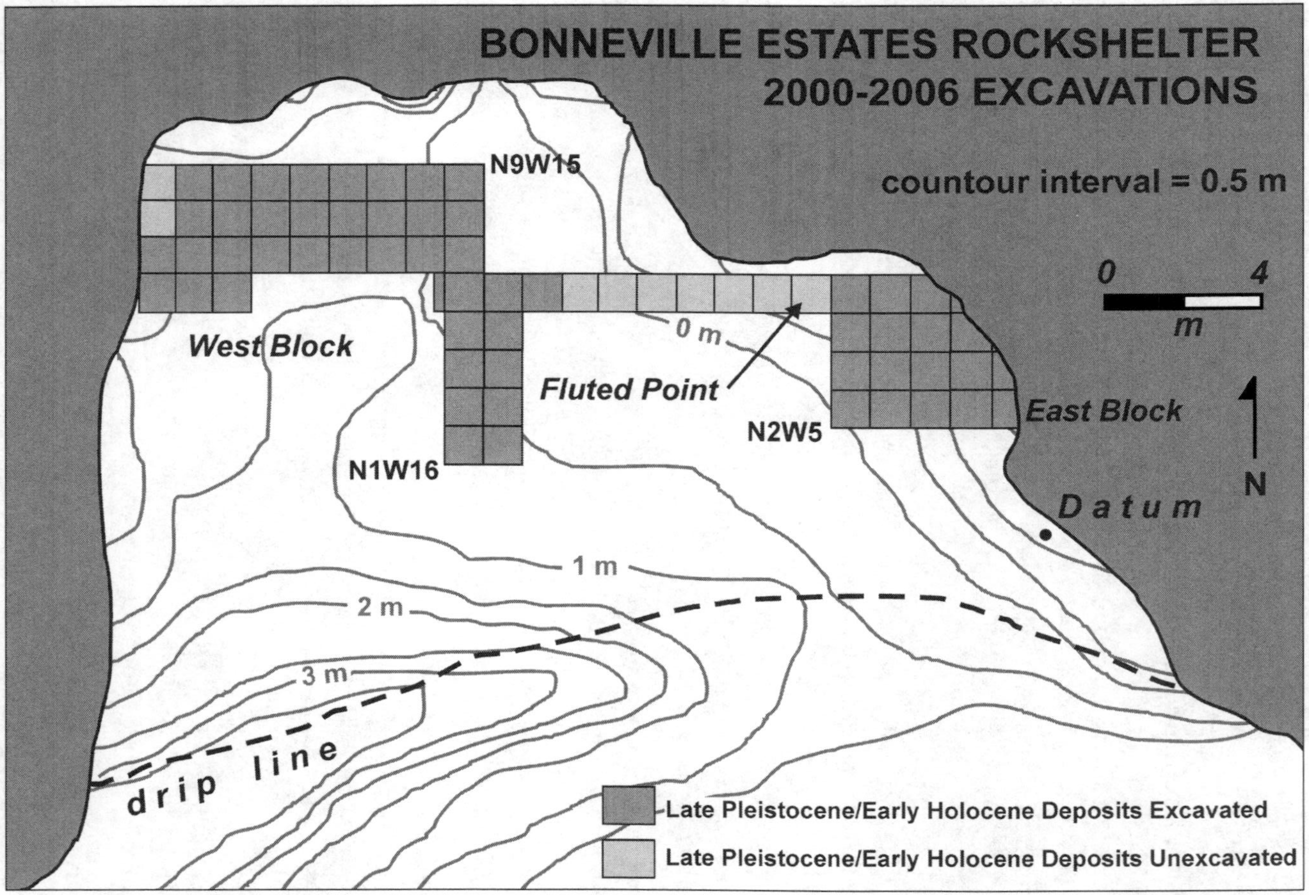

FIGURE 9.1. Excavation map for Bonneville Estates Rockshelter showing locations of Clovis fluted point find and the East and West Blocks.

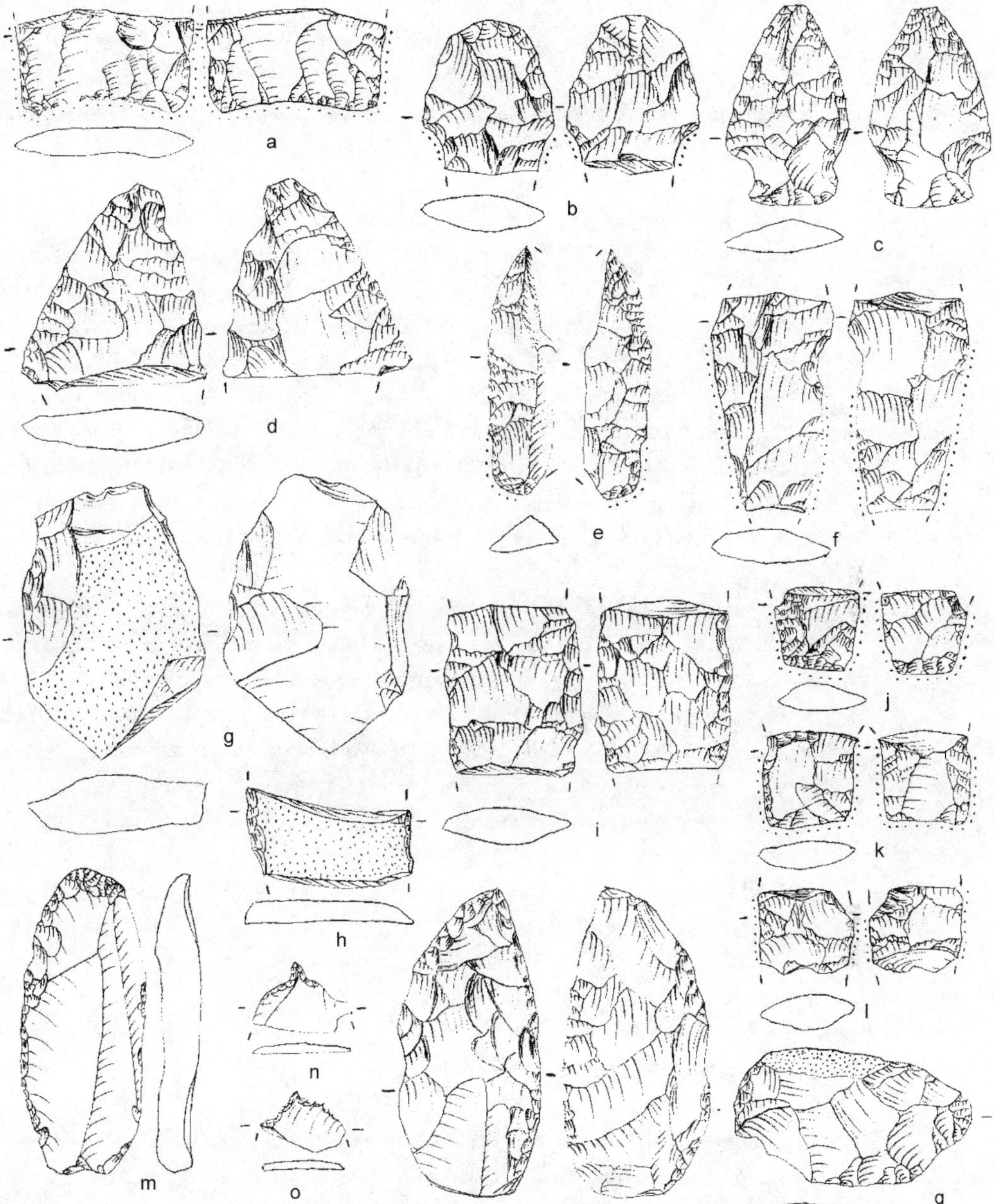

FIGURE 9.2. Lithic artifacts from the Pre-Archaic strata (Component 6) of Bonneville Estates Rockshelter: *a*, fluted point fragment; *b–f, i, j–l*, stemmed point fragments; *g*, edged biface; *h, q*, side scrapers; *m*, end scraper; *n, o*, gravers; *p*, bifacial preform (*a*, Stratum 1 [trench]; *b, m*, Stratum 10 [East Block]; *c*, Stratum 9 [East Block]; *g, p*, Stratum 18b [West Block]; *e, h, i, n, o, q*, Stratum 18a [West Block]; *f, k*, Stratum 17b' [West Block]; *d, l, j*, Strata 18a–18b [West Block]).

The East Block's Pre-Archaic Lithic Assemblage

Through 2006, 21 lithic artifacts had been recovered from the early-period components of Bonneville Estate Rockshelter's East Block. These include 18 debitage pieces and three tools. Since the debitage sample is still very small, I treat it as a single assemblage despite the fact that these materials came from two strata, 12 and 10. Raw materials include cryptocrystalline silicates (CCS) (n = 13), fine-grained volcanics (FGV; i.e., basalt) (n = 3), and obsidian (n = 2). Debitage types include biface thinning flakes (6), core reduction flakes (2), a retouch chip, cortical spall, flake fragments (7), and a retouch chip fragment. The majority of type-able debitage pieces clearly relate to biface reduction.

Two of the three tools from the East Block came from Stratum 12. These include the midsection-to-tip fragment of a stemmed bifacial point made on FGV (Figure 9.2b) and a complete end scraper made on a white CCS prismatic blade (Figure 9.2m). The third tool, from Stratum 10, is apparently a finished but unhafted stemmed point made on gray CCS (Figure 9.2c). No perishable artifacts have been found in the lower layers of the East Block.

The West Block's Pre-Archaic Lithic Assemblage

Through the 2006 field season, a total of 1,063 lithic artifacts had been recovered from the early-period components of the West Block. These include 1,031 debitage pieces and 29 tools. Of these, 853 can be precisely attributed to either Stratum 19, 18b, 18a, or 17b', while 178 have less precise stratigraphic provenience (due to differences in the preservation of these strata). Of the latter 178 artifacts, 154 come from Stratum 18 (a combination of 18a and 18b), and 53 come from either 18 or 17b'. In addition, there are three artifacts from Stratum 20. Each of these lithic assemblages is described in detail below, starting from the bottom of the stratigraphic profile.

STRATUM 20

Three lithic artifacts have been recovered from Stratum 20, the lowest stratigraphic layer yet exposed in the West Block of the rockshelter. These include a CCS flake fragment, a CCS biface thinning flake, and an FGV primary cortical spall, all of which were probably post-depositionally introduced into this layer, which predates 12,300 ^{14}C BP (15,000 cal BP).

STRATUM 19

The basal cultural occupation of the West Block (tentatively dated to between 12,300 and 10,900 ^{14}C BP [15,000 and 13,000 cal BP]) has produced 32 debitage pieces (Table 9.1). Lithic raw materials include CCS (53.1 percent), FGV (28.1 percent), and obsidian (18.8 percent) (Table 9.2). Only one of the six obsidian pieces is of sufficient size to be characterized elementally using X-ray fluorescence (XRF) analysis; it originated from the Malad source, southeast Idaho. The debitage assemblage itself is dominated by small biface thinning flakes (25.0 percent), retouch chips (i.e., pressure flakes) (18.8 percent), and retouch

TABLE 9.1. Debitage Categories by Stratigraphic Layer, Pre-Archaic

Debitage Category	Stratigraphic Layer							Total
	17b'	18a	18b	18a–18b[a]	17b'–18b[b]	19	20	
Flake fragment	24	71	56	52	9	7	1	219
Complete flake	9	29	26	13	4	1	—	82
Bipolar flake	—	5	—	1	—	—	—	6
Primary cortical spall	2	5	1	1	—	1	1	11
Secondary cortical spall	2	6	6	2	—	—	—	16
Retouch chip fragment	26	77	60	28	12	8	—	211
Retouch chip	4	31	25	6	6	6	—	78
Biface thinning flake	60	109	111	39	21	8	1	349
Angular shatter	6	32	10	7	1	1	—	57
Unworked cobble	—	1	—	—	—	—	—	1
Initially flaked cobble	—	1	—	—	—	—	—	1
Unidirectional flake core	—	1	—	1	—	—	—	2
Bidirectional flake core	—	1	—	—	—	—	—	1
Total	133	369	295	150	53	32	3	1,034

[a] Artifacts from Stratum 18, where 18a and 18b could not be discerned stratigraphically.
[b] Artifacts from Pre-Archaic strata, but where 17b', 18, and 18b could not be discerned stratigraphically.

TABLE 9.2. Raw Materials Present in the Pre-Archaic Assemblages

Raw Material	Stratigraphic Layer							Total
	17b'	18a	18b	18a–18b[a]	17b'–18b[b]	19	20	
Cryptocrystalline silicates	37	162	193	50	10	17	2	471
Obsidian	42	85	33	72	11	6	—	249
Fine-grained volcanics	56	128	79	32	32	9	1	336
Quartzite	—	4	1	—	—	—	—	5
Granite	—	1	—	—	—	—	—	1
Quartz	—	1	—	—	—	—	—	1
Total	135	381	306	154	53	32	3	1,063

[a] Artifacts from Stratum 18, where 18a and 18b could not be discerned stratigraphically.
[b] Artifacts from Pre-Archaic strata, but where 17b', 18, and 18b could not be discerned stratigraphically.

chip fragments (25.0 percent), while flake fragments (21.9 percent), complete flakes (3.1 percent), primary cortical spalls (3.1 percent), and angular shatter fragments (3.1 percent) are less common.

Only one (3.1 percent) debitage piece in the Stratum 19 assemblage bears cortex (Table 9.3), and debitage sizes are either tiny (< 1 cm^2) (64.5 percent) or small (1–3 cm^2) (35.5 percent). Of platform types, 50 percent are smooth, and 50 percent are complex (Table 9.4).

The high proportions of biface thinning flakes and retouch chips, many with complex platforms, obviously represent the resharpening and rejuvenation of lithic tools, chiefly bifaces. This is supported by the overall lack of cortex in the

TABLE 9.3. Amount of Cortex Preserved on Dorsal Faces of Artifacts, Pre-Archaic

Amount of Cortex	Stratigraphic Layer							
	17b'	18a	18b	18a–18b[a]	17b'–18b[b]	19	20	Total
0%	130	363	297	149	53	31	2	1,024
< 10%	2	4	4	2	—	—	—	12
10–49%	1	6	3	—	—	—	—	10
50–90%	1	2	2	—	—	1	—	6
> 90%	1	5	—	2	—	—	1	9
Total	135	380	306	153	53	32	3	1,061

[a] Artifacts from Stratum 18, where 18a and 18b could not be discerned stratigraphically.
[b] Artifacts from Pre-Archaic strata, but where 17b', 18, and 18b could not be discerned stratigraphically.

TABLE 9.4. Platform Surface Preparation, Pre-Archaic

Platform Category	Stratigraphic Layer							
	17b'	18a	18b	18a–18b[a]	17b'–18b[b]	19	20	Total
Cortical	1	4	4	—	—	—	1	10
Smooth	13	61	45	9	12	8	—	148
Complex	61	111	114	39	20	8	1	354
Crushed	—	5	—	2	1	—	—	8
Total	75	181	163	50	33	16	2	520

[a] Artifacts from Stratum 18, where 18a and 18b could not be discerned stratigraphically.
[b] Artifacts from Pre-Archaic strata, but where 17b', 18, and 18b could not be discerned stratigraphically.

assemblage, as well as the large number of very small debitage pieces and complex platforms.

Stratum 18b

For Stratum 18b (10,800–10,500 ^{14}C BP [13,000–12,400 cal BP]), there are 295 debitage pieces and 11 tools. Lithic raw materials are dominated by CCS (63.1 percent), while FGV (25.8 percent) and obsidian (10.8 percent) are less common (Table 9.2). XRF analysis of one obsidian biface thinning flake from Stratum 18b indicates that its source is Malad, Idaho. The type-able debitage assemblage is dominated by biface thinning flakes (37.6 percent), while retouch chip fragments (20.3 percent), retouch chips (pressure flakes) (8.5 percent), flakes (8.8 percent), and cortical spalls (2.4 percent) occur less frequently. Debitage fragments (flake fragments and angular shatter) together make up 22.4 percent of the assemblage. No cores have been recovered from Stratum 18b (Table 9.1).

Only nine (2.9 percent) artifacts bear cortex on their dorsal surfaces (Table 9.3). Most debitage sizes are very small (53.4 percent) or small (45.3 percent), and only two (0.7 percent) are of medium size (3–5 cm^2). Platforms are dominated by complex, multifaceted surfaces (69.9 percent), while smooth surfaces

TABLE 9.5. Tool Forms, Pre-Archaic

	STRATIGRAPHIC LAYER							
TOOL CATEGORY	17B'	18A	18B	18A–18B[a]	17B'–18B[b]	19	20	TOTAL
Bifacial points								
Parman Stemmed	—	1	—	—	—	—	—	1
Haskett Stemmed	1	—	1	—	—	—	—	2
Windust Stemmed	1	—	—	—	—	—	—	1
Unident. stemmed	—	—	—	3[d]	—	—	—	3
Fragment	—	1	1	—	—	—	—	2
Bifaces								
Edged bifaces (Stage 2)	—	—	2	—	—	—	—	2
Finished bifaces (Stage 5)	—	—	2	—	—	—	—	2
Fragment	—	—	1	1	—	—	—	2
Retouched flake	1	4	1	—	—	—	—	6
Side scraper	—	2	2	—	—	—	—	4
Graver	—	2	—	—	—	—	—	2
Denticulate	—	1	—	—	—	—	—	1
Combination tool[c]	—	—	1	—	—	—	—	1
Total	3	11	11	4	0	0	0	29

[a] Artifacts from Stratum 18, where 18a and 18b could not be discerned stratigraphically.
[b] Artifacts from Pre-Archaic strata, but where 17b', 18, and 18b could not be discerned stratigraphically.
[c] Side scraper/biface knife.
[d] Rhode et al. (2005) assigned these to 18a, but in this part of the excavation it was impossible to differentiate 18a from 18b.

occur infrequently (27.6 percent), and cortical surfaces are rare (2.5 percent) (Table 9.4).

Lithic tools include one basal fragment of an FGV (basalt) Haskett point, one CCS bifacial point tip fragment, two edged (Stage 2) CCS biface fragments (Figure 9.2g), two finished (Stage 5) but unhafted bifaces (one CCS and one FGV [Figure 9.2p]), one unidentifiable CCS biface fragment, one CCS retouched flake, two side scrapers (one on CCS and one on FGV), and one FGV combination tool (Table 9.5). The combination tool is a side scraper/biface knife (possibly a fragmented biface recycled into a side scraper).

Taken together, the high proportions of biface thinning flakes, retouch chips, and retouch chip fragments indicate that the chief technological activity carried out during the Stratum 18b occupation was the secondary reduction of bifaces and/or bifacial points. This is supported by the overall lack of cortex on the debitage and tools, the high percentage of complex platforms, and the small sizes of most of the debitage pieces, as well as the fact that seven of the 11 tools recovered are bifacial points, bifaces, or biface fragments.

STRATUM 18A

The Stratum 18a assemblage (10,100 ^{14}C BP [11,600 cal BP]) includes 369 debitage pieces and 12 tools. Among raw materials, no individual toolstone category clearly dominates; instead, toolstones used include CCS (42.5 percent), FGV (33.6 percent), and obsidian (22.3 percent) (Table 9.2). In addition, there are four quartzite pieces (including a primary cortical spall, unworked cobble, and initially flaked cobble), one quartz retouch chip fragment, and one granite secondary cortical spall. Three debitage pieces from Stratum 18a have been XRF analyzed; two are from Browns Bench, and one is from Ferguson Wash, Nevada. Chief among debitage pieces are biface thinning flakes (29.5 percent) (Table 9.1); less common are retouch chip fragments (20.9 percent), retouch chips (8.4 percent), flakes (7.9 percent), cortical spalls (3.0 percent), and bipolar flakes (1.4 percent). Flake fragments and angular shatter make up 27.9 percent of the assemblage. In addition to these debitage pieces, there are two cores and two cobbles. The cores include one small unidirectional flake core (maximum linear dimension [MLD] measures 46.15 mm) made on FGV and one large bidirectional flake core (MLD = 80.63 mm) made on CCS.

Only 16 artifacts (4.5 percent) bear cortex on their dorsal surfaces (Table 9.3). Most debitage sizes are very small (54.1 percent) or small (44.5 percent), although there are four (1.1 percent) medium-sized and one (0.3 percent) large-sized (>5 cm²) pieces. Among platforms, complex surfaces are most common (61.3 percent), while smooth (33.7 percent), cortical (2.2 percent), and crushed (2.8 percent) surfaces are less common (Table 9.4).

The tool assemblage consists of two bifacial point fragments and nine flake tools (Table 9.5). The two point fragments include a longitudinally fragmented Parman Stemmed point made on obsidian (Figure 9.2e) and a midsection fragment of a large, thick obsidian point that cannot be typed (Figure 9.2i). The Parman point fragment is weakly ground around its base. Among the flake tools are four retouched flakes (two CCS, one FGV, one obsidian), two side scrapers (one CCS [Figure 9.2q] and one FGV [Figure 9.2h]), two gravers (both CCS [Figure 9.2n, o), and one denticulate (CCS).

Thus, the debitage assemblage of Stratum 18a is characterized by numerous small-sized biface thinning flakes and retouch chips, a low frequency of debitage pieces with cortex, and a high frequency of complex platforms; however, cores, cobbles, and bipolar flakes are also present, and flake fragments and angular shatter occur in relatively high frequencies. These details clearly indicate that technological activities focused on the resharpening of finished bifacial points that had been manufactured elsewhere, as well as the production and resharpening of small unifacial flake tools. The bifacial points are clearly stemmed and therefore indicate that Stratum 18a can be ascribed to the Great Basin Stemmed point complex.

STRATUM 17B'

Stratum 17b' (9500 ^{14}C BP [10,700 cal BP]) is represented by 133 debitage pieces and three tools (Tables 9.1 and 9.5). Of these, raw materials are dominated by FGV (41.5 percent), while obsidian (31.1 percent) and CCS (27.4 percent) occur less frequently (Table 9.2). XRF analysis of three debitage pieces from Stratum

17b' reveal that two come from Browns Bench obsidian (Nevada), and one is from an unknown source. The majority of type-able debitage pieces are biface thinning flakes (45.1 percent), while retouch chip fragments (19.5 percent), flakes (6.8 percent), and retouch chips (3.0 percent) are less common, and cortical spalls are rare (3.0 percent). Flake fragments and angular shatter together make up 22.6 percent of the debitage assemblage.

Just 3.7 percent of the artifacts have cortex on their dorsal faces (Table 9.3). The vast majority of debitage sizes are very small (49.2 percent) or small (48.4 percent), while only three (2.4 percent) are medium in size. Platform surfaces are mostly complex (81.3 percent), while smooth surfaces (17.3 percent) occur much less frequently (Table 9.4).

Tools include two bifacial point fragments and a retouched flake. Both point fragments are made on FGV; they include a midsection fragment of a Haskett Stemmed point (Figure 9.2f) and basal fragment of a Windust Stemmed point (Figure 9.2k).

The debitage assemblage of Stratum 17b' is characterized chiefly by very small debitage pieces, most of which are biface thinning flakes with complex platforms. The lack of cortical pieces and larger-sized debitage suggest that most technological activities during this occupation centered on resharpening of bifacial points that were manufactured elsewhere, away from Bonneville Estates Rockshelter.

Strata 18a–18b

In the eastern part of the West Block, Strata 18a and 18b were difficult to distinguish from each other, so some artifacts could only be attributed generally to Stratum 18. These include 150 debitage pieces and four tools. Of these, raw materials are chiefly obsidian (46.8 percent), while CCS (32.7 percent) and FGV (20.8 percent) are less common (Table 9.2). Of the 17 obsidian artifacts subjected to XRF analysis from Strata 18a–18b, 15 originated from Browns Bench, one from Topaz Mountain, Utah, and one from an unknown source. Debitage categories include biface thinning flakes (26.0 percent), retouch chip fragments (18.7 percent), retouch chips (4.0 percent), flakes (8.7 percent), and cortical spalls (2.0 percent) (Table 9.1). Flake fragments and angular shatter make up 39.3 percent of the debitage assemblage.

Only 2.6 percent of the assemblage is made up of artifacts bearing cortex on their dorsal surfaces (Table 9.3). Debitage sizes are either very small (51.8 percent) or small (48.2 percent). Platform surfaces are predominantly complex (78.0 percent), while smooth (18.0 percent) and cortical (4.0 percent) surfaces occur less frequently (Table 9.4).

Of the four tools assignable to Strata 18a–18b, three are bifacial point fragments, and one is a biface fragment (Table 9.5). The three point fragments are made on obsidian and are fragments of stemmed points. The first is a basal fragment of an edge-ground stemmed point (Figure 9.2j), the second is a midsection fragment of an edge-ground stemmed point (Figure 9.2l), and the third is a tip fragment that still bears one shoulder of a stemmed point (Figure 9.2d). The biface fragment is an edge fragment of a large obsidian biface. Like the long edge fragment from Stratum 18a described above, it may also be a fragment of a stemmed point.

STRATA 17B'–18B

Fifty-three debitage pieces cannot be attributed to any specific Pre-Archaic stratum but are known to have originated from below Stratum 17b. Of these, 60.4 percent are made on FGV, 20.8 percent on obsidian, and 18.9 percent on CCS (Table 9.2). One obsidian flake from Strata 17b'–18b was subjected to XRF analysis; its source is Topaz Mountain. Debitage categories are dominated by biface thinning flakes (39.6 percent), while retouch chip fragments (22.6 percent) and retouch chips (11.3 percent) and flakes (7.5 percent) are less common (Table 9.1). None of these bears cortex, and debitage sizes are either very small (63.8 percent) or small (36.1 percent) (Table 9.3). Platform surfaces are chiefly complex (60.6 percent), while smooth surfaces are less common (36.4 percent), and crushed surfaces are rare (3.0 percent) (Table 9.4).

Pre-Archaic Raw Material Procurement and Technological Activities

Relative proportions of raw materials transported to Bonneville Estates Rockshelter vary between the three main Pre-Archaic layers (Strata 18b, 18a, and 17b'). The Stratum 18b assemblage is clearly dominated by CCS, the Stratum 18a assemblage is dominated by CCS and FGV, and the Stratum 17b' assemblage is dominated by FGV and obsidian. These differences are statistically significant (X2 = 63.904; df = 6; p <0.001).

Cortex is very rare in all of these assemblages (X^2 = 1.113; df = 2; p = 0.573), suggesting that most raw material packages were initially worked at some other place and then transported to the rockshelter. The only strong evidence for primary working of lithic materials comes from Stratum 18a, which contains two cores, four bipolar flakes, and a significantly higher than expected proportion of angular shatter among debitage categories (X^2 = 26.614; df = 12; p = 0.009), perhaps the product of hard-hammer core bashing or block-on-block reduction. Overall, though, these materials are quite rare, and their near absence from the Pre-Archaic strata suggests that primary reduction activities rarely occurred within the rockshelter.

Evidence for secondary working is much more substantial. Debitage pieces from all three Pre-Archaic strata are very small or small (a Kruskal-Wallis test comparing mean debitage size ranks failed to identify a significant difference between the assemblages [X^2 = 1.205; df = 2; p = 0.548]), and all of the Pre-Archaic debitage assemblages are dominated by biface thinning flakes, retouch chips, and retouch chip fragments (66.4 percent for 18b, 58.8 percent for 18a, and 67.7 percent for 17b') as well as complex platforms (69.9 percent for 18b, 61.3 percent for 18a, and 81.3 percent for 17b'). Clearly, the chief technological activities carried out in the rockshelter related to the reduction and rejuvenation of tools, primarily bifaces. Although not scored metrically, most of the flakes, biface thinning flakes, and retouch chips from the Pre-Archaic strata have small platforms and weak bulbs of percussion, suggesting the use of soft-hammer percussion or pressure flaking in the retouching of both the bifaces and unifaces.

Interestingly, Stratum 18a seems to differ from 18b and 17b' in terms of secondary reduction activities. Stratum 18a has fewer biface thinning flakes and more smooth platforms than expected, indicating that unifacial as well as bifacial tools were being retouched and resharpened. Not surprisingly, Stratum 18a also has the highest frequency of flake tools of any of the Pre-Archaic assemblages. These differences indicate that a greater variety of technological activities occurred during the Stratum 18a occupation, and that this assemblage may represent a slightly longer period of occupation than those associated with Strata 18b and 17b'.

Pre-Archaic Techno-Typological Affinities

Strata 18b, 18a, and 17b' have all yielded stemmed bifacial points, as have Strata 12 and 10 in the rockshelter's East Block. Together, these points indicate that at Bonneville Estates Rockshelter the Great Basin's Stemmed point complex (i.e., the Mount Moriah Complex of Bryan [1979, 1988] or Stemmed Point Tradition of Bryan [1980]) spans from about 10,800 to 9500 ^{14}C BP (13,000 to 10,700 cal BP). Sample sizes are still small, however, so we cannot reliably define any chrono-stratigraphic succession of stemmed point styles within the Bonneville Estates record. We do have two Parman points from deposits predating 10,000 ^{14}C BP (11,300 cal BP) (in the east and West Blocks), Haskett points from deposits dating to about 10,800–10,500 and 9500 ^{14}C BP (13,000–12,400 and 10,700 cal BP) (in just the West Block), and a Windust point only from deposits dating to about 9500 ^{14}C BP (10,700 cal BP) (in the West Block).

Because Stratum 19 has yet to produce any diagnostic bifacial points, it is not yet possible to determine whether this occupation (if it is an occupation at all) relates to the Great Basin's stemmed point complex or to an earlier fluted point complex (e.g., Clovis). Continued excavations in the rockshelter may shed more light on this issue.

BONNEVILLE ESTATES ROCKSHELTER'S EARLY ARCHAIC LITHIC ASSEMBLAGES

Through the 2005 excavations at Bonneville Estates Rockshelter, the Early Archaic assemblages consisted of materials from the West Block ranging in age from about 7400 to 6000 ^{14}C BP (8300 to 7100 cal BP), and materials from the East Block postdating 5400 ^{14}C BP (6300 cal BP) (Graf, this volume; Hockett, this volume).

TABLE 9.6. Debitage Categories by Stratigraphic Layer, Early Archaic

	Stratigraphic Layer							
Debitage Category	14	15	16	17a	17b	14–17b[a]	PM[b]	Total
Flake fragment	321	6	69	16	3	10	15	440
Complete flake	197	2	28	9	4	1	2	243
Bipolar flake	55	—	5	1	—	1	—	62
Cortical spall fragment	8	—	2	—	—	—	—	10
Primary cortical spall	33	—	9	—	—	2	1	45
Secondary cortical spall	80	—	10	4	1	2	1	98
Retouch chip fragment	370	4	60	9	2	6	15	466
Retouch chip	200	1	17	1	1	4	2	226
Biface thinning flake	713	4	119	29	10	13	42	930
Angular shatter	375	3	47	15	—	7	1	448
Unworked cobble	2	—	—	—	—	—	—	2
Initially flaked cobble	—	—	—	—	—	—	1	1
Unidirectional flake core	—	1	—	—	—	—	—	1
Bidirectional flake core	—	—	—	—	—	1	—	1
Bipolar flake core	1	—	—	—	—	—	—	1
Total	2,356	20	366	84	21	47	80	2,974

[a] Artifacts from Early Archaic strata, but where 14, 15, 16, 17a, and/or 17b could not be discerned stratigraphically.
[b] Artifacts from packrat midden deposits between Strata 17a and 16.

The West Block's Early Archaic Lithic Assemblage

Only the materials from the West Block are included in the lithic analysis presented here, a total of 2,974 debitage pieces and 117 tools (Table 9.6). Of these, 2,847 can be precisely attributed to either Stratum 17b, 17a, 16, 15, or 14, while 80 come from a fossil wood rat midden (PM2 and PM1) stratigraphically occurring between Strata 17a and 16. Forty-seven debitage pieces have less precise stratigraphic provenience (due to differences in the preservation of these strata) and are assigned to Strata 17b–14. Each of these lithic assemblages is described in detail below, starting with the lowest stratigraphic layer.

STRATUM 17B

Only 22 lithic artifacts have been recovered from Stratum 17b, the silt and rubble zone stratigraphically separating the Pre-Archaic layers from the later Early Archaic layers. Of these, 54.5 percent are made on CCS, 22.7 percent on obsidian, and 22.7 percent on FGV (Table 9.7). Debitage categories are dominated by biface thinning flakes (47.6 percent) and flakes (19.0 percent), while retouch chip fragments (9.5 percent), retouch chips (4.8 percent), and cortical spalls (4.8 percent) are less common (Table 9.6). Two (9.1 percent) artifacts bear cortex (Table 9.8), and debitage sizes are small (71.4 percent) or very small (28.6 percent). Platform surfaces are chiefly complex (66.7 percent), while smooth

TABLE 9.7. Raw Material, Early Archaic

	Stratigraphic Layer							
Raw Material	14	15	16	17a	17b	14–17b[a]	PM[b]	Total
Cryptocrystalline silicates	1,613	16	230	47	12	22	18	1,958
Obsidian	423	3	84	19	5	13	28	575
Fine-grained volcanics	395	3	63	18	5	14	34	532
Quartzite	6	—	1	—	—	—	—	7
Granite	2	—	—	—	—	—	—	2
Quartz	10	—	—	—	—	—	—	10
Limestone/dolomite	—	—	2	—	—	—	—	2
Conglomerate	4	1	—	—	—	—	—	5
Total	2,453	23	380	84	22	49	80	3,091

[a] Artifacts from Early Archaic strata, but where 14, 15, 16, 17a, and/or 17b could not be discerned stratigraphically.
[b] Artifacts from packrat midden deposits between Strata 17a and 16.

TABLE 9.8. Amount of Cortex on Dorsal Face, Early Archaic

	Stratigraphic Layer							
Amount of Cortex	14	15	16	17a	17b	14–17b[a]	PM[b]	Total
0%	2,310	22	356	80	20	44	77	2,909
< 10%	43	—	6	3	2	3	—	57
10\-49%	47	—	9	1	—	—	1	58
50\-90%	29	—	5	—	—	1	1	36
> 90%	22	1	4	—	—	1	—	28
Total	2,451	23	380	84	22	49	79	3,088

[a] Artifacts from Early Archaic strata, but where 14, 15, 16, 17a, and/or 17b could not be discerned stratigraphically.
[b] Artifacts from packrat midden deposits between Strata 17a and 16.

surfaces are less common (33.3 percent) (Table 9.9). The single tool from Stratum 17b is a marginally retouched flake made on FGV (Table 9.10).

Stratum 17a

Eighty-four debitage pieces have been recovered from Stratum 17a. Most of these were associated with an unlined hearth ^{14}C dated to about 7400 ^{14}C BP (8300 cal BP). Raw materials include CCS (56.0 percent), obsidian (22.6 percent), and FGV (21.4 percent) (Table 9.7). Debitage categories are dominated by biface thinning flakes (34.5 percent) and angular shatter (17.9 percent), while flakes (10.7 percent), retouch chip fragments (10.7 percent), and cortical spalls (4.8 percent) are less common, and retouch chips (1.2 percent) and bipolar flakes (1.2 percent) are rare (Table 9.6). Cortex occurs on 4.8 percent of the debitage (Table 9.8), and debitage sizes are medium (2.4 percent), small (63.1 percent), and very small (34.5 percent). Platform surfaces are chiefly complex (68.9

TABLE 9.9. Platform Surface Preparation, Early Archaic

Platform Category	Stratigraphic Layer							Total
	14	15	16	17A	17B	14–17B[a]	PM[b]	
Cortical	20	—	8	—	—	1	—	29
Smooth	426	3	51	11	5	8	5	509
Complex	748	4	121	31	10	13	41	968
Crushed	101	—	12	3	—	—	—	116
Total	1,295	7	192	45	15	22	46	1,622

[a] Artifacts from Early Archaic strata, but where 14, 15, 16, 17a, and/or 17b could not be discerned stratigraphically.
[b] Artifacts from packrat midden deposits between Strata 17a and 16.

percent) or smooth (24.4 percent), but crushed platforms (6.7 percent) also occur (Table 9.9).

Stratum PM

An assemblage of 80 debitage pieces has been recovered from Strata PM2 and PM1, the fossil wood rat midden deposit stratigraphically situated between Strata 17a and 16. Raw materials include FGV (42.5 percent), obsidian (35.0 percent), and CCS (22.5 percent) (Table 9.7). Debitage pieces are chiefly biface thinning flakes (52.5 percent) or retouch chip fragments (18.8 percent); however, retouch chips (2.5 percent), complete flakes (2.5 percent), cortical spalls (2.5 percent), an angular shatter fragment (1.3 percent), and an initially flaked cobble (1.3 percent) also occur (Table 9.6). Only 2.5 percent of these debitage pieces bear cortex (Table 9.8), and most are very small (60.9 percent) or small (39.1 percent). Virtually all platforms are complex (89.1 percent) (Table 9.9).

Stratum 16

Stratum 16 is a well-preserved layer with numerous plant macrofossils and perishable artifacts, including the base of a coiled basket directly ^{14}C dated to 7200 ^{14}C BP (8100 cal BP) (Graf, this volume). Among the lithic artifacts recovered from this stratum are 366 debitage pieces and 14 tools. In this assemblage, CCS is most common (60.5 percent), while obsidian (22.1 percent), FGV (16.6 percent), limestone/dolomite (0.5 percent), and quartzite (0.3 percent) occur less frequently (Table 9.7). Eleven obsidian artifacts have been characterized using XRF analysis; the majority (seven) appear to have originated from the local Ferguson Wash source, while three originated from Browns Bench (Nevada) and one from Topaz Mountain (Utah). The debitage assemblage contains numerous biface thinning flakes (32.5 percent), retouch chip fragments (16.4 percent), and angular shatter fragments (12.8 percent) (Table 9.6). Less common are complete flakes (7.7 percent), cortical spalls (5.7 percent), retouch chips (4.6 percent), and bipolar flakes (1.4 percent).

Interestingly, 6.3 percent of the artifacts from Stratum 16 bear cortex on their dorsal surfaces (Table 9.8). Debitage sizes are typically small (56.1 percent) or very small (42.5 percent), but medium-sized debitage pieces also occur (1.4 percent). Most platform surfaces are complex (63.0 percent), but smooth (26.6 percent), crushed (6.3 percent), and cortical (4.2 percent) platforms are also present (Table 9.9).

Of the 14 tools from Stratum 16, one is a bifacial point, five are bifaces, six are flake tools, and two are ground stone tools (Figure 9.3, Table 9.10). The bifacial point is an unidentifiable FGV fragment. All of the bifaces are made on CCS, and they include two unidentifiable fragments, one edged (Stage 2) biface (Figure 9.3i), one preform (Stage 4), and one finished but unhafted biface (Stage 5). The finished biface is a midsection fragment that has one notch; however, it is not clear whether this was a notch formed from the side or corner of the point (Figure 9.3p). Flake tools include three retouched flakes (two obsidian, one CCS), two side scrapers (both CCS) (Figure 9.3m), and one end scraper (CCS) (Figure 9.3n). The ground stone tools consist of two mano fragments, one made on quartzite and the other on limestone/dolomite.

Stratum 15

From Stratum 15 comes a small assemblage of 20 debitage pieces and three tools. Stratum 15 is a thin layer of silt that separates Stratum 16 from 14, and its artifacts are probably from one or the other of these two layers. Raw materials are chiefly CCS (69.6 percent), while obsidian (13.0 percent), FGV (13.0 percent), and a conglomerate (4.3 percent) are uncommon (Table 9.7). Debitage categories consist of biface thinning flakes (20.0 percent), retouch chip fragments (20.0 percent), angular shatter (15.0 percent), and complete flakes (10.0 percent) (Table 9.6). One artifact has cortex (4.3 percent), and debitage sizes are small (60.0 percent) and very small (40.0 percent) (Table 9.8). Platforms are complex (57.1 percent) or smooth (42.9 percent) (Table 9.9). The three tools include one edged (Stage 2) biface (CCS), one thinned (Stage 3) biface (CCS) (Figure 9.3o), and one metate fragment (conglomerate) (Table 9.10).

Stratum 14

This Early Archaic stratum represents the most extensive early occupation uncovered in Bonneville Estates Rockshelter; through the 2005 field season, 2,356 debitage pieces and 97 stone tools had been recovered. Associated with these lithic remains were well-preserved hearths, faunal remains (Hockett, this volume), macrobotanical remains (Rhode, this volume), and an extensive assemblage of human coprolites, cordage, and basket fragments.

The lithic assemblage is made up primarily of CCS (65.8 percent), FGV (16.1 percent), and obsidian (17.2 percent), but quartz (0.4 percent), quartzite (0.2 percent), and granite (0.1 percent) artifacts also occur (Table 9.7). Thirteen obsidian artifacts have been subjected to XRF analysis. Results suggest that five pieces originated from the local source, Ferguson Wash, while four pieces came from Topaz Mountain (Utah) and four came from Wild Horse Canyon (Utah). Among debitage pieces, biface thinning flakes dominate (30.3 percent) (Table

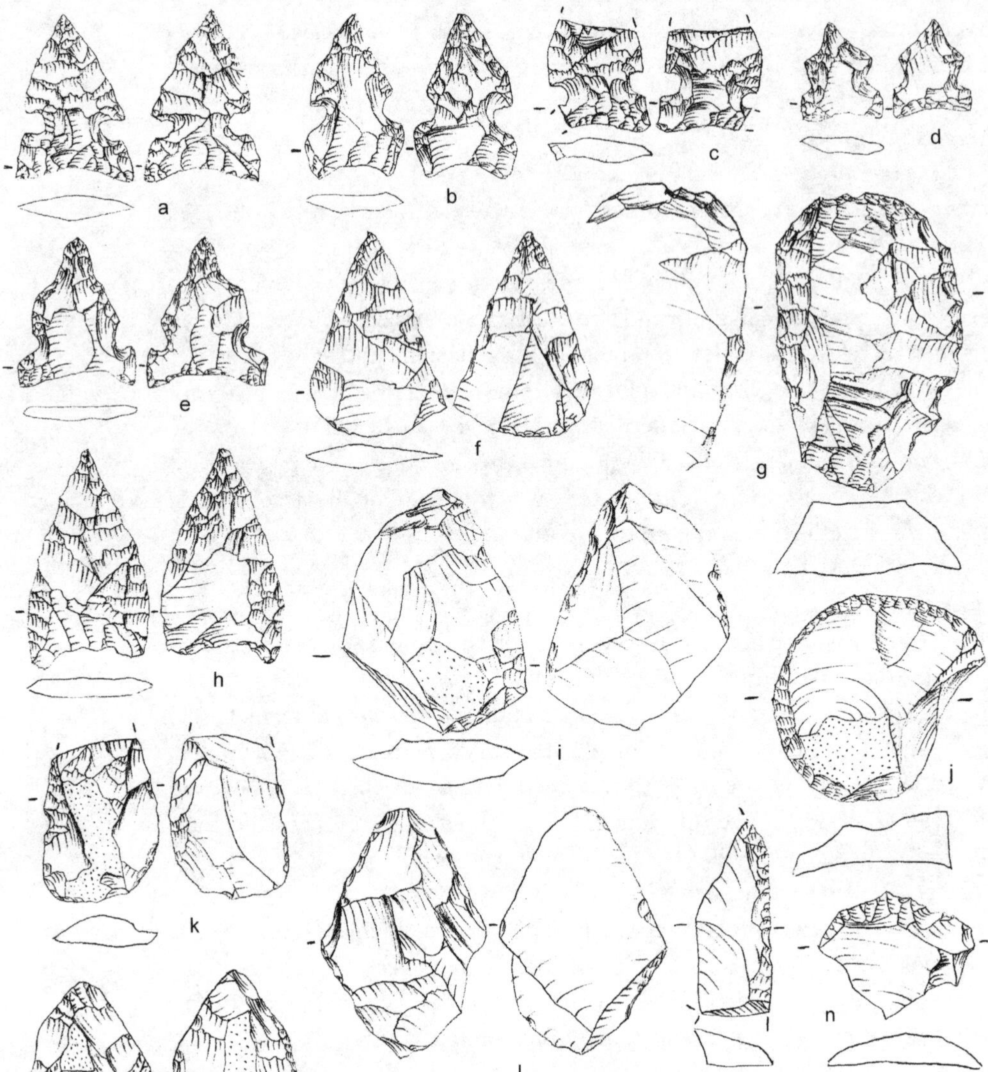

FIGURE 9.3. Lithic artifacts from the Early Archaic strata (Component 5) of Bonneville Estates Rockshelter: *a–e*, large side-notched points; *f, h*, biface preforms; *g, j, n*, end scrapers; *i, k, l, o*, edged bifaces; *m*, side scraper fragment; *p*, notched point fragment (*a–h, j–l*, Stratum 14; *o*, Stratum 15; *i, m, n, p*, Stratum 16).

9.6). Angular shatter fragments (15.9 percent), retouch chip fragments (15.7 percent), retouch chips (8.5 percent), complete flakes (8.4 percent), and cortical spalls (5.1 percent) are common, and bipolar flakes (2.3 percent), unworked cobbles (0.1 percent), and cores (0.1 percent) are rare. The cores include one unidirectional flake core made on FGV and one bipolar core made on obsidian (MLD = 34.65 mm).

TABLE 9.10. Tool Forms, Early Archaic

	Stratigraphic Layer							
Tool Category	14	15	16	17a	17b	14–17b[a]	PM[b]	Total
Bifacial points								
Large side-notched	12	—	—	—	—	—	—	12
Concave based	1	—	—	—	—	—	—	1
Corner-notched	3	—	—	—	—	—	—	3
Fragment	10	—	1	—	—	1	—	12
Biface								
Blank (Stage 1)	3	—	—	—	—	—	—	3
Edged biface (Stage 2)	28	1	1	—	—	—	—	30
Thinned biface (Stage 3)	4	1	—	—	—	—	—	5
Preform (Stage 4)	3	—	1	—	—	—	—	4
Finished biface (Stage 5)	1	—	—	—	—	—	—	1
Fragment	5	—	2	—	—	1	—	8
Retouched flake	13	—	3	—	1	—	—	17
End scraper	3	—	1	—	—	—	—	4
Side scraper	1	—	2	—	—	—	—	3
Metate fragment	6	1	—	—	—	—	—	7
Mano fragment	5	—	2	—	—	—	—	7
Total	97	3	14	0	1	2	—	117

[a] Artifacts from Early Archaic strata, but where 14, 15, 16, 17a, and/or 17b could not be discerned stratigraphically.
[b] Artifacts from packrat midden deposits between Strata 17a and 16.

Stratum 14 yielded 141 artifacts (5.8 percent) that bear cortex on their dorsal surfaces (Table 9.8). Most debitage pieces are very small (50.9 percent) or small (47.9 percent), although there are 26 (1.1 percent) medium and two (0.1 percent) large (> 5 cm^2) pieces. Among platforms, complex surfaces are most common (57.8 percent), while smooth (32.9 percent), crushed (7.8 percent), and cortical (1.5 percent) surfaces are less common (Table 9.9).

Retouched tools from Stratum 14 include 26 bifacial points (26.8 percent), 43 bifaces (44.3 percent), 17 flake tools (17.5 percent), and 11 ground stone artifacts (11.3 percent) (Table 9.10). Among the type-able projectile points, large side-notched points (93.8 percent) are most common (Figure 9.3a–e), but a concave-based point (Figure 9.3h) also occurs. The concave-based point may be an unnotched preform of a large side-notched point. Twenty-eight (73.7 percent) of the 38 type-able bifaces are edged (Stage 2) bifaces (Figure 9.3f, k–l). Other biface types (i.e., blanks, thinned bifaces, and preforms) occur in much lower frequencies. Most flake tools are marginally retouched flakes (76.5 percent), but end scrapers (17.6 percent) (Figure 9.3j) and a side scraper (5.9 percent) (Figure 9.3g) are also present. In addition to these flaked stone tools, there are 11 ground stone artifacts in the Stratum 14 assemblage. Among these are six metate fragments and five mano fragments.

STRATA 17B–14

Among the Early Archaic artifacts from Bonneville Estates Rockshelter, 47 debitage pieces and two tools cannot be unequivocally assigned to a specific stratum. Of these, 44.9 percent are on CCS, 28.6 percent on FGV, and 26.5 percent on obsidian (Table 9.7). Six obsidian artifacts have been subjected to XRF analysis; significantly, all of them come from the nearby Ferguson Wash source. The majority of debitage pieces are biface thinning flakes (27.7 percent), angular shatter fragments (14.9 percent), and retouch chip fragments (12.8 percent); also present in low frequencies are retouch chips (8.5 percent), cortical spalls (8.5 percent), a complete flake (2.1 percent), a bipolar flake (2.1 percent), and a bidirectional flake core (2.1 percent) (Table 9.6). Five artifacts (10.2 percent) bear cortex (Table 9.8), while 48.9 percent of the artifacts are very small, 48.9 percent are small, and 2.1 percent are medium-sized. Platforms are mostly complex (59.1 percent), but smooth (36.4 percent) and cortical (4.5 percent) platforms are also present (Table 9.9). The two tools from Strata 17b–14 are a nondiagnostic bifacial point fragment and unidentifiable biface fragment (Table 9.10).

Early Archaic Raw Material Procurement and Technological Activities

In terms of raw materials, the three main Early Archaic layers (Strata 17a, 16, and 14) are similar in that all are dominated by CCS and have moderate amounts of obsidian and FGV ($X2 = 9.969$; $df = 6$; $p = 0.126$).

All three Early Archaic assemblages are dominated by raw materials that were probably procured short distances from the rockshelter (less than 10 km). Although cortex is relatively rare in all of the Early Archaic assemblages ($X^2 = 0.361$; $df = 2$; $p = 0.835$), ranging from just 4.8 percent to 6.3 percent of each assemblage, there are numerous signs that primary reduction activities were taking place within the rockshelter. For example, Ferguson Wash obsidian and a gray chert appear to have been procured locally and carried to the rockshelter in nodule form and preliminarily worked there. This is further suggested by the occurrence, in similar proportions, of complete flakes (many of which are probably core reduction flakes), bipolar flakes, and angular shatter fragments in all three of the Early Archaic assemblages ($X^2 = 17.865$; $df = 12$; $p = 0.120$), as well as the occurrence of crushed platforms (a result of bipolar reduction) in similar proportions in all three assemblages ($X^2 = 12.222$; $df = 6$; $p = 0.057$). From these data, we can interpret that flakes were being manufactured inside the rockshelter, and that these were detached from unprepared or simply prepared nodules through direct percussion, core bashing, and the block-on-block bipolar technique.

Secondary reduction activities are also represented by articles recovered from the three Early Archaic strata. Biface thinning flakes and retouch chip fragments occur in similar proportions throughout, as do complex platforms. In fact, these are the most frequently occurring debitage types and platform types in all of the Early Archaic assemblages. Interestingly, many of the complex platforms observed on larger-sized biface thinning flakes are wide platforms resulting from hard-hammer percussion of early- to middle-stage bifaces. Mean debitage sizes are relatively small, especially for Stratum 14. A Kruskal-Wallis test comparing debitage sizes for the three assemblages indicates that debitage pieces in Stratum 14 are significantly smaller than those in Strata 16 and 17a ($X^2 = 16.425$; df = 2; $p < 0.001$). Perhaps this suggests that although secondary reduction activities were occurring throughout all three Early Archaic occupations, some bifaces and bifacial points were being worked to a further extent during the later, Stratum 14 occupation. This is supported by the preponderance of finished bifacial points in the Stratum 14 assemblage; many of these were manufactured on what appear to be locally procured raw materials.

Tool assemblages from Strata 16 and 14 (Stratum 17a has not yielded any tools) are similar in that both have examples of bifacial points, bifaces, flake tools, and ground stone artifacts, but are different in terms of the relative proportions of these tools. Stratum 16 is chiefly made up of bifaces and flake tools, while Stratum 14 has numerous bifaces and bifacial points. There are also many edged bifaces in the Stratum 14 assemblage. Some of these appear to have been purposefully manufactured just to this stage for utilization as a kind of expedient, unhafted bifacial knife. Thinned bifaces and preforms also occur, however, further indicating that some of these bifaces were on their way to becoming hafted bifacial points when they failed and broke apart. Among flake tools, both Stratum 16 and 14 have expediently retouched flakes as well as small numbers of more formal end and side scrapers. Ground stone artifacts are also common in both strata, with Stratum 14 having both metate and mano fragments, and Stratum 16 having just mano fragments. Together, these artifact forms suggest that the Early Archaic occupants of the rockshelter were grinding some plant foods, probably seeds.

Early Archaic Techno-Typological Affinities

Of all the Early Archaic assemblages described above, diagnostic bifacial points occur only in Stratum 14. All of these are large side-notched points that bear notches high upon their lateral margins (what Holmer [1986:104] refers to as "high" side-notched points). Throughout the northern Great Basin and Snake River basin of southern Idaho, these point forms consistently have been found in deposits dating to between 7500 and 4500 ^{14}C BP (8400 and 5200 cal BP) and thus are diagnostic of the Early Archaic period (Elston 2005; Holmer 1986; O'Connell and Inoway 1994). Some of the side-notched points from Stratum 14 have distinctive "comma-shaped" notches that are characteristic of Northern Side-notched points (Gruhn 1961; Holmer 1986). The samples from Strata 17a and 16 analyzed here have not yielded any diagnostic bifacial points, but Stratum 16 does have one fragment of a notched point, indicating that, like Stratum 14, it can be assigned generally to the Early Archaic. In addition, during the 2006 field season at Bonneville Estates, several large side-notched points were found in the Stratum 16 deposit, indicating that these were being used by the rockshelter's occupants as early as 7200 ^{14}C BP (8100 cal BP).

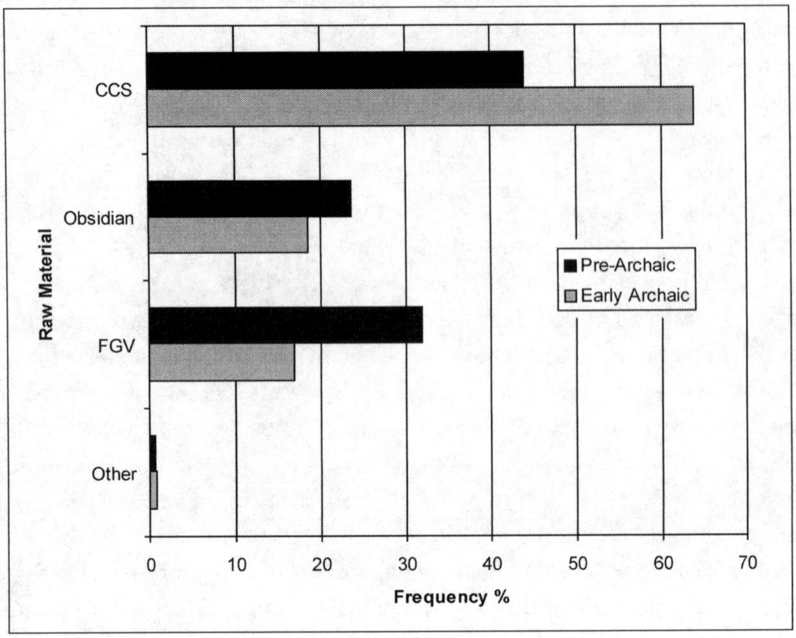

FIGURE 9.4. Proportions of lithic raw materials represented in the Pre-Archaic and Early Archaic components, Bonneville Estates Rockshelter ($X^2 = 149.842$; df = 3; $p < 0.001$). CCS = cryptocrystalline silicate; FGV = fine-grained volcanic.

PRE-ARCHAIC AND EARLY ARCHAIC COMPARISONS

Here I combine the lithic assemblages from Strata 19 through 17b' into a single Pre-Archaic component (11,000–9500 ^{14}C BP [13,000–10,700 cal BP]) and the lithic assemblages from Strata 17b through 14 into a single Early Archaic component (7400–6000 ^{14}C BP [8300–7100 cal BP]) in order to facilitate comparisons between the two groups of assemblages. Comparisons are made in terms of (1) raw material procurement, (2) technological activities carried out in the rockshelter, and (3) tool production. In analyzing the two components in this way, my goal is to determine whether the Pre-Archaic component is better characterized as Paleoindian or Paleoarchaic.

Raw Material Procurement

In Bonneville Estates Rockshelter's Early Archaic assemblage, CCS is by far the most prevalent raw material present, while in the Pre-Archaic lithic assemblage no single raw material type dominates (Figure 9.4). This difference in proportions of raw materials is significant (X2 = 149.842; df = 3; $p < 0.001$), suggesting that Early Archaic occupants preferred CCS as a toolstone, whereas Pre-Archaic occupants did not necessarily have such a predilection.

The Early Archaic component is characterized by an overwhelming reliance on locally procured raw materials, while the Pre-Archaic component is not. Specifically (although not presented in detail above), 36 percent of the Early Archaic assemblage is a poor-quality, gray-colored chert that was procured locally, probably in the Lead Mine Hills within 10 km of the rockshelter. This gray chert also occurs in the Pre-Archaic assemblage, but at a much lower

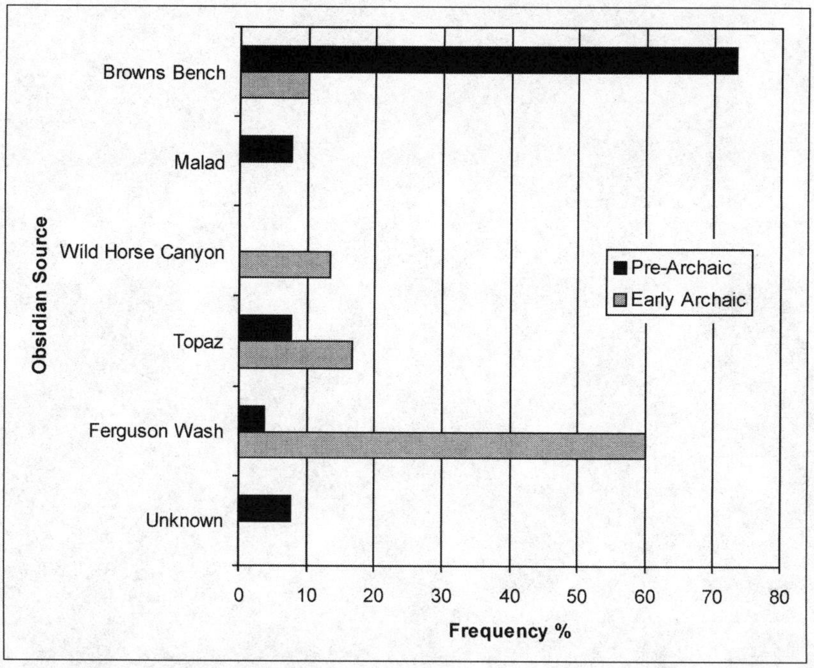

FIGURE 9.5. Proportions of obsidian sources represented in the Pre-Archaic and Early Archaic components, Bonneville Estates Rockshelter.

proportion (14 percent). Further, there is a major difference in the proportions of local vs. nonlocal obsidian sources represented in the two assemblages (Figure 9.5). In the Pre-Archaic assemblage, 19 of 26 (73.1 percent) of the obsidian artifacts analyzed have a chemical signature consistent with the Browns Bench source, located 180 km north-northwest of Bonneville Estates Rockshelter (Figure 9.6). In the Early Archaic assemblage, 18 of 30 (60.0 percent) of the obsidian artifacts appear to have originated from the Ferguson Wash obsidian source, located just 6 km south of the rockshelter, and five (16.7 percent) came from Topaz Mountain, located 105 km southeast of the rockshelter.

Differences in the degree of local procurement of raw materials also can be seen in the proportions of cortex present in the two assemblages (Figure 9.7). Specifically, 5.8 percent of the Early Archaic assemblage is made up of debitage pieces bearing cortex, while only 3.4 percent of the Pre-Archaic assemblage has cortex. This difference is significant ($X^2 = 10.752$; df = 4; p = 0.029). Thus, there are multiple lines of evidence suggesting a greater propensity toward local procurement of lithic raw materials during the Early Archaic period. This difference confirms an earlier assertion (Jones et al. 2003) that Pre-Archaic lithic conveyance zones in east-central Nevada were quite far-reaching and followed north-south-trending valleys, while later conveyance zones became more localized and east-west-trending. At Bonneville Estates Rockshelter, Pre-Archaic foragers made frequent moves beyond the Bonneville Basin, whereas Early Archaic foragers appear to have restricted their movements to within the southern Bonneville Basin.

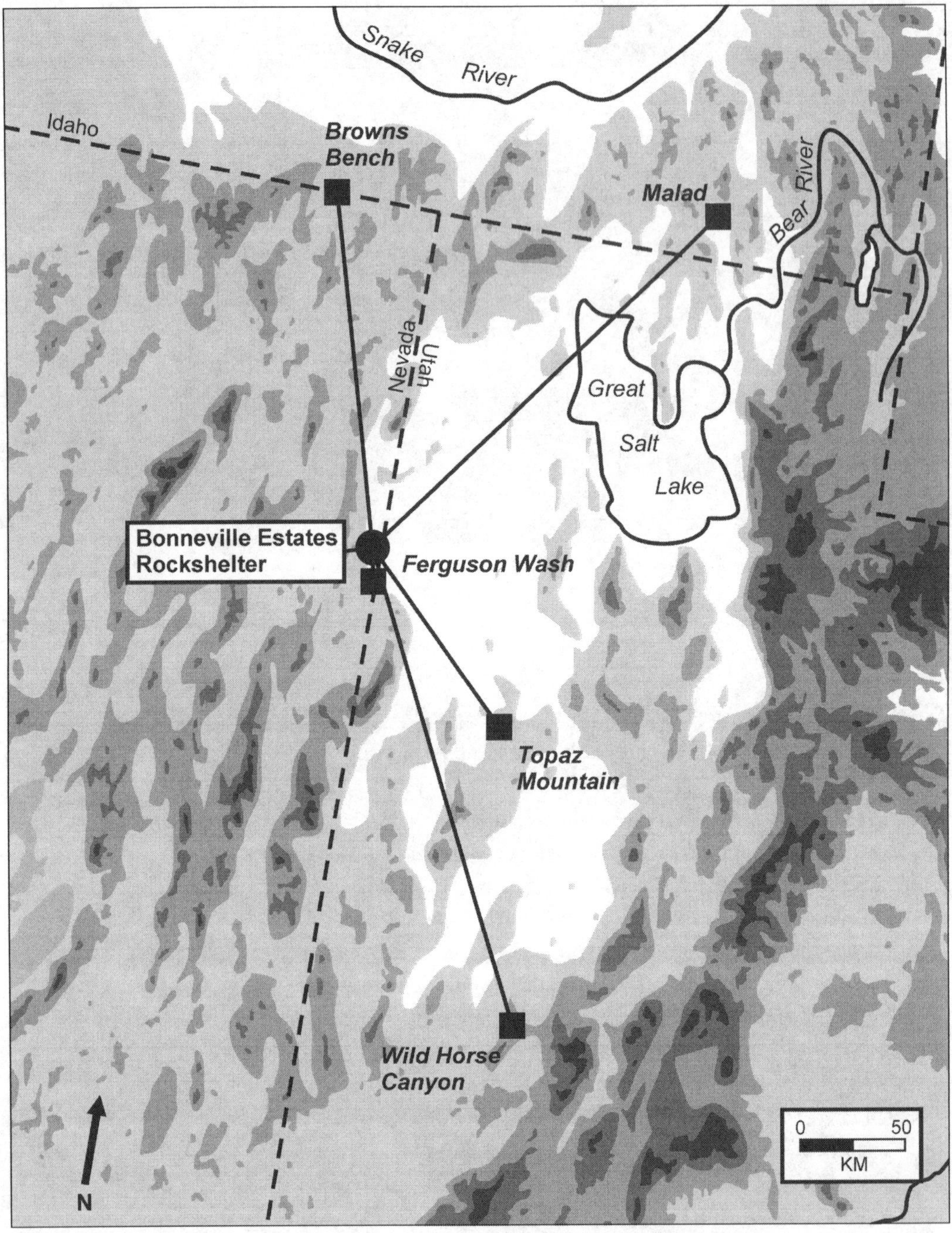

FIGURE 9.6. Obsidian source localities represented in the archaeological assemblages of Bonneville Estates Rockshelter. The Early Archaic component is dominated by obsidian from Ferguson Wash, Topaz Mountain, and Wild Horse Canyon, while the Pre-Archaic component is dominated by obsidian from Browns Bench.

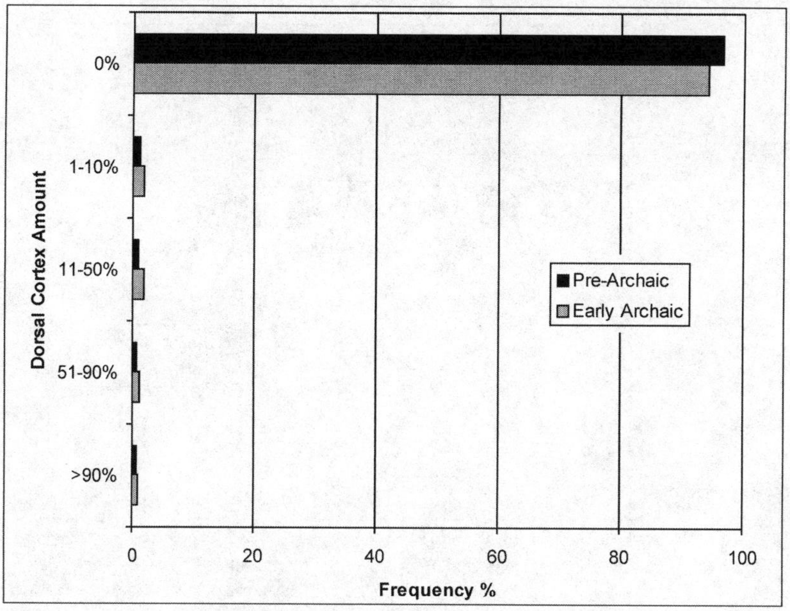

FIGURE 9.7. Proportions of dorsal cortex amount represented in the Pre-Archaic and Early Archaic components, Bonneville Estates Rockshelter (X^2 = 10.752; df = 4; p = 0.029).

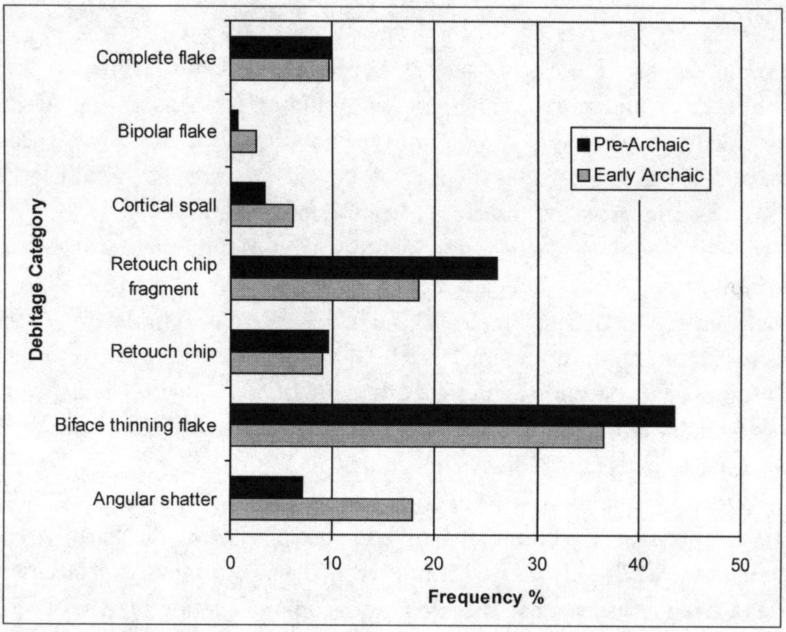

FIGURE 9.8. Proportions of debitage categories represented in the Pre-Archaic and Early Archaic components, Bonneville Estates Rockshelter (X^2 = 92.566; df = 6; p < 0.001).

Technological Activities

Major differences exist between the assemblages in terms of debitage category, debitage size, and platform preparation, suggesting that Early Archaic and Pre-Archaic technological activities in the rockshelter varied.

Frequencies of debitage categories in the Early Archaic component are significantly different from those in the Pre-Archaic component ($X^2 = 92.566$; df = 6; $p < 0.001$) (Figure 9.8). The Early Archaic assemblage has fewer than

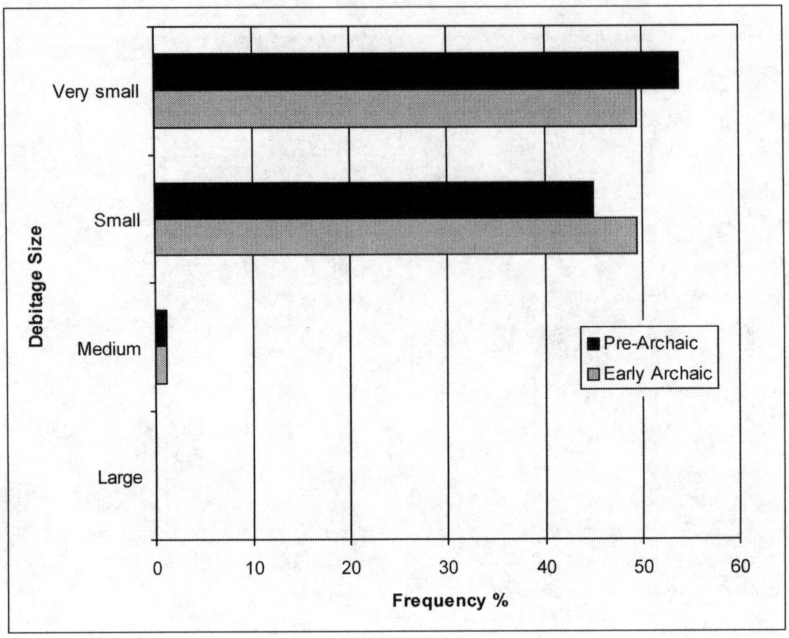

FIGURE 9.9. Proportions of debitage size ranks represented in the Pre-Archaic and Early Archaic components, Bonneville Estates Rockshelter: very small, < 1 cm^2; small, 1–3 cm^2; medium, 3–5 cm^2; large, > 5 cm^2 ($U = 1{,}314{,}043$; $p = 0.019$).

expected debitage pieces produced by secondary reduction activities (biface thinning flakes and retouch chip fragments), and a greater than expected number of debitage pieces produced by primary reduction (bipolar flakes, cortical spalls, and angular shatter fragments). For the Pre-Archaic assemblage, the opposite is true: products of secondary reduction occur more.

To analyze debitage size, a nonparametric Mann-Whitney test of mean ranks was conducted. Results indicate that although virtually all debitage pieces in both assemblages are very small and small, debitage pieces in the Pre-Archaic assemblage are significantly smaller than those in the Early Archaic assemblage ($U = 1314043$; $p = 0.019$) (Figure 9.9). This difference is further evidence that more primary reduction activities were taking place in the rockshelter during the Early Archaic than in Pre-Archaic times.

Platform preparation also differs by component (Figure 9.10). Although both assemblages are dominated by complex platforms, the Early Archaic assemblage contains more smooth and crushed platforms than expected, while the Pre-Archaic assemblage contains more complex platforms than expected ($X^2 = 28.987$; $df = 3$; $p < 0.001$).

Taken together, these data suggest that biface production and/or rejuvenation was the most common technological activity conducted in Bonneville Estates Rockshelter during the Early Archaic and Pre-Archaic occupations. This, however, is where the similarities end. During the Early Archaic occupation, primary reduction activities (e.g., cobble decortication, bipolar reduction, and blank production) also occurred frequently. In other words, many Early Archaic tools were manufactured within the rockshelter, and only a few finished tools were transported to Bonneville Estates from some other place.

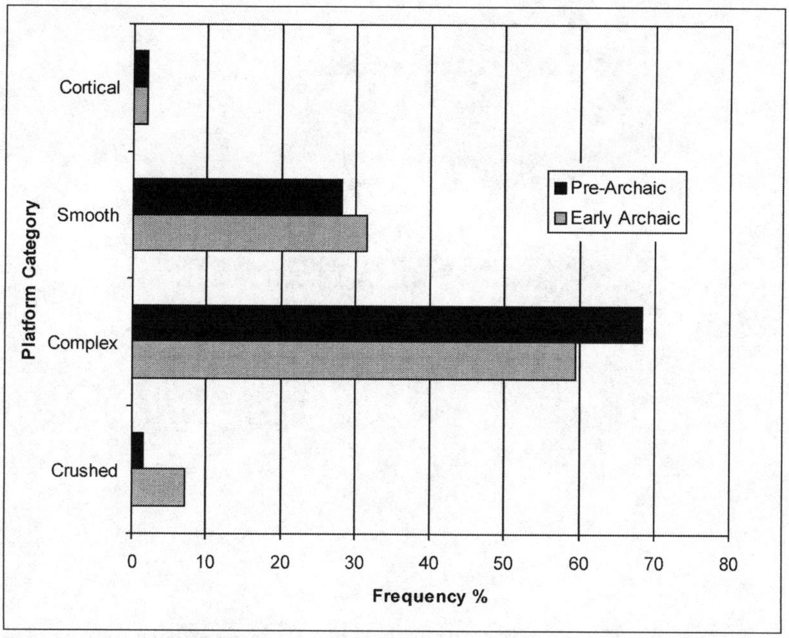

FIGURE 9.10. Proportions of platform categories represented in the Pre-Archaic and Early Archaic components, Bonneville Estates Rockshelter (X^2 = 28.987; df = 3; p < 0.001).

During the Pre-Archaic occupation, there are fewer indications of primary reduction activities, suggesting that most of the tools, especially bifacial points and bifaces, had been produced at some distance from the rockshelter and then transported there for use.

Tool Production

Both the Early Archaic and Pre-Archaic assemblages contain very few tools (3.8 percent and 2.6 percent, respectively). Although these proportions are not significantly different (X2 = 3.060; df = 1; p = 0.080), there are some important differences in tool production between the two components. The Early Archaic tool assemblage contains more unhafted bifaces than expected, whereas the Pre-Archaic tool assemblage contains more bifacial points and flake tools than expected (Figure 9.11). Further, there are 14 ground stone tools in the Early Archaic assemblage, but none in the Pre-Archaic assemblage. Together, these differences are statistically significant (X2 = 14.115; df = 3; p = 0.003).

Variation in production of bifacially worked tools is further highlighted in Figure 9.12. As shown in this graph, all biface stages are present in the Early Archaic assemblage, especially edged bifaces (Stage 2), while virtually no stages other than finished bifaces (Stage 5) are present in the Pre-Archaic assemblage. As discussed above, this is a clear indication that most of the Early Archaic bifacial points were being produced within the rockshelter, while most of the Pre-Archaic bifacial points were transported to the rockshelter from some other place on the landscape. More than this, there are some important differences in the manner in which these bifaces were resharpened. Early Archaic bifaces and points show clear signs of hard-hammer percussion, while Pre-Archaic bifaces

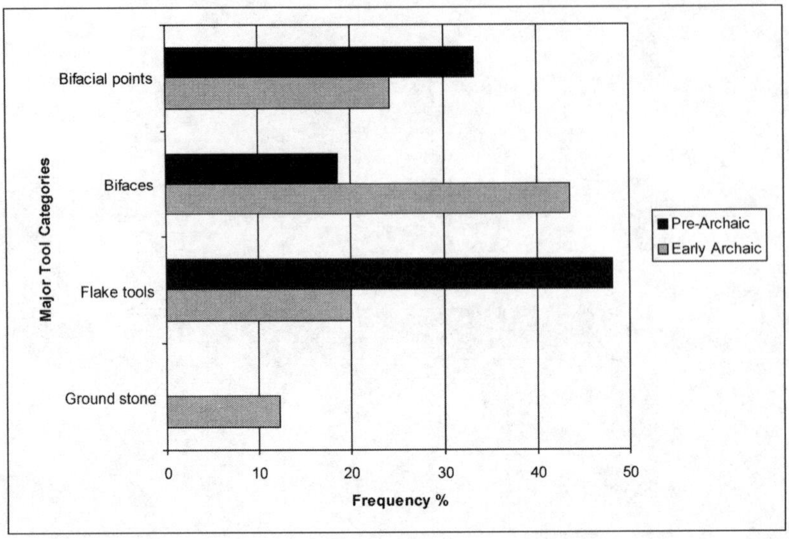

FIGURE 9.11. Proportions of major tool categories represented in the Pre-Archaic and Early Archaic components, Bonneville Estates Rockshelter ($X^2 = 14.315$; $df = 3$; $p = 0.003$).

and points show signs of soft-hammer percussion or pressure flaking. In terms of tool production, then, Early Archaic technology was much more expedient than Pre-Archaic technology, even in terms of biface production.

CONCLUSIONS: THE PRE-ARCHAIC OCCUPATION OF BONNEVILLE ESTATES ROCKSHELTER—PALEOINDIAN OR PALEOARCHAIC?

During Early Archaic times, hunter-gatherers transported stone nodules to the rockshelter for tool production. Most of these raw materials (cherts and obsidians) were procured locally, probably from outcrops as well as alluvial and lacustrine deposits in the Lead Mine Hills area, within 10 km of Bonneville Estates Rockshelter. Few raw materials were transported from greater distances; these include obsidians from Topaz Mountain and Wild Horse Canyon (Utah) as well as some fine-grained volcanic rocks (basalts) from some unknown source(s). Given these data, I conclude that the Early Archaic occupants of the rockshelter relied chiefly on the local lithic landscape for their toolstones, but also had access (through direct procurement or trade) to more distant obsidian sources in the southern Bonneville Basin.

Most of the lithic tools in the Early Archaic assemblage appear to have been manufactured within the rockshelter. Raw material packages were reduced through either the block-on-block technique or simple direct hard-hammer percussion. Many of the bifaces were manufactured only to the preliminary edged-biface stage (Stage 2), and these were coarsely reduced through the use of a large, hard hammer. Perhaps these served as expedient hand-held knives. Other expedient tools in the Early Archaic assemblage include marginally retouched flakes and a denticulate. In fact, virtually the only formal flaked stone

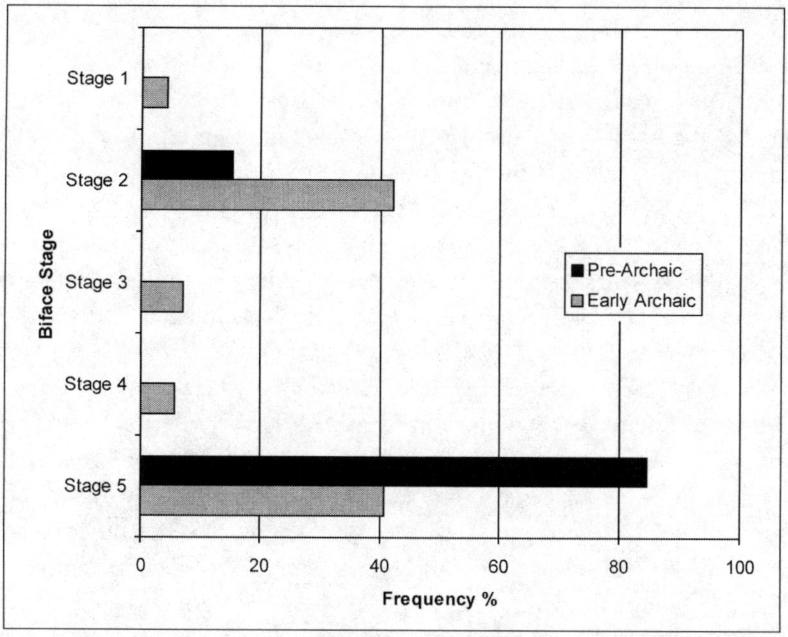

FIGURE 9.12. Proportions of biface stages represented in the Pre-Archaic and Early Archaic components, Bonneville Estates Rockshelter: Stage 1, blank; Stage 2, edged biface; Stage 3, thinned biface; Stage 4, preform; Stage 5, finished biface.

tools appear to be finished bifacial points. Thus, the local procurement of toolstone, its primary and secondary reduction within the rockshelter, and the use of expediently prepared lithic tools (bifaces and unifaces) all suggest that Early Archaic foragers practiced a tool production strategy focusing on the provisioning of places (Kuhn 1995). This implies relatively long stays in the rockshelter or nearby area and low levels of residential mobility during the Early Archaic. Even Stratum 17a, which certainly represents the shortest occupation of the rockshelter during the Early Archaic, is characterized by signs of expedient tool manufacture and place-provisioning. The occurrence of both flaked stone and ground stone tools in the Early Archaic assemblages, finally, implies significant diet breadth: specifically, that these middle Holocene foragers were incorporating small-sized seeds (Rhode and Louderback, this volume).

Pre-Archaic technological organization was decidedly different. The Pre-Archaic assemblage is characterized by a preponderance of nonlocal lithic raw materials. Obsidians are primarily from Browns Bench and Malad, two sources that lie nearly 200 km north of the rockshelter and outside of the Bonneville Basin. None of the bifacial points in the assemblage appear to have been manufactured within the rockshelter; instead they were transported to the rockshelter in finished form. The same can be said for some of the bifaces and flake tools, especially those produced on obsidian and fine-grained volcanic rocks. These were initially prepared at some other place on the landscape and then transported to the rockshelter as either blanks or finished tools. Based on these interpretations I conclude that Pre-Archaic lithic conveyance zones were much more extensive than those described above for the Early Archaic, and that most tools in the assemblage were prepared well in advance of use in the rockshelter. Further, the Pre-Archaic inhabitants of the rockshelter do not appear to have utilized ground stone technologies, perhaps indicating that their diets excluded the hard-to-process plant foods that later became so important to Early Archaic foragers of the Great Basin. Pre-Archaic technology at Bonneville Estates Rockshelter, then, was organized to facilitate relatively short and focused stays in the rockshelter, to take advantage of an extensive lithic conveyance system reaching beyond the Bonneville Basin, and to permit the provisioning of individuals with formal tools and blanks that could be easily transported across this large territory.

In terms of technological organization, then, the lithic data suggest that the Pre-Archaic inhabitants of Bonneville Estates Rockshelter were not behaving in the same manner as those of the Early Archaic. Instead, their inferred actions are much more reminiscent of late Paleoindian foragers known from other regions of North America (Goodyear 1989; Kelly 1996; Kelly and Todd 1988; Kuhn 1995). The extensive lithic conveyance zone, long-distance transport of finished tools, and formal character of the Pre-Archaic lithic assemblage imply elevated levels of mobility on par with those inferred for late Paleoindian foragers of the Great Plains and Midwest of North America (Goodyear 1989; Pitblado 2003). The locally oriented, place-provisioning behavior more typical of the Archaic of the Great Basin appeared much later at Bonneville Estates Rockshelter, not until close to 7400 ^{14}C BP (8300 cal BP). Thus, in terms of technological and settlement organization, the Pre-Archaic inhabitants of Bonneville Estates Rockshelter are better characterized as Paleoindian, not Paleoarchaic.

ACKNOWLEDGMENTS

Research at Bonneville Estates Rockshelter has been funded primarily by the National Science Foundation Archaeology Program (grant numbers BCS-0313813 and BCS-0514504), the U.S. Department of the Interior Bureau of Land Management, and the Sundance Archaeological Research Fund of the University of Nevada, Reno. Time and money were also provided by the many undergraduate and graduate students who enrolled in our archaeological field schools at Bonneville Estates Rockshelter during the 2000–2004 seasons. Special thanks to Craig Skinner, who conducted the XRF analysis of obsidian artifacts presented here. Kelly Graf and Bryan Hockett provided helpful comments on an early draft of this chapter.

PART IV

Subsistence Organization

10

Artiodactyl Use and Adaptive Discontinuity Across the Paleoarchaic/Archaic Transition in the Northern Great Basin

Ariane Pinson

In the northern Great Basin (Figure 10.1), the Paleoarchaic/Archaic transition (ca. 8000 ^{14}C BP [9000 cal BP]) coincides with a transition from the cooler, moister climates of the late Pleistocene and early Holocene to the generally warmer and more arid climates of subsequent time periods. This transition was accompanied by a significant reduction in the availability and quality of surface water across the landscape, significant shifts in vegetation distributions, and changes in the availability of food resources for humans who occupied that landscape (Madsen 1999; Wigand and Rhode 2002).

Culturally, the Paleoarchaic/Archaic transition is characterized by a shift from Great Basin Stemmed series projectile points to smaller, notched projectile points (e.g., Elko series). Somewhat later, circa 5300 ^{14}C BP (6000 cal BP), there is evidence for the development of more-sedentary lifeways in the form of stone houses and storage pits (see Jenkins et al. 2004). Ground stone tools, present even in Paleoarchaic sites (e.g., the Tucker site [35LK3227] [Pinson 2004]), become more elaborate and formal over time. Rock art and fiber arts (sandals, baskets) are present throughout the sequence (Connolly and Barker 2004; Connolly et al. 1998; Ricks and Cannon 1998).

Adaptation across the Paleoarchaic/Archaic transition in the northern Great Basin has been modeled as a period of increasing reliance on small game and plant resources as a natural consequence of the shift from mesic to xeric climates. This climate change resulted in a reduction in precipitation and the consequent loss or reduction of many surface water sources, as well as vegetation changes that reduced forage preferred by artiodactyls (e.g., Bedwell 1973; Elston 1982; Grayson 1993).

More recently, archaeologists have recognized that Paleoarchaic adaptations may have been much more generalized than earlier researchers had thought (e.g., Jenkins et al. 2004; Willig 1988, 1989, 1991). These researchers essentially argue that Paleoarchaic foragers followed a broad-spectrum adaptation similar

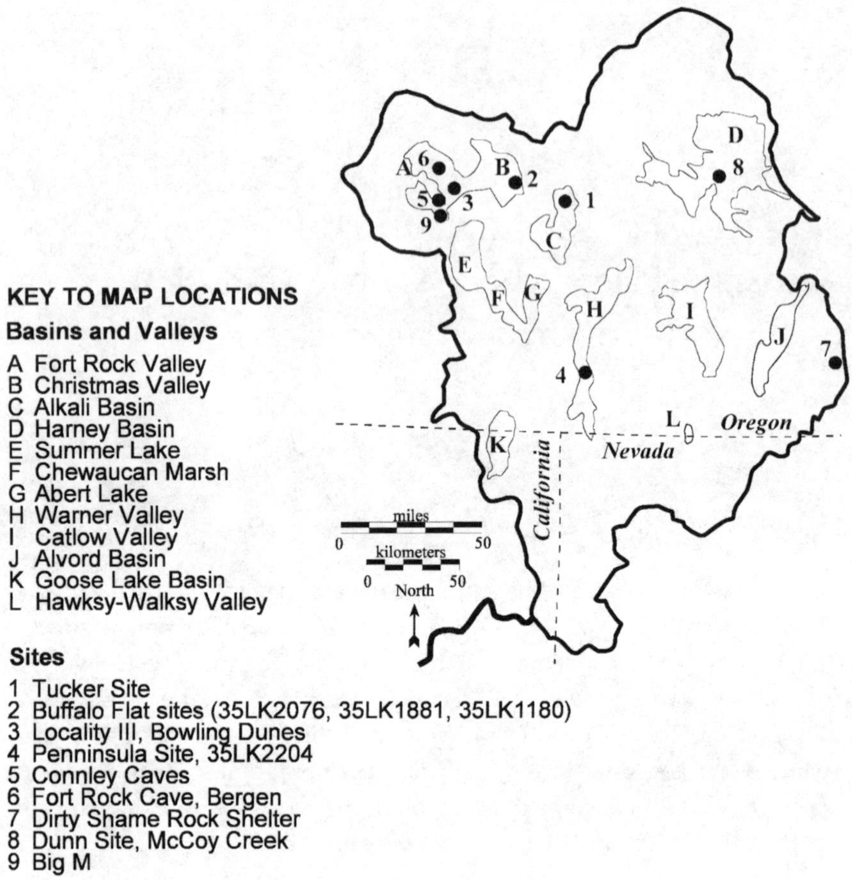

FIGURE 10.1. Map of the northern Great Basin showing sites and localities mentioned in the text.

to that in later time periods, with a diet that included a wide range of fish, small game, and plant resources. Thus, these researchers emphasize continuity of adaptation, arguing that dietary differences may not have been one of kind (wholesale shift in the types of resources used) but one of degree (shifting patterns of use in a relatively static resource set). However, even these researchers point to declining artiodactyl abundances due to climate change as one factor leading to greater diet breadth.

This chapter examines the degree of continuity and change in diet and adaptation across the Paleoarchaic/Archaic transition. Specifically, it investigates whether there were any changes in small and large game animal use in the northern Great Basin between Paleoarchaic and later populations. Published data on faunal assemblages from firmly dated, clearly cultural components of archaeological sites in the northern Great Basin are analyzed, and data on the abundance of lagomorph and artiodactyls extracted. These abundances are compared over time in order to better understand the shifts in terrestrial mammal use across the Paleoarchaic/Archaic transition. Changes in land use patterns and technology during the transition are also discussed in relation to the faunal assemblages. Finally, the implications of these data for foraging breadth across this transition are explored.

METHODS

The data used in this chapter consist of published counts of lagomorph and artiodactyl bones from 12 archaeological sites in the northern Great Basin. Sites from all occupation periods were included (Tables 10.1, 10.2, and 10.3). The 23 assemblages selected for analysis came from identifiably cultural components (e.g., pits, floors, hearths, houses) of archaeological sites to minimize mixing with fauna introduced to assemblages by noncultural means. Most of the selected assemblages were also unambiguously associated with one or more radiocarbon dates so that they could be firmly assigned to cultural periods for analysis. Radiocarbon dates were averaged using the method of Long and Rippeteau (1974) and calibrated to calendar dates using CalPal 2005 ver. 1.4 (CalPal 2006). A few of the assemblages were not directly dated, but were recovered from well-dated strata. Assemblages were excluded if they contained fewer than five identifiable specimens (NISP lagomorph + NISP artiodactyls = < 5).

The contribution of artiodactyls to the diet relative to lagomorphs is calculated using the artiodactyl index. This index is simply the ratio of artiodactyl bones to the sum of the artiodactyl and lagomorph bones:

$$\frac{NISP_{artio}}{NISP_{artio} + NISP_{lagomorph}}$$

Like all faunal quantification techniques, the results are sensitive to aggregation, fragmentation, and sample size effects (Grayson 1984). To minimize aggregation effects, the analysis includes only faunal remains from identified cultural contexts that could be assumed to be relatively short-term behavioral units, rather than averaging the fauna over an arbitrary number of levels. To account for fragmentation effects, which would be expected to reduce the rate of identification of one species relative to another, the artiodactyl index is calculated based on both NISP and on the size classes as identified by Thomas (1969). If the same results occur using both the NISP and size class data, differential fragmentation is probably not significantly influencing the outcome of the analysis.

Sample size is not correlated with mean radiocarbon date for either the NISP data set (Spearman's rho = -0.289, p-value = 0.181, alpha = 0.05) or the size class data set (Spearman's rho = -0.176, p-value = 0.421, alpha = 0.05).

The faunal data from the 23 assemblages in the northern Great Basin were evenly distributed among four cultural periods (Table 10.2, Figure 10.1): Boulder Village (2900 ^{14}C BP [3000 cal BP] to historic), Bergen (5300 to 2900 ^{14}C BP [6000 to 3000 cal BP]), Lunette Lake (8000 to 5300 ^{14}C BP [9000 to 6000 cal BP]), and Fort Rock (10,200 to 8000 ^{14}C BP [12,000 to 9000 cal BP]). No cultural assemblages date to the Paisley period (prior to 10,200 ^{14}C BP [12,000 cal

TABLE 10.1. Cultural Chronology of the Northern Great Basin

DATES— CAL BP (^{14}C BP)	CULTURAL PERIOD	DOMINANT CULTURAL PATTERNS, SITES, AND ARTIFACTS
3000 BP to historic period (2900 ^{14}C BP to historic period)	Boulder Village Period	Climatic changes make the basin wetlands less predictable (dry periods increasingly punctuate wet intervals); spring habitation sites appear in the uplands where more-predictable root crops are abundant, and lowland villages are positioned to provide access to both uplands and basin floor environments; some evidence of defensive considerations in village location and configuration (Boulder Village, Carlon Village).
6000 to 3000 BP (5300 to 2900 ^{14}C BP)	Bergen Period	Relatively wet period when basin lakes and marshes are rejuvenated, becoming very productive resource areas. Long-term occupations at lake margins with substantial houses, large storage pits; fish and other wetland resources most important; elaborate artifacts present, as well as exotic shell and stone beads providing evidence for long-distance contacts (Bergen, Big M, GP2, and DJ Ranch sites).
9000 to 6000 BP (8000 to 5300 ^{14}C BP)	Lunette Lake Period (Mount Mazama eruption ~7600 cal BP)	After ~7600 cal BP, Cascade style and Northern Side-notched projectile points; Multiple Warp and Spiral Weft sandals; decorated twined basketry; no evidence of residential structures, storage features, etc. Before ~7600 cal BP, a relatively dry period during which many lakes disappear, punctuated by brief wet intervals; temporary hunting and foraging camps, primarily on the margins of permanent lakes (Paulina Lake site) and intermittently active/seasonal lakes and ponds (Locality III site); Cascade-style projectile points.
12000 to ~9000 BP (10,200 to 8000 ^{14}C BP)	Fort Rock Period	Winter residential use of caves around the edges of substantial but receding lakes (Fort Rock Cave, Connley Caves, Cougar Mountain Cave); summer upland residential site at Paulina Lake; mass harvesting of occasionally abundant foods (Buffalo Flat); opportunistic hunting/foraging around receding lakes; preferential and repeated use of areas of localized abundance (ecotone areas) where stream channels enter lakes and both adjoin uplands (sites in North Alkali Basin, Chewaucan Marsh, and Alvord Basins); compared to later periods, less dietary emphasis on large game and grain resources (Tucker site); artifacts include Fort Rock-style sandals; Great Basin Stemmed series (GBSS) points, Great Basin Fluted (GBF) points; crescents; undecorated twined basketry; simple ground stone (manos and metates).
Prior to 12,000 BP (Prior to 10,200 ^{14}C BP)	Paisley Period (Clovis Horizon ~13,000 cal BP [~11,000 RCYBP])	GBF point (aka "Clovis") at Dietz site and occurring elsewhere as isolates. Associated with Glacier Peak tephra at East Wenatchee site (Washington). GBSS dated to 11,300–11,400 RCYBP at Cooper's Ferry site (Idaho), possibly other sites in and immediately adjacent to the Great Basin proper. Early radiocarbon and obsidian hydration ages from Connley, Paisley, and Fort Rock caves with possible cultural associations; ephemeral evidence: possible associations of cultural materials with the remains of mastodon, camel, horse, and bison at Fossil Lake, Paisley Caves, and Lower Klamath Lake.

Note: This chronology is modified from Jenkins et al. 2004.

TABLE 10.2. Description of Faunal Assemblages from Valley Floor Sites

Site Name (Feature Number)	^{14}C BP/ $\bar{X}$ ^{14}C BP	Spec. No.	Cal BP	Period	Environment During Occupation	Description of Assemblage	Data Source
Peninsula Site/ 35LK2579	$\bar{X}$=374±39 240±65 440±80 450±80 625±90	AA-19784 WSU-4197 WSU-4195 WSU-4194	347–484 136–403 360–532 366–541 556–664	Boulder Village	Sagebrush flat at the base of Hart Mountain, close to Hart Lake, in Warner Valley	Bones come from two house pits with overlapping radiocarbon dates, which were combined in the published data into a single assemblage.	Eiselt 1997a
McCoy Creek 35LK1263 (Feature 11)	480±70	Beta-31736	450–591	Boulder Village	Sagebrush flat bordering Diamond Swamp, in Harney Basin	Bones come from Feature 11, identified floor of a Wickiup.	Greenspan 1995b
McCoy Creek 35LK1263 (House Floor H2, Feature 7)	$\bar{X}$=1,190±39 990±90 1140±60 1340±60	Beta-28037 Beta-35150 Beta-31735	1,071–1,171 805–996 993–1146 1198–1306	Boulder Village	Sagebrush flat bordering Diamond Swamp, in Harney Basin	Bones come from Feature 7, a storage pit dug into Feature H2, identified as a house floor.	Greenspan 1995b
McCoy Creek 35LK1263 (House Floor H1, Feature 3)	1270±70	Beta-31734	1107–1265	Boulder Village	Sagebrush flat bordering Diamond Swamp, in Harney Basin	Bones come from Feature 3, a rock-lined hearth dug into Feature H1, identified as a house floor.	Greenspan 1995b
35LK2204 (Locus 9)	2070±110	not avail.	1926–2217	Boulder Village	Flagstaff Dune, a lunette dune proximate to Flagstaff Lake, in Warner Valley	Bones are from Locus 9, one of three charcoal concentrations observed at the site.	Moore 1993
Bowling Dune (Feature 2)	2890±70	Beta-75078	2939–3150	Boulder Village	Lunette dune adjacent to active paleo channel, in Fort Rock Basin	Bones come from Feature 2, a house floor.	Singer 2004
Bowling Dune (Feature 9)	3060±60	Beta-75083	3186–3343	Bergen	Lunette dune adjacent to active paleo channel, in Fort Rock Basin	Bones come from Feature 9, an area of darkly stained anthropic soil identified as a living floor.	Singer 2004
Dunn Site (Feature 7)	3255±65	Beta-31537; ETH-5660	3420–3564	Bergen	Sagebrush flat bordering Diamond Swamp	Bones come from Feature 7, a living surface laterally adjacent to a house pit containing a hearth and storage pit (Features 1–6), with which it is assumed to be contemporaneous.	Greenspan 1995a

TABLE 10.2. (continued) Description of Faunal Assemblages from Valley Floor Sites

SITE NAME (FEATURE NUMBER)	^{14}C BP/ $\bar{X}$ ^{14}C BP	SPEC. NO.	CAL BP	PERIOD	ENVIRONMENT DURING OCCUPATION	DESCRIPTION OF ASSEMBLAGE	DATA SOURCE
Dunn Site House (Features 1–6)	3255±65	Beta-31537; ETH-5660	3420–3564	Bergen	Sagebrush flat bordering Diamond Swamp	Bones come from Features 1-6, a house pit, hearth, and storage pit.	Greenspan 1995a
Bowling Dune (Feature 8)	4460±160	Beta-67523	4904–5313	Bergen	Lunette dune adjacent to active paleo channel, in Fort Rock Basin	Bones come from Feature 8, a storage pit associated with several other pit and hearth features collectively identified as being part of a living floor or house floor.	Singer 2004
Big M (Structure 3)	4910±60	Beta-57007	5609–5717	Bergen	Sagebrush flat adjacent to active paleo channel, in Fort Rock Basin	Bones come from stained sediment identified as Structure 3, which also contained a hearth (Feature 2).	Dean 1994
Bergen House (Quad A, Unit 3, Levels 9–13)	5190±40	Beta-153979	5925–5988	Bergen	Lunette dune adjacent to paleo Lake Beasely, in Fort Rock Basin	Bones come from Unit 3, Quad A, Levels 11–13, a portion of a house floor immediately adjacent to Feature 3, a hearth.	O'Grady 2004
Bowling Dune (Feature 18)	5380±70	Beta-75076	6052–6257	Lunette Lake	Lunette dune adjacent to active paleo channel, in Fort Rock Basin	Bones come from Feature 18, a living floor, also producing lithics, ground stone, and seeds.	Singer 2004
Bowling Dune (Feature 12)	6040±120	Beta-75080	6761–7086	Lunette Lake	Lunette dune adjacent to active paleo channel, in Fort Rock Basin	Bones come from Feature 12, a large (2.3 m diameter) hearth.	Singer 2004
Locality III (Unit B-5, Hearth EB-2)	6370±190	WSU-4864	7036–7436	Lunette Lake	Lunette dune adjacent to active paleo channel, in Fort Rock Basin	Bones from Levels 8, 9, and 11 in Hearth EB-2 in Structure B-5, representing 95% of B-5 faunal assemblage.	O'Grady 2004
Locality III (Unit B-7, Hearth EB-8)	7040±135	WSU-4866	7738–7988	Lunette Lake	Lunette dune adjacent to active paleo channel, in Fort Rock Basin	Bones from levels 11–14, in Hearth EB-8, in Unit B-7, representing ~83% of assemblage from B-7.	O'Grady 2004
Locality III (Unit B-2, Hearth EB-3)	$\bar{X}$=7155±81 7445±190 7010±90	WSU-4863 Beta 990020	7897–8076 8070–8428 7747–7927	Lunette Lake	Lunette dune adjacent to active paleo channel, in Fort Rock Basin	Bones from levels 6–7 from Hearth EB-3 in Unit B-2, representing 87% of fauna in B-2.	O'Grady 2004

TABLE 10.2. (continued) Description of Faunal Assemblages from Valley Floor Sites

SITE NAME (FEATURE NUMBER)	^{14}C BP/ $\bar{X}$ ^{14}C BP	SPEC. NO.	CAL BP	PERIOD	ENVIRONMENT DURING OCCUPATION	DESCRIPTION OF ASSEMBLAGE	DATA SOURCE
Bowling Dune (Feature 22)	ca. < 6850 Undated, but stratigraphically immed. post-Mazama	n/a	ca. < 7600	Lunette Lake	Lunette dune adjacent to active paleo channel, in Fort Rock Basin	Bones come from Feature 22, a storage pit approx. 90 cm in diameter and 30 cm deep, also containing charcoal and lithic cultural materials.	O'Grady 2004
Bowling Dune (Feature 22)	ca. < 6850 Undated, but stratigraphically immed. post-Mazama	n/a	ca. < 7600	Lunette Lake	Lunette dune adjacent to active paleo channel, in Fort Rock Basin	Bones come from Feature 15, a thin deposit consisting of scattered patches of charcoal-stained soil, bones, chipped stone, and ground stone, collectively identified as camp debris.	Singer 2004
35LK2076 (Feature 4)	$\bar{X}$=8810±103 8780±120 8870±200		9699–10,087 9657–10,073 9688–10,189	Fort Rock	Sagebrush/shadscale flat, in Fort Rock Basin	Bones come from Feature 4.	Oetting 1993; Greenspan 1993
35LK1881 (pit-like feature)	$\bar{X}$=8983±69 8880±120 8950±120 9120±120		9971–10,208 9759–10,138 9836–10,195 10,181–10,463	Fort Rock	Sagebrush/shadscale flat, in Fort Rock Basin	Bones from a pit feature also containing charcoal, tools, flakes, and cores.	Oetting 1993; Greenspan 1993
35LK1180	$\bar{X}$=9327±276 8220±430 9880±360		10,239–11,001 8655–9692 10,852–12,032	Fort Rock	Sagebrush/shadscale flat, in Fort Rock Basin	Bones from horizon containing flakes, tools and charcoal.	Oetting 1993; Greenspan 1993
Tucker 35LK3227 (Block 4)	9430±80	Beta-102259	10,581–10,869	Fort Rock	Lunette dune adjacent to active playa, in Alkali	Faunal remains come from Block 4, Basin which contained a large, diffuse charcoal stain (Feature 3) associated with burned bones, flakes, and metate fragments. Artifacts and bones were noticeably concentrated in the stained portion of the block, and rare outside it.	Pinson 1999
Tucker 35LK3227 (Block 5)	$\bar{X}$=9490±40 9480±50 9520±60	Beta-102260 Beta-102259	10,700–10,993 10,671–10,989 10,732–11,036	Fort Rock	Lunette dune adjacent to active playa, in Alkali Basin	Bones come from Block 5, which contained Feature 4, a large, diffuse charcoal stain associated with burned bone and flakes. Artifacts and bones were noticeably concentrated in the stained portion of the block, and rare outside it.	Pinson 1999

TABLE 10.3. Faunal Assemblage Data from Valley Floor Sites

SITE NAME (FEATURE NUMBER)	^{14}C BP/ $\bar{X}$ ^{14}C BP	NISP LAGOMORPH	NISP ARTIODACTYL	ARTIODACTYL INDEX-NISP	N CLASS III	N CLASS V	ARTIODACTYL INDEX–SIZE CLASS
Peninsula 35LK2579	374±39	641	440	0.4070	641	440	0.4070
McCoy Creek 35LK1263 (Feature 11)	480±70	199	27	0.1195	235	501	0.6807
McCoy Creek 35LK1263 (House Floor H2, Feature 7)	1190±39	15	0	0.0000	15	4	0.2105
McCoy Creek 35LK1263 (House Floor H1, Feature 3)	1270±70	49	1	0.0200	50	11	0.1803
35LK2204 (Locus 9)	2070±110	244	48	0.1644	244	48	0.1644
Bowling Dune (Feature 2)	2890±70	155	9	0.0549	196	17	0.0798
Bowling Dune (Feature 9)	3060±60	17	8	0.3200	26	145	0.8480
Dunn (Feature 7)	3255±65	24	2	0.0769	159	106	0.4000
Dunn House (Features 1–6)	3255±65	38	27	0.4154	437	275	0.3862
Bowling Dune (Feature 8)	4460±160	60	1	0.0164	122	16	0.1159
Big M (Structure 3)	4910±60	193	16	0.0766	193	16	0.0766
Bergen House (Quad A, Unit 3, Levels 9–13)	5190±40	93	4	0.0412	93	4	0.0412
Bowling Dune (Feature 18)	5380±70	4	0	0.0000	10	5	0.3333
Bowling Dune (Feature 12)	6040±120	24	38	0.6129	49	328	0.8700
Locality III (Unit B-5, Hearth EB-2)	6370±190	20	0	0.0000	20	0	0.0000
Locality III (Unit B-7, Hearth EB-8)	7040±123	19	0	0.0000	23	0	0.0000
Locality III (Unit B-2, Hearth EB-3)	7155±81	11	0	0.0000	12	2	0.1429
Bowling Dune (Feature 22)	ca.<6850	11	0	0.0000	27	0	0.0000
Bowling Dune (Feature 15)	ca.>6850	19	3	0.1364	35	23	0.3966
35LK2076 (Feature 4)	8810±103	247	247	0.0000	0	0	0.0000
35LK1881 (pit-like feature)	8983±69	3843	3843	0.0000	0	0	0.0000
35LK1180	9327±276	795	0	0.0000	796	1	0.0013
Tucker 35LK3227 (Block 4)	9430±80	5	0	0.0000	70	1	0.0141
Tucker 35LK3227 (Block 5)	9490±40	11	0	0.0000	88	2	0.0222

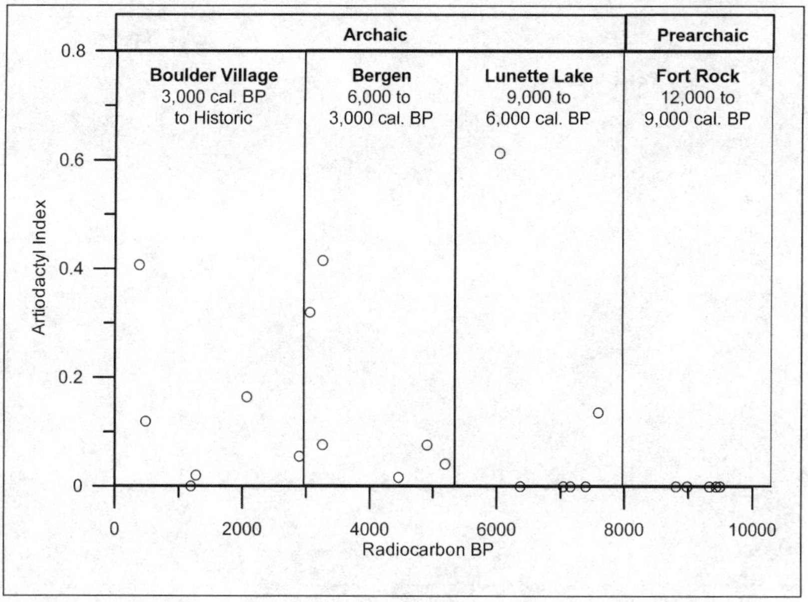

FIGURE 10.2. Artiodactyl index vs. time for northern Great Basin cultural assemblages using NISP data.

BP]). Sites from the Boulder Village, Bergen, and Lunette Lake periods were collectively considered to belong to the Archaic period; sites dating to the Fort Rock period were further classified as Paleoarchaic. All sites considered in this analysis are from open-air contexts; there is no patterning with respect to paleoenvironments and time.

RESULTS

As can be seen in Figure 10.2, Paleoarchaic assemblages consistently have artiodactyl index values close to 0.0, while the artiodactyl index values of Archaic assemblages vary from 0.0 to about 0.60. This graph highlights the unexpected rarity of artiodactyls in the Paleoarchaic diet, contrary to the notion that this was a time of relatively greater reliance on them. The difference between Paleoarchaic diets, on the one hand, and those of later time periods is statistically significant (Mann Whitney U = 12.50, p-value = 0.012, alpha = 0.05, n = 23). The small number of samples precludes more-detailed statistical comparisons.

To check if differential fragmentation is having an effect on the data, the analysis can be re-done using size class data, where Class 3 equals small game such as lagomorphs and Classes 5 and 6 equal large game, such as artiodactyls (Figure 10.3). Two things happen when we do this: the first is that artiodactyls do show up as part of the Paleoarchaic diet inventory; the second is that the fundamental relationship remains unchanged. Archaic period site assemblages are characterized by a much greater abundance of large game species than are Paleoarchaic site assemblages (U = 12.000, p-value = 0.012, alpha = 0.05, n = 23).

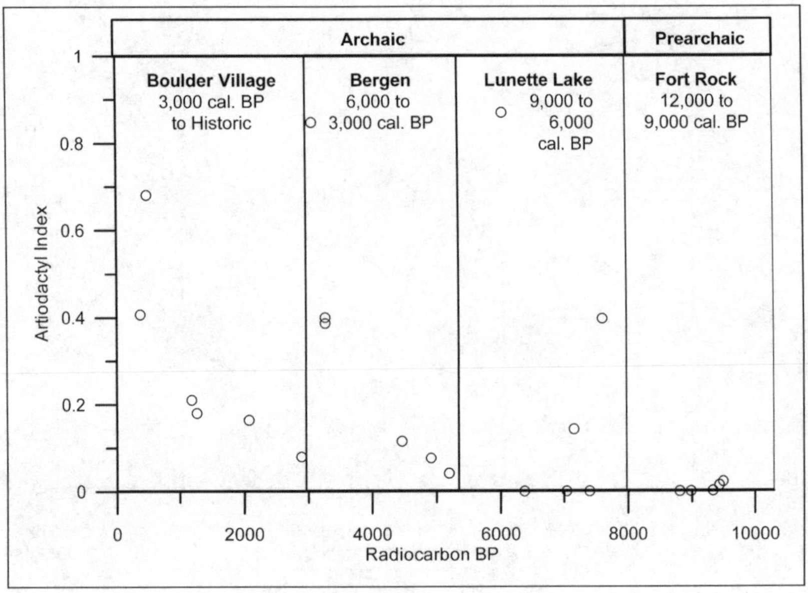

FIGURE 10.3. Artiodactyl index vs. time for northern Great Basin cultural assemblages using size class data.

In sum, both NISP and size class data show that, contrary to expectations, the relative abundance of artiodactyls to lagomorphs in the diet increases over time. This increase occurs in tandem with reductions in projectile point sizes and a shift to more xeric environments.

The observed increasing importance of artiodactyls in the diet over time is contrary to what has generally been assumed; that is, that large game consumption decreased relative to small game over the Paleoarchaic/Archaic transition. Furthermore, the lower abundance of large game species in Paleoarchaic assemblages compared to later assemblages would appear to contradict predictions from foraging theory for small populations of hunter-gatherers experiencing a shift from mesic environments comparatively rich in large game to more-xeric environments. Foraging theory robustly predicts a dietary emphasis on larger terrestrial mammals during the Paleoarchaic, particularly compared to later foragers occupying more-xeric environments (Elston 1982; Elston and Zeanah 2002; Elston et al. 1995; Simms 1985, 1987, 1988). The same theory predicts that as climate becomes drier and large game animals became scarcer, diet breadth expands to include a larger number of small game animals (e.g., lagomorphs) and a wider range of plant resources. The pattern of artiodactyl use over time does not match theoretical predictions.

DISCUSSION OF POTENTIAL DATA BIASES

Because the faunal assemblage data from the northern Great Basin contradicts what has historically been assumed for diet changes over time, and contradicts predictions made based on foraging theory, it is important to determine whether other biases are affecting the study results. Three factors must be considered in evaluating the observed results:

1. It is possible that the observed trend is an artifact of the small number of Paleoarchaic sites excavated to date, and the poor quality of the faunal assemblages derived from these sites.
2. It is possible that Paleoarchaic foragers were much more residentially mobile than their later counterparts: Paleoarchaic foragers might have hunted and consumed artiodactyls in upland settings, whereas later foragers might have chosen to return to a central place in the lowlands to distribute and consume their kill. Thus, the fact that all the sites included in this analysis are low elevation sites might bias the data set against the recovery of Paleoarchaic large faunal remains.
3. Finally, while projectile point size is not necessarily a good proxy for target game size, it has always at least been assumed that large projectile points were used on large game animals rather than on small game animals. The emphasis on large projectile points and the absence of smaller ones in Paleoarchaic lithic assemblages would appear to logically contradict the observed low levels of artiodactyl resource use in the Paleoarchaic.

Each of these concerns is addressed below.

Site Abundance and Data Quality

The number of samples available for this analysis are indeed few, particularly given the long range of time spanned by the data. In addition, sediments in the northern Great Basin are very alkaline (pH > 8.0), resulting in comparatively poor organic matter preservation. The faunal assemblages are thus highly fragmentary, a factor that this analysis addressed by comparing the assemblages using NISP and size class data. Finally, this analysis does not include either fish or avifauna, both of which were likely significant resources throughout northern Great Basin prehistory. Consequently, although this analysis used all available published faunal data from the northern Great Basin, the analyses described in this chapter must be considered preliminary, awaiting more and better data from a larger number of excavations.

Land Use and Mobility Biases

As suggested above, one possible reason for the differences in the abundance of artiodactyl bones in lowland assemblages over time might be due to differences in land use patterns over the same time period. Since all of the sites included in the analysis are low elevation, open-air sites, it is reasonable to ask if temporal changes in settlement patterns are biasing the results. Is it possible that Paleoarchaic hunter-foragers hunted and consumed artiodactyls primarily in upland hunting settings, whereas Archaic foragers were more likely to transport hunted game to village sites in the lowlands for consumption?

To assess whether or not land use and mobility patterns are biasing the subsistence data, two other data sets were analyzed:

1. Faunal assemblage data from valley margin cave sites, which may have served as loci for central place foraging in adjacent upland areas, might reveal

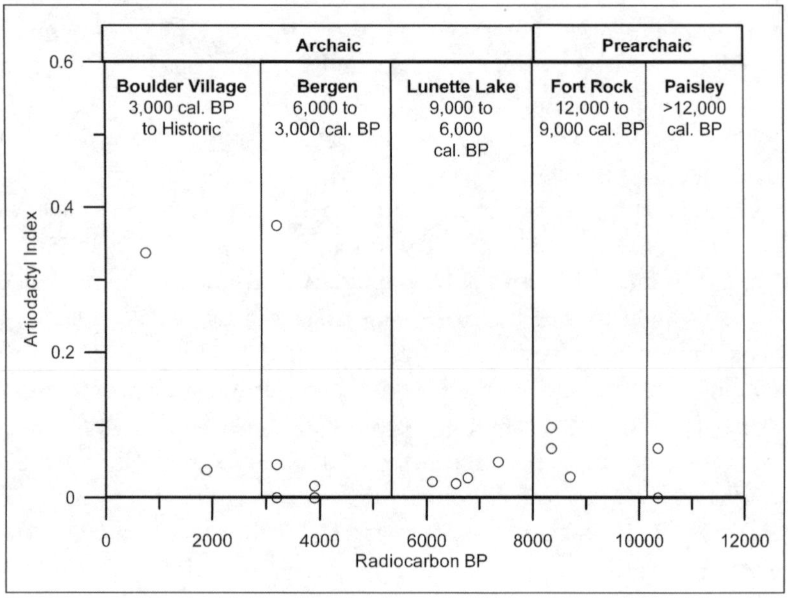

FIGURE 10.4. Artiodactyl index vs. time for northern Great Basin valley margin cave assemblages using NISP data.

that Paleoarchaic foragers preferentially used these settings for obtaining and consuming artiodactyls compared to later foragers. Thus, artiodactyl remains in Paleoarchaic lowland sites might be less abundant than in later lowland sites because Paleoarchaic large game hunting and consumption occurred at different locations on the landscape than those used by later foragers.

2. Distributional data might show that changes in upland land use over time account for the observed differences in lowland artiodactyl use-rates. Specifically, distributional data might show that Paleoarchaic site abundance in upland areas was much more common than later use of these same areas, suggesting that Paleoarchaic foragers were hunting and camping (and presumably consuming artiodactyls) in upland settings, whereas later Archaic foragers were practicing central place foraging centered on lowland residential localities.

Each of these possibilities is considered below.

To assess whether or not Paleoarchaic foragers relied more heavily on valley margin cave sites for artiodactyl procurement and consumption, the faunal analysis used above was repeated using faunal assemblages from valley margin cave sites (Table 10.4, Figure 10.4). These caves contain some of the earliest evidence of human occupation in the northern Great Basin, with episodic occupations dating from the Paisley period through the later Archaic. If Paleoarchaic foragers were consuming game closer to the site of procurement than did later foragers, it would be profitable to compare Paleoarchaic faunal material with later Archaic faunal assemblages from cave sites.

The faunal assemblages in these caves accumulated through a mix of

Table 10.4. Faunal Assemblage Data from Valley Margin Cave Sites

Site Name (Feature/ Strat.)	Est. Median ^{14}C Date (cal BP)	Period	Location	NISP Lagomorphs	NISP Artiodactyls	Artiodactyl Index–NISP	Data Source
Dirty Shame Rockshelter Zone I	750 (685)	Boulder Village	Owyhee Uplands	358	182	0.3370	Grayson 1977
Dirty Shame Rockshelter Zone II	1900 (1855)	Boulder Village	Owyhee Uplands	276	11	0.0383	Grayson 1977
Connley Cave 3 Stratum 1	3200 (3425)	Bergen	Fort Rock Basin	42	2	0.0455	Grayson 1979
Connley Cave 4 Stratum 1	3200 (3425)	Bergen	Fort Rock Basin	20	0	0.0000	Grayson 1979
Connley Cave 5 Stratum 1	3200 (3425)	Bergen	Fort Rock Basin	22	0	0.0000	Grayson 1979
Connley Cave 6 Stratum 1	3200 (3425)	Bergen	Fort Rock Basin	5	3	0.3750	Grayson 1979
Connley Cave 4 Stratum 2	3900 (4360)	Bergen	Fort Rock Basin	9	0	0.0000	Grayson 1979
Connley Cave 6 Stratum 2	3900 (4360)	Bergen	Fort Rock Basin	123	2	0.0160	Grayson 1979
Dirty Shame Rockshelter Zone III	6100 (6970)	Lunette Lake	Owyhee Uplands	574	13	0.0221	Grayson 1977
Dirty Shame Rockshelter Zone IV	6550 (7450)	Lunette Lake	Owyhee Uplands	555	11	0.0194	Grayson 1977
Fort Rock Cave Stratum 2	6775 (7630)	Lunette Lake	Fort Rock Basin	71	2	0.0274	Grayson 1979
Dirty Shame Rockshelter Zone V	7350 (8180)	Lunette Lake	Owyhee Uplands	270	14	0.0493	Grayson 1977
Connley Cave 4 Stratum 3	8350 (9380)	Fort Rock	Fort Rock Basin	382	28	0.0683	Grayson 1979
Connley Cave 5 Stratum 3	8350 (9380)	Fort Rock	Fort Rock Basin	176	19	0.0974	Grayson 1979
Dirty Shame Rockshelter Zone VI	8700 (9640)	Fort Rock	Owyhee Uplands	236	7	0.0288	Grayson 1977
Connley Cave 4 Stratum 4	10,350 (12,270)	Paisley	Fort Rock Basin	461	34	0.0687	Grayson 1979
Connley Cave 6 Stratum 4	10,350 (12,270)	Paisley	Fort Rock Basin	9	0	0.0000	Grayson 1979

processes—many natural, some cultural. The "assemblages" consist of faunal remains from one or more stratigraphic levels grouped into time periods in the original published analysis. The assemblages do not meet the rigorous criteria for acceptance used in selecting cultural assemblages for analysis, so the data from cave assemblages were excluded from that analysis. Analysis of these assemblages can only address whether or not there are differences over time in the factors influencing the abundance of animals in the assemblage, but it cannot account for these differences. The sites used in this analysis are Connley Caves (Grayson 1979), Fort Rock Cave (Grayson 1979), and Dirty Shame Rock Shelter (Grayson 1977). Assemblages were excluded if the combined NISP of lagomorphs and artiodactyls was less than five, or if the date of the assemblage was not secure. Only NISP data were available for analysis.

The graph of cave assemblages shows that there are no significant differences in the processes by which large and small game were added to the cave assemblages across the Paleoarchaic/Archaic transition ($U = 21.5$, $p = 0.383$, $alpha = 0.05$, $n = 17$). There do not appear to be the kinds of differences through time in artiodactyl abundance that might be expected had Paleoarchaic foragers been using the caves as base camps for upland artiodactyl hunting more frequently than did later, Archaic foragers.

If Paleoarchaic hunters do not appear to have been consuming many artiodactyls in the lowlands, and there is no evidence for a Paleoarchaic valley margin preference compared to later foragers, it is still possible that Paleoarchaic foragers were hunting and consuming artiodactyls almost exclusively in upland settings, and preferentially consuming lagomorphs in lowland settings. Although there are no upland faunal assemblages available for analysis, this proposition can be evaluated in a preliminary way through the use of land use data (a detailed treatment of this land use analysis can be found in Pinson 1999). If Paleoarchaic foragers were in the uplands procuring and consuming artiodactyls and other upland resources, a relatively high abundance of Paleoarchaic sites in the uplands would be expected, perhaps even at a rate higher than during later periods when permanent villages became established in the lowlands.

Land use patterns were investigated by dividing the landscape into three classes: upland basalt plateaus, lowland valley bottoms, and ecotone areas. *Ecotone areas* are defined as those parts of the landscape where a wide range of resource types are available in a small geographic space; these areas generally consist of a point on the landscape where a stream or streams enter a lake, usually resulting in the development of a small marshland. Ecotone areas also include upland areas immediately adjacent to the lowland stream/marsh/lake complex.

Land use patterns through time in the different classes were tracked using projectile points. Projectile points are an imprecise measure of time because of the considerable overlap in size and form among the recognized types, as well as overlap between types in time and space. However imprecise, projectile points remain the primary means of assessing land use patterns through time and provide at least an approximate time-averaged understanding of such patterns.

For this analysis, Great Basin Stemmed series and fluted points are considered indicative of the Paleoarchaic period (the Fort Rock and Paisley periods, respectively). Large side-notched, large corner-notched, and large split-stemmed projectile points (e.g., Elko series) are indicative of the Archaic period (Lunette Lake and Bergen periods combined). Small arrowheads of the Rosegate, Desert Side-notched, Cottonwood Triangular, and other series are indicative of Boulder Village period foraging.

The land use data show that Paleoarchaic sites are concentrated in lowland ecotone areas and are virtually absent in adjacent lowland areas and in the uplands. For instance, in Alkali Basin, Paleoarchaic sites are concentrated in the vicinity of Dietz Basin at the head of pluvial Lake Alkali, in a region that probably had at least one permanent stream and wetland area (Pinson 1999). This ecotone is close to the Horse Mountain obsidian source, and uplands occur within a few miles of the site. Paleoarchaic sites are rare almost everywhere else in Alkali Basin. Archaic sites have a much wider distribution across the landscape, including both the upland and lowland areas.

The same Paleoarchaic land use pattern is observed elsewhere in the northern Great Basin. Surveys in the Hawksy-Walksy Valley (Christian 1997) and Beattys Butte (Moore 1999) areas also showed that Paleoarchaic sites occur almost exclusively in the lowlands, and primarily in lowland ecotone areas. Later land use is more evenly distributed between uplands and lowlands, with lowland use not concentrated in the ecotone areas (Pinson 1999).

In the Alvord Basin (Beck 1984; Pettigrew 1984), land use is also restricted to lowland areas during the Paleoarchaic period. Distributional maps (Pettigrew 1984) show that within the lowlands, Paleoarchaic sites occur in what could be classified as ecotone areas. Although later foragers also emphasized the lowlands, these same distributional maps show that Archaic foragers did not concentrate their lowland use in ecotone areas. Instead, they made heavy use of areas away from the ecotone spots, including significantly greater use of the upland areas. Elsewhere, I have argued that at least some of this difference in land use resulted from differences in resource predictability across the Paleoarchaic/Archaic transition (see Pinson 1999).

The land use data show that although Paleoarchaic foragers used upland areas, they did so at a rate substantially lower than subsequent foragers: upland Paleoarchaic sites are extremely rare compared to later period upland sites. Archaic sites in upland areas are abundant and widely distributed, indicating that Archaic period foragers made extensive use of the uplands. Furthermore, the Archaic period upland sites are not strictly temporary hunting scatters, but through their highly diverse lithic assemblages represent a wide range of residential and nonresidential activities (Fagan 1988; Jenkins 1994; Jenkins et al. 2004; Oetting 1989). The distributional data provide no support for the argument that upland Paleoarchaic central place foraging accounts for the relatively low abundance of artiodactyl use evident in lowland Paleoarchaic sites.

Projectile Points and Preservation Bias

Paleoarchaic projectile points are indisputably large, and it is unlikely that they were used to hunt lagomorphs. Rather, it is likely that they were used on large game animals and as weapons against other groups of humans. But, frustratingly for archaeologists, such projectile points are relatively rare in the archaeological record compared to later point types. Certainly, reuse of Paleoarchaic points as toolstone sources by later individuals, site formation processes, and modern collecting have all had a negative effect on the abundance of these projectile points, particularly in surface assemblages.

The rarity of these points may also, however, be a factor of diet. If Paleoarchaic peoples were emphasizing small game animals (lagomorphs, avifauna, fish), then the suite of hunting tools would almost certainly have emphasized the technologies used for their capture: nets, clubs, and fire. We know from analysis of sandals that weaving and cordage technologies were already sophisticated by 9200 ^{14}C BP (10,500 cal BP) (Connolly and Barker 2004), making it more than likely that Paleoarchaic foragers could have made and maintained nets for small game capture. It is possible that the comparative rarity of Paleoarchaic projectile points in the archaeological record may be due, at least in part, to their relatively lesser importance in a Paleoarchaic toolkit focused more on hunting small game than large game.

One factor that may be driving the rarity of artiodactyls in the Paleoarchaic diet might be the efficiency of hunting technologies. Although the more mesic environments of the Paleoarchaic period likely led to higher artiodactyl abundances on the landscape compared to later periods, the hunting technology may have been less efficient. The change from socket-hafted to notched projectile points at the end of the Paleoarchaic may well have been accompanied by other technological changes that increased projectile velocity, force, and accuracy. Improved projectile technologies would have increased both the return rate for hunting artiodactyls and the chances of killing an animal during any particular hunt. Thus, technological changes accompanying the shift to notched projectile points may have significantly increased the odds of a successful artiodactyl hunt, making the undertaking more attractive to a hunter with competing foraging opportunities. Such changes in foraging return may have more than compensated for reductions in artiodactyl availability in the more xeric Archaic landscapes.

CONCLUSION

In the foregoing analysis, it has been shown that Paleoarchaic faunal assemblages are poorer in artiodactyl bones than later Archaic assemblages. The inference from these data is that, contrary to expectations, Paleoarchaic foragers consumed artiodactyls at a much lower rate than later foragers in the same region. Dependence on artiodactyls appears to have increased over time despite increasingly xeric environments and decreasing projectile point sizes. Furthermore, there is no evidence that the conclusions are biased due to differences in land use patterns and mobility over time.

The faunal data, taken in conjunction with the evidence for land use patterns, strongly suggest that Paleoarchaic and Archaic foragers used the land in very different ways. Northern Great Basin Paleoarchaic foragers appear to have focused their foraging in lowland ecotone areas, where they made extensive use of lagomorphs and other small game resources (avifauna, fish). It is likely that they did so using perishable technologies, particularly nets and wooden clubs. Large projectile points were likely used to hunt artiodactyls, but the efficiency of this method compared to later Archaic methods may have led to comparatively lower consumption rates of artiodactyls than during subsequent periods.

The Archaic pattern in the northern Great Basin features a broad diet in terms of greater use of large game resources and a greater reliance on a wider range of plant resources (Jenkins et al. 2004). Humans exploited a wider range of landscape elements and were at least seasonally resident in lowland villages near lakes, marshes, and rivers. The increased reliance on artiodactyl resources may in part be a result of more efficient large game hunting technology.

The changes in foraging behavior that appear evident in the archaeological data from the northern Great Basin do not inform on the foraging choices made by Paleoarchaic peoples outside the northern Great Basin. In areas with different geographies and different biogeographic histories, Paleoarchaic foragers may have found it advantageous to focus more heavily on artiodactyls. This preference may be evident in the faunal assemblages of these regions.

The archaeological data cited in this chapter, however, support the argument for adaptive discontinuity across the Paleoarchaic/Archaic transition in the northern Great Basin. A shift in foraging strategies appears to have occurred about 8000 ^{14}C BP (9000 cal BP), coincident with significant environmental changes and changes in hunting technology. The changes in strategy resulted in significant differences in land use and settlement patterns between the Paleoarchaic and Archaic periods that are observable in the archaeological record of the northern Great Basin.

ACKNOWLEDGMENTS

I wish to thank the archaeologists who contributed to faunal analyses at the Tucker Site (35LK3227) as well as the archaeologists who generously shared data with me during the first phases of this research: Ruth Greenspan, Shannon Jung, Dennis Jenkins, and Patrick O'Grady. I would also like to thank the editors of this volume for inviting me to contribute this chapter.

A preliminary version of this analysis was presented at the Twenty-seventh Great Basin Anthropological Conference, Ogden, Utah, October 6–8, 2000.

11

Nutritional Ecology of Late Pleistocene to Middle Holocene Subsistence in the Great Basin

Zooarchaeological Evidence from Bonneville Estates Rockshelter

Bryan Hockett

BONNEVILLE ESTATES ROCKSHELTER (BER) PRESERVES A RECORD OF human subsistence in the west Bonneville Basin for at least the past 13,000 calendar years (Table 11.1). The modern excavation techniques employed at BER (e.g., fine-grained stratigraphic excavation and three-dimensional piece plotting of lithic artifacts and faunal specimens), together with the multitude of accelerator mass spectrometry (AMS) radiocarbon dates has allowed us to achieve a precision in matching artifacts and features with relatively short-term occupation sequences that was not obtained at other nearby caves excavated 40 to 50 years ago, such as Danger and Hogup caves. BER is also unique because human foragers began occupying the shelter at least 700 years before Danger Cave was first occupied, and 2,500 years prior to the initial occupation of Hogup Cave.

The faunal remains recovered from most of the occupation surfaces in BER are exceptionally well preserved, including those from the late Pleistocene and early Holocene layers. And because our excavations occurred more than 20 years after the seminal taphonomic studies of Binford (1981), Brain (1981), Haynes (1980), and Behrensmeyer and Hill (1980), we can also tease out the relative contributions of human and nonhuman predators to the creation of that faunal record—for both the large and small faunal components. As a result, these specimens serve as an excellent sample to test models that are designed to explain changes in forager subsistence preferences (e.g., Hockett 2005).

In recent publications (Haws and Hockett 2004; Hockett and Haws 2003, 2005), we have attempted to revitalize a nutritional and demographic-based model to interpret prehistoric foraging choices during the middle and upper Paleolithic of Europe. In this chapter, I focus on the late Pleistocene, early Holocene, and middle Holocene faunal records recovered through the 2006 field season at BER using this same nutritional ecology perspective. I discuss three principal items: (1) late Pleistocene to middle Holocene climate in the

TABLE 11.1. Late Pleistocene, Early Holocene, and Middle Holocene Occupation Zones (Simplified) from Bonneville Estates Rockshelter by Radiocarbon Years BP and Their Associated Calendar Dates

STRATA	^{14}C YEARS BP	CAL BP[a]	NOTES
West Block: 18b (base)	11,000 BP	12,900 BP	Hearth; no associated fauna
West Block: 18b (top)	10,500 BP	12,600 BP	Several hearths and associated fauna
West Block: 18a East Block: 12	10,100 BP	12,200 BP	Numerous hearths and associated fauna
West Block: 17b' East Block: 10	9500 BP	10,700 BP	Numerous hearths and associated fauna
East Block: 9 (base)	8800 BP	9900 BP	Hearth; no associated fauna
West Block: 16	7200 BP	8000 BP	Hearth and associated thin organic lens with fauna
West Block: 14	6100 BP	7000 BP	Hearths and associated thick organic lens with fauna
East Block: 9 (upper)	5200 BP	6000 BP	Hearths associated with thick, deep organic pits with fauna

[a] Calibrated dates after Reimer 2004.

west Bonneville Basin; (2) assumptions of the nutritional ecology model; and (3) the subsistence and settlement choices of late Pleistocene/early Holocene and middle Holocene foragers at BER and related sites in the Great Basin, and whether these data suggest that they were influenced by climate change, the underpinnings of the nutritional ecology model, or both.

LATE PLEISTOCENE TO MIDDLE HOLOCENE CLIMATE CHANGE IN THE WEST BONNEVILLE BASIN

Madsen and Schmitt (2005) recently reviewed Holocene environmental change in the Bonneville Basin. My review here generally follows theirs for the early and middle Holocene, although additional paleoclimatic studies pertinent to understanding the BER cultural record are included below. Correlations between major shifts in climate and the occupation of BER are also highlighted.

Younger Dryas/Latest Pleistocene, 10,800–10,100 ^{14}C BP (12,850–11,650 cal BP)

Between the Last Glacial Maximum (ca. 18,000 ^{14}C BP [21,400 cal BP]) and the end of the Pleistocene epoch (traditionally set about ca. 10,100 ^{14}C BP [11,700 cal BP]), global climate witnessed a general warming trend that was offset by a millennial-scale cooling known as the Younger Dryas (e.g., Gosse et al. 1995).

At this time, the western shoreline of Lake Bonneville (Gilbert) would have been situated about 4,250 ft (1,289 m) asl, about the same elevation as the modern lakes and marshes that form Blue Lakes, located some 10 km (5 mi) east of BER (see Graf, this volume). Any drop in lake level from this Gilbert highstand would have facilitated the creation of the Blue Lakes marshes east of BER because lake levels at or above the 4,250 ft threshold would have essentially drowned the marshes under deeper water.

The Younger Dryas was a relatively cool period in the Great Basin (e.g., Bischoff and Cummins 2001), with cool and mesic-adapted small mammals occupying low-level elevations in the northern Great Basin (Grayson 1998, 2000a, 2000b; Schmitt et al. 2002b). Mature stands of sagebrush must have existed near the rockshelter during this time given that pygmy rabbit (*Brachylagus idahoensis*) and sage grouse (*Centrocercus urophasianus*) were among the most commonly identified animals in the Younger Dryas–aged sediments. Human foragers repeatedly occupied BER during this time, signaling the earliest population pulse in the western Bonneville Basin.

Early Holocene, 10,100–8300 ^{14}C BP (11,650–9400 cal BP)

Early Holocene climate appears to have warmed compared to the previous latest Pleistocene (Younger Dryas) conditions, but it was still relatively cool compared to the succeeding middle Holocene. Caves such as Homestead Cave and Camels Back Cave in the northern Bonneville Basin (Grayson 1998, 2000a, 2000b; Schmitt et al. 2002b; Schmitt and Madsen 2005) and Pintwater Cave in the northern Mojave Desert of southern Nevada (Hockett 2000) show that mesic small mammals continued to occupy low elevation settings throughout the Great Basin. However, while cool and wet conditions prevailed in the northern Great Basin at this time (Grayson 2000a), the southern Great Basin experienced greater warming but still relatively wet conditions (Hockett 2000). Near BER, sagebrush-adapted species such as pygmy rabbit and sage grouse continued to be present, and human foragers continued to repeatedly occupy the rockshelter.

Early Middle Holocene, 8300–7200 ^{14}C BP (9400–8000 cal BP)

Numerous records document a relatively rapid deterioration of climate throughout western North America ca. 9400 cal BP (Grayson 2000b; Hockett 2000; Overpeck et al. 1996; Madsen and Schmitt 2005; Schmitt et al. 2002b). This deterioration resulted in both warmer and drier conditions in the northern Great Basin, although the southern Great Basin probably continued to experience rather predictable summer precipitation (Hockett 2000). Mesic-adapted small mammals such as pika (*Ochotona princeps*), marmot (*Marmota flaviventris*), pygmy rabbit (*Brachylagus idahoensis*), sage vole (*Lemmiscus curtatus*) and other voles of the genus Microtus, bushytailed wood rat (*Neotoma cinerea*), and western harvest mouse (*Reithrodontomys megalotis*) were either extirpated from or experienced significant population declines in many low elevation settings in the northern Great Basin (Grayson 2000a, 2000b; Grayson and Madsen 2000; Schmitt et al. 2002b). Near BER, neither pygmy rabbits nor sage grouse were found in deposits that date between 8800 and 6000 ^{14}C BP (9900–6800 cal BP), signaling the dismantling of the sagebrush habitat. Humans abandoned BER beginning in the late early Holocene (8800 ^{14}C BP [9900 cal BP]), and this settlement strategy continued throughout the early middle Holocene.

The 7200 ^{14}C BP (8000 cal BP) Event

Around 8000 cal BP, ancient Lake Agassiz catastrophically emptied into, and caused a short-term freshening of, the North Atlantic (Alley and Agústsdóttir 2005). This occurrence—tentatively referred to as the "Finse," "Goldilocks," or "8k" event (Alley and Agústsdóttir 2005; Schmidt and LeGrande 2005)—caused a short-term (perhaps no more than a decade or two) cooling in the North Atlantic and subsequent landscape-specific changes in climate on a near-global scale. Foragers reoccupied BER for two very brief periods (radiocarbon dated at or slightly earlier than ca. 8000 cal BP) after a hiatus of nearly 2,000 years. Bison (Bison bison) appears in the faunal record of BER for the first time in the 7200 ^{14}C BP (8000 cal BP) zone. Neither sage grouse nor pygmy rabbits, however, appear in the shelter deposits associated with this climatic event. This suggests that increases in precipitation may have occurred primarily during the summer rather than winter months in the west Bonneville Basin. The brief reoccupations of the shelter by human foragers appear to correspond closely with this 8k event, which probably brought more mesic (but warm) conditions to the west Bonneville Basin for one to several decades in one or more mesic spikes.

The 6100 ^{14}C BP (7000 cal BP) Event

Following the brief 8k climate event, conditions in the northern Great Basin appear to have returned to the warm and arid state of the early-middle Holocene. However, another relatively mesic event centered around 6100 ^{14}C BP (7000 cal BP) is recorded in the Great Basin paleoenvironmental record (Madsen and Currey 1979; Madsen and Schmitt 2005; Schmitt and Madsen 2005). At BER, human foragers once again returned to the shelter after a 1,000-year absence following the 8k event. Sage grouse bones return to the shelter deposits after an approximately 3,500-year absence. Interestingly, neither bison nor pygmy rabbits are found in the shelter deposits of this age. This may suggest that the "7k" event was cooler than the 8k event, but still relatively brief. Based on the thickness of the occupation zones dating to this time period, human foragers either repeatedly occupied BER or occupied the shelter for longer periods of time, or both, during the 7k event compared to the occupational intensity seen during the 8k event.

The 5200 ^{14}C BP (6000 cal BP) Event?

Following the 7k event, conditions once again turned warm and arid in the Great Basin (e.g., Schmitt and Madsen 2005). However, there are indications that a third middle Holocene mesic spike may have occurred in the Great Basin around 5200–5300 ^{14}C BP (6000 cal BP) (reviewed in Madsen and Schmitt 2005). BER may once again record this climatic event in the form of human occupation of the rockshelter. Following another hiatus of 1,000 years following the 7k event, foragers briefly returned to BER during this "6k" event. Artiodactyls were heavily exploited during this time. However, no sage grouse, pygmy rabbit, or bison bones were found in the shelter. The event generally corresponds to the initial occupation of many important cave and rockshelter sites across the central Great Basin (e.g., Hockett 2005), signaling the beginning of an occupational intensity that rivaled that of the late Pleistocene/early Holocene 6,000 to 7,000 years earlier.

NUTRITIONAL ECOLOGY

Nutritional ecology models the health consequences of prehistoric dietary choice. A more detailed overview of the nutritional ecology approach can be found in Hockett and Haws 2003 and 2005. Here I note that it has only been in the past two decades that nutrition scientists have realized the powerful links between micronutrient diversity, fetal development, and infant and maternal mortality rates (e.g., Allen 2005; Ashworth and Antipatis 2001; Bendich 2001; Fall et al. 2003; Gambling and McArdle 2004; Keen et al. 2003; Ramakrishnan et al. 1999; Solomons 2000; Underwood 2002; Viteri and Gonzalez 2002). Although many excellent publications with a nutritional ecology perspective have been published over the past several decades (e.g., Schwarcz and Schoeninger 1991), it was our belief that most of them were idiosyncratic in the sense that they lacked a broader theoretical framework to focus attention on one of the main truisms of modern nutrition science: the essentiality of nutrient diversity to human health and its impact on demographic trends (Harper 1999).

Briefly, nutritional impacts to human demographic patterns center around the health and well-being of the mother and child. These impacts will vary depending on maternal consumption of the essential nutrients necessary for proper fetal-to-neonatal growth, as well as for maternal body maintenance during and immediately following pregnancy (e.g., Fall et al. 2003; Ramakrishnan et al. 1999). There are three principal reasons why micronutrients must be brought into models concerned with nutritional impacts to human mortality—and hence to human selective fitness.

First, ethnographic research supports the idea that most foragers maintained caloric intake levels above chronic energy deficiency (Lee and Devore 1968). As a result, ancient foragers who obtained calories more efficiently than others were not likely to gain a selective advantage unless chronic energy deficiency was a leading factor in infant and maternal mortality rates. It is possible, indeed likely, that such occurrences impacted human demographic trends in specific places and times. However, many short-term and much of the long-term mortality patterns in human populations are likely to have been primarily influenced by overall nutrient diversity. Further, in higher primates, low calorie diets reduce the rate of cell damage and result in extended life expectancy rates (e.g., Yoshida et al. 1999).

Second, many micronutrients are essential to fetal development and the survival of neonates. It would be difficult for most foragers to consistently consume these critical micronutrients within a context of restricted dietary diversity. In other words, the broader the diet (note that broad or diverse diets in this context are defined by diversity of essential nutrient intake rather than by N-Species consumed; see Hockett and Haws 2005:26, Figure 2), the more likely it is that human females consume the diversity of micronutrients necessary for proper fetal development. Hence, in general, diverse diets should be associated with other evidence suggesting population growth and/or improved health conditions, while more restricted diets should be associated with periods of demographic downturns and/or deteriorating health patterns. An excellent example of this principle is presented by Lockett et al. (2000:196), who reported on the health consequences of pastoralists who either did or did not consume relatively low calorie but micronutrient-dense wild plant resources during times of subsistence stress:

> During 7 years of inadequate rainfall between 1965–72, agro-pastoral societies in the eastern Kalahari of Botswana thrived and remained well nourished.... In contrast, agro-pastoral societies who occupied the Sahel of West Africa, a zone similar to the Kalahari in climate and geography, suffered from terrible famine and hardship.... One primary difference between the two areas was a more diversified diet in the Kalahari characterized by an abundance of edible wild plants. Elsewhere in Africa uncultivated indigenous plants were protective during the 1984–85 Darfur famine in the Sudan where malnutrition-related mortality rates were lower among groups who used edible wild species compared to communities where such species were infrequently utilized.

Third, proteins, carbohydrates, and fats can be converted to useful energy only in the presence of specific micronutrients. Historically, this may be the most underappreciated fact in foraging theory in both the biological and anthropological sciences. But this is beginning to change. Within biology, new megacatylst theories are being developed to explain biogeographic variability based on differences in the availability of micronutrients to ruminants (e.g., Milewski and Diamond 2000). At their core, these theories are based on the understanding that body size in some mammals, including humans, is partially determined by the availability of specific micronutrients, not simply by the availability of energy (Lagiou et al. 2005). In human evolution, researchers are determining the nutrients required for proper brain development and function, and then comparing those nutrients with those available to early hominids in order to narrow the subsistence components that would have been required for the evolution of the human brain (e.g., Broadhurst et al. 2002).

These studies are based on the essentiality of both micronutrients and macronutrients to mammalian development and maintenance. They can be further linked up the theoretical chain to the roles that nutritional factors play in the selection of populations. We may state the general proposition that selection operates on differential rates of fertility and mortality (Ward and Weiss 1976), but in order to gain interpretive power from this proposition, archaeologists must isolate and operationalize those factors that influence fertility and mortality patterns in human populations (see Hockett and Haws 2005:25, Figure 1). Nutritional ecology seeks to interpret, among other things, nutritional impacts to human mortality, and then apply those interpretations to the archaeological record. And it is within this general context that I apply the nutritional ecology model to the ecofacts recovered from BER.

THE BER LATE PLEISTOCENE, EARLY HOLOCENE, AND MIDDLE HOLOCENE FAUNA

The faunal remains described below date to two general time frames: (1) the late Pleistocene and early Holocene, represented by Strata 17b', 18a, and 18b in the West Block and Strata 10, 11, and 12 in the East Block (Component 6); and (2) the middle Holocene, represented by Strata 14, 15, and 16 in the West Block (Stratum 17a contained lithic artifacts and a hearth but few faunal remains) and Stratum 9 in the East Block (Components 5a, 5b, 5c) (Table 11.1; also see Graf, this volume). Stratum 19, which dates to about 12,300 ^{14}C BP (14,100 cal BP), also contains an abundance of well-preserved fauna. However, because the possibility of human occupation in BER prior to 11,000 ^{14}C BP (13,000 cal BP) has not been demonstrated, the faunal remains from Stratum 19—as well as those recovered from below this level (Stratum 20)—will be discussed in an upcoming paper.

As noted above, the late Pleistocene/Component 6 occupations generally correlate with the Younger Dryas cooling, although human foragers continued to repeatedly occupy BER until ca. 8800 ^{14}C BP (9900 cal BP). The three to four "pulses" of middle Holocene/Component 5 occupations generally correspond to the 8k, 7k, and 6k cal BP climatic events, with each separated by at least a 1,000-year occupational hiatus. As a result, these three middle Holocene occupations may be compared and contrasted with one another, as well as combined together for a general picture of middle Holocene subsistence in the west Bonneville Basin. These latter trends can then be contrasted with the late Pleistocene/early Holocene occupations.

TABLE 11.2. Number of Identified Specimens of Faunal Remains Deposited Principally by Human Foragers During the Late Pleistocene to Middle Holocene at Bonneville Estates Rockshelter

	COMPONENT 6 12,600–10,700 BP	COMPONENT 5A 8000 BP	COMPONENT 5B 7000 BP	COMPONENT 5C 6000 BP
Bison	—	1 (1)	—	—
Mountain sheep	3 (1)	2 (1)	—	1 (1)
Pronghorn	7 (1)	—	11 (1)	1 (1)
Deer	1 (1)	1 (1)	—	1 (1)
Artiodactyl	14	1	15	4
Large mammal	1,314	284	595	346
Black bear	1 (1)	—	—	—
Hare	169 (9)	13 (2)	77 (4)	2 (1)
Small mammal (burned)	261	12	58	22
Sage grouse	502 (18)	—	16 (1)	—
Grasshoppers	18 (18)	—	—	—
Small game index	.93	.40	.83	.25

Note: Minimum number of individuals are included in parentheses; dates are reported in calendar years before present.

Late Pleistocene/Early Holocene Animal Procurement

The excavations of the late Pleistocene and early Holocene hearths and the sediments surrounding these features have produced the largest set of radiocarbon dates, lithic artifacts, and faunal remains yet recovered from this early period of the Great Basin. These bones represent a mixture of those deposited by humans and those accumulated by nonhuman agents (Tables 11.2 and 11.3). Taphonomic analysis can tease these components apart to various levels of certainty.

As I describe in further detail below, humans appear to have deposited the majority of artiodactyl, sage grouse, bear, and hare bones (as well as grasshoppers) in the shelter. However, most of the cottontail remains, all of the pygmy rabbit and rodent bones, and perhaps all of the waterfowl bones appear to derive from degraded raptor pellets and, to a lesser extent, from carnivore scats. The cottontail bones are biased toward young individuals, forelimb bones outnumber hindlimb bones, and mandibles were relatively common—all characteristics that generally indicate raptor pellet accumulations (Hockett 1991). A number of these bones also display single puncture marks. Many of the waterfowl bones appear similar (broken hind limb bones; evidence of slight gastric juice corrosion) to the collection of Pleistocene-aged teal and pintail bones deposited in raptor pellets in Pintwater Cave (Hockett 2000).

In contrast, there is ample evidence for the systematic butchering of sage grouse carcasses, as well as the butchering and extraction of marrow from both

TABLE 11.3. Faunal Remains (NISP) Deposited Principally by Raptors and Carnivores During the Late Pleistocene to Middle Holocene at Bonneville Estates Rockshelter

	COMPONENT 6 12,600–10,700 BP	COMPONENT 5A 8000 BP	COMPONENT 5B 7000 BP	COMPONENT 5C 6000 BP
Cottontail	239	2	33	2
Pygmy rabbit	149	—	—	—
Leporidae	685	24	73	10
Small carnivores	2	—	1	2
Raptors	10	—	—	—
Waterfowl	16	—	—	2
Passerine birds	38	—	—	2
Fish	1	—	2	—

Note: Dates are reported in calendar years before present.

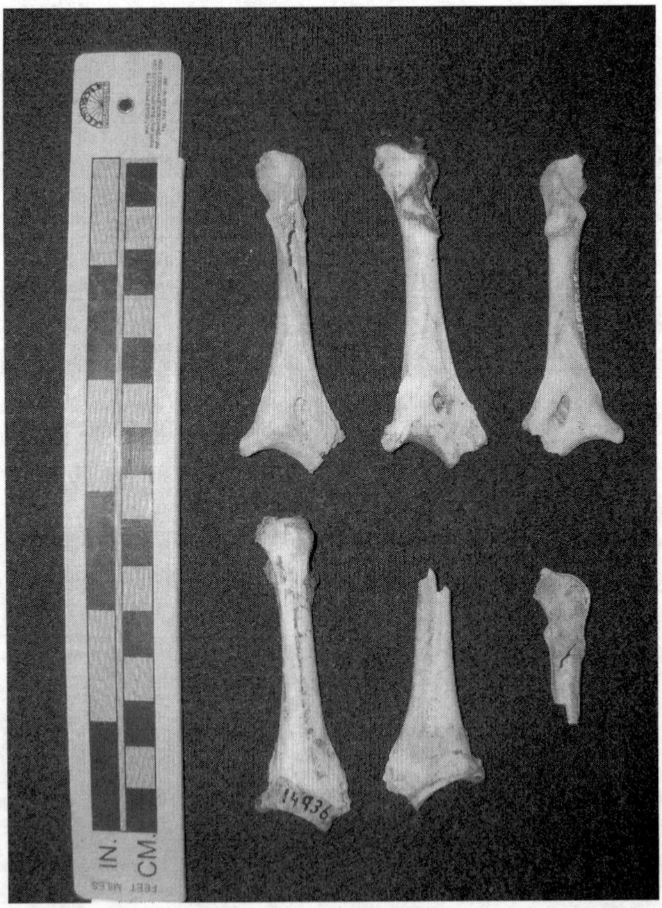

FIGURE 11.1. Late Pleistocene/early Holocene sage grouse coracoids recovered from Bonneville Estates Rockshelter.

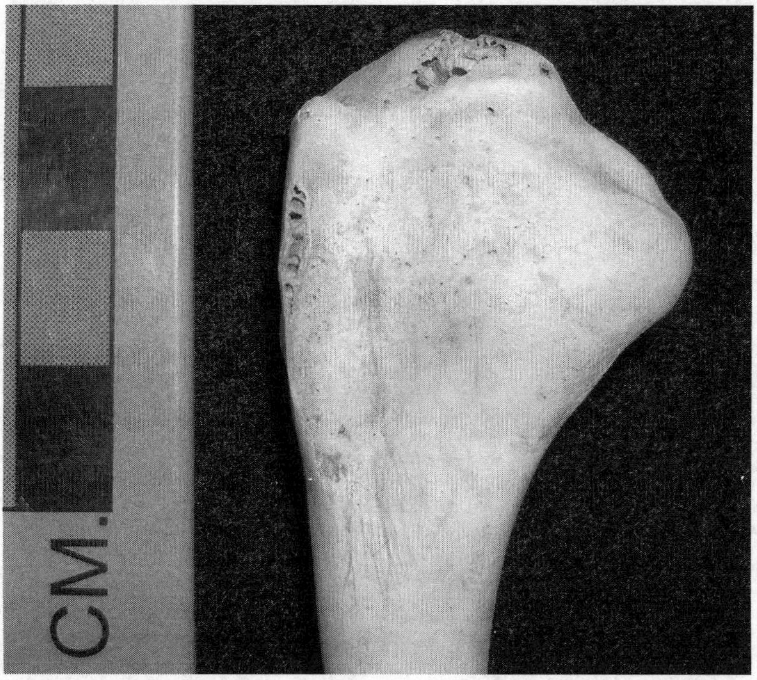

FIGURE 11.2. Late Pleistocene/early Holocene sage grouse proximal humerus with numerous cut marks recovered from Bonneville Estates Rockshelter.

TABLE 11.4. Sage Grouse Deposited by Human Foragers During the Late Pleistocene to Middle Holocene at Bonneville Estates Rockshelter

	COMPONENT 6 12,600–10,700 BP	COMPONENT 5A 8000 BP	COMPONENT 5B 7000 BP	COMPONENT 5C 6000 BP
Total	502	—	16	—
Unburned	489	—	15	—
Burned	13	—	1	—
Cut	33	—	4	—
% Burned and Cut	.09	.00	.31	.00

Note: Dates are reported in calendar years before present.

artiodactyl and hare carcasses. Over 500 sage grouse bones have been identified to date (Table 11.4), some with ligaments still attached after 13,000 calendar years (Figure 11.1). These bones represent at least 16 individual birds. Three percent of the grouse bones are charred, and about 7 percent of them display stone tool cut marks (Figure 11.2). The preservation and taphonomic analysis of these bones leave no doubt that the carcasses were systematically butchered and then left behind directly surrounding the hearths dated from these levels.

All portions of the grouse carcass were recovered. The butchering or dismemberment patterns are as follows: (1) the wing joint between the proximal

TABLE 11.5. Large Mammal (Artiodactyl) Fragments Deposited Principally by Human Foragers During the Late Pleistocene to Middle Holocene at Bonneville Estates Rockshelter

	COMPONENT 6 12,600–10,700 BP	COMPONENT 5A 8000 BP	COMPONENT 5B 7000 BP	COMPONENT 5C 6000 BP
Total	1,314	284	595	346
Unburned	877	247	519	283
Burned	437	37	76	63
Cut	2	7	45	3
% Burned and Cut	.33	.15	.20	.19

Note: Dates are reported in calendar years before present.

humerus and the scapula was separated with a stone tool of some kind; this left many proximal humeri with cut marks on the posterior side of the bone and also on the proximal blade of the scapula (Figure 11.2); (2) stone tools were also used to separate the upper femur from the pelvis; thus, several femora display slicing marks on their proximal ends (Figure 11.3); (3) the wings were further reduced in size by snapping the joints just below the proximal ends of the ulna and radius, and again just above the distal ends of these bones; this left almost every ulna and radius as diaphyses cylinders (Figure 11.4); (4) the lower leg was snapped just below the knee joint and again just above the ankle, leaving almost every tibiotarsus as a diaphysis cylinder (Figure 11.5); and (5) the feet were also snapped just below the ankle and again above the tarsals, leaving every tarsometatarsus as a midsection (Figure 11.6). Similar long bone cylinders created by human processing of bird carcasses have been identified in late Holocene contexts from the Great Basin, such as at the Stillwater Marsh sites in western Nevada (Schmitt 1988).

Although sage grouse sterna, vertebrae, clavicles, phalanges, and skull and mandible fragments were recovered, their relative frequency values were less than those of the wing and leg portions. Thus, the axial skeleton was either processed more intensively (which rendered these bones unidentifiable), discarded in a portion of the shelter not yet excavated, or differentially transported away from the shelter.

The artiodactyl bones identified belong primarily to pronghorn (*Antilocapra americana*) but also include mountain sheep (*Ovis canadensis*) and deer (*Odocoileus hemionus*). These bones were extensively broken and form the typical pattern of relatively small long bone shaft fragments created through marrow extraction. Approximately 34 percent of these artiodactyl fragments were burned, and two fragments retained stone tool cut marks (Table 11.5).

Regarding the hares, 26 tibiae, femora, and humeri cylinders were recovered, six of which displayed stone tool cut marks (Figure 11.7). The cylinders were created by human foragers snapping or biting off the proximal and distal

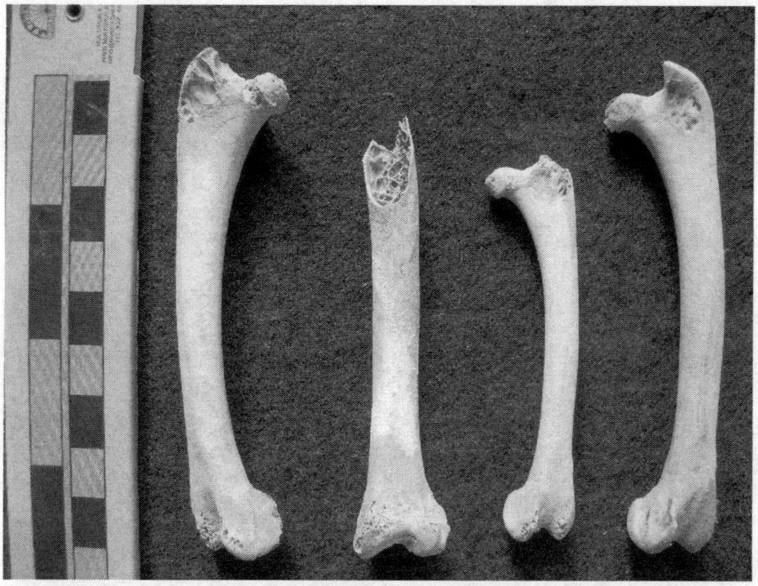

FIGURE 11.3.
Late Pleistocene/early Holocene sage grouse femora recovered from Bonneville Estates Rockshelter.

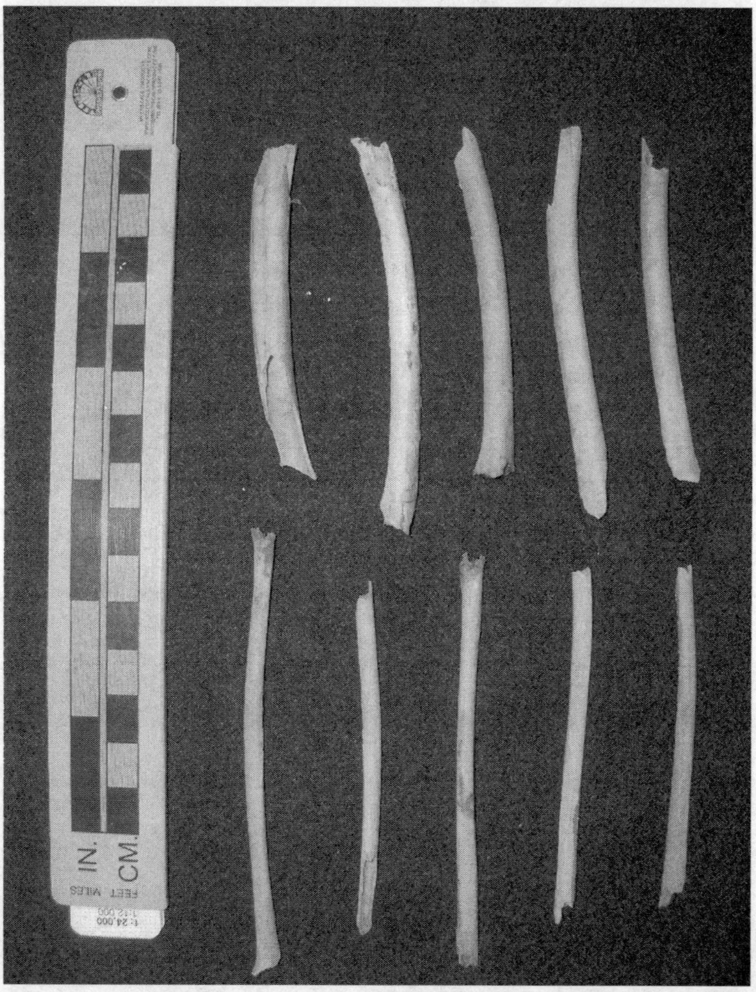

FIGURE 11.4.
Late Pleistocene/early Holocene sage grouse ulnae (*top*) and radii (*bottom*) diaphyses recovered from Bonneville Estates Rockshelter.

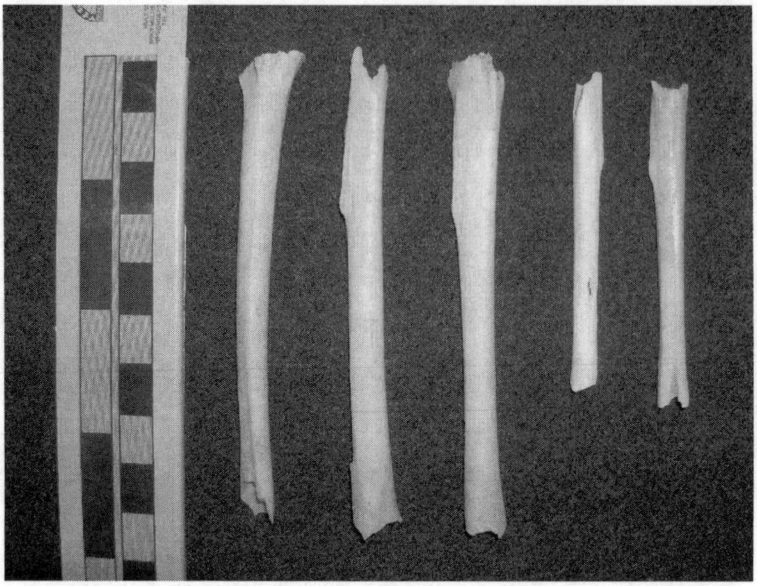

FIGURE 11.5.
Late Pleistocene/early Holocene sage grouse tibiotarsii diaphyses cylinders recovered from Bonneville Estates Rockshelter.

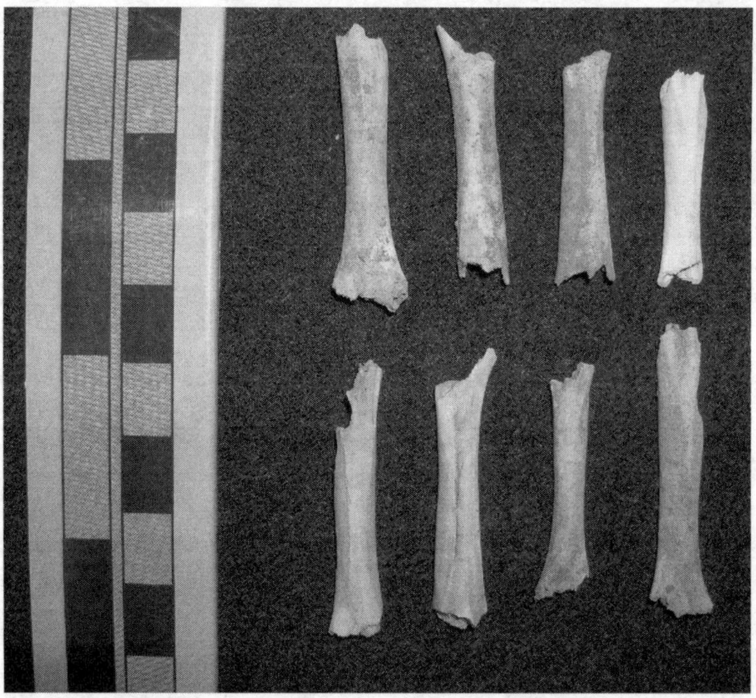

FIGURE 11.6.
Late Pleistocene/early Holocene sage grouse tarsometatarsii diaphyses recovered from Bonneville Estates Rockshelter.

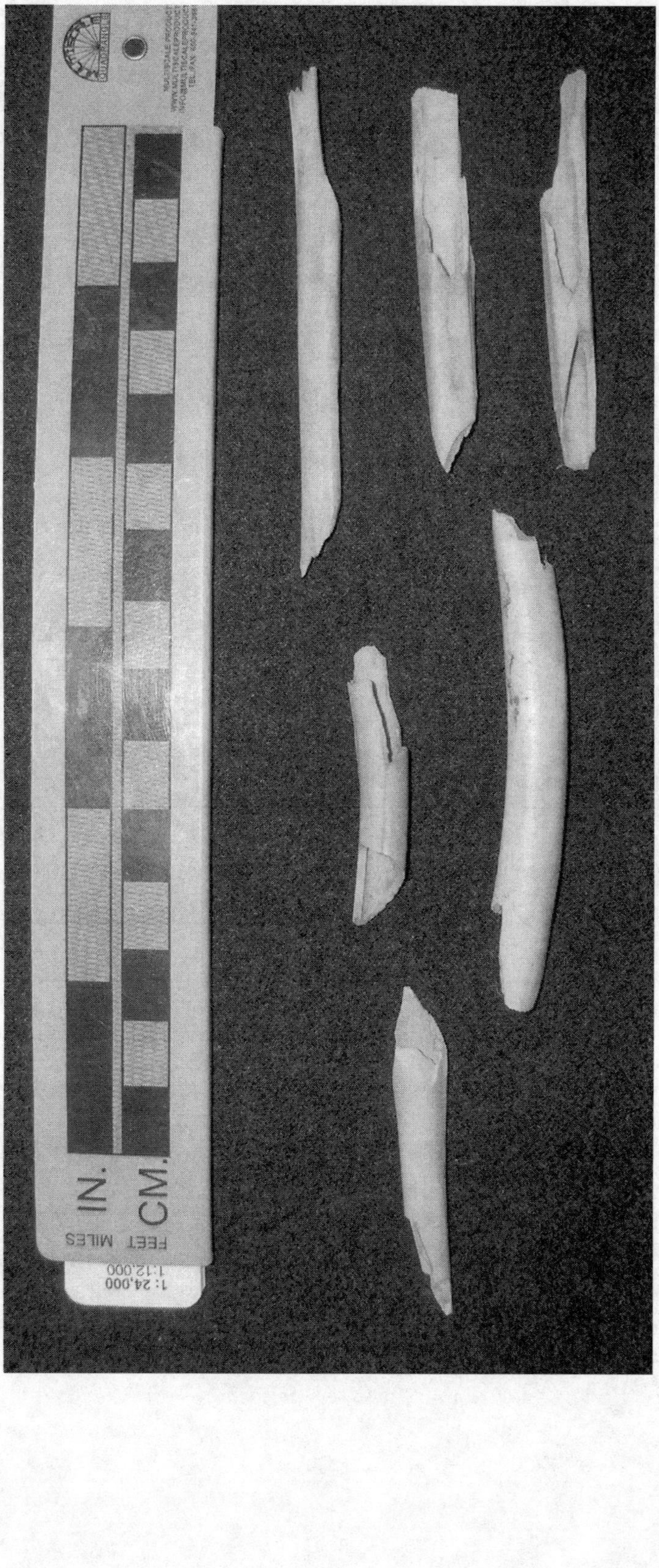

FIGURE 11.7.
Late Pleistocene/early Holocene hare diaphyses cylinders recovered from Bonneville Estates Rockshelter.

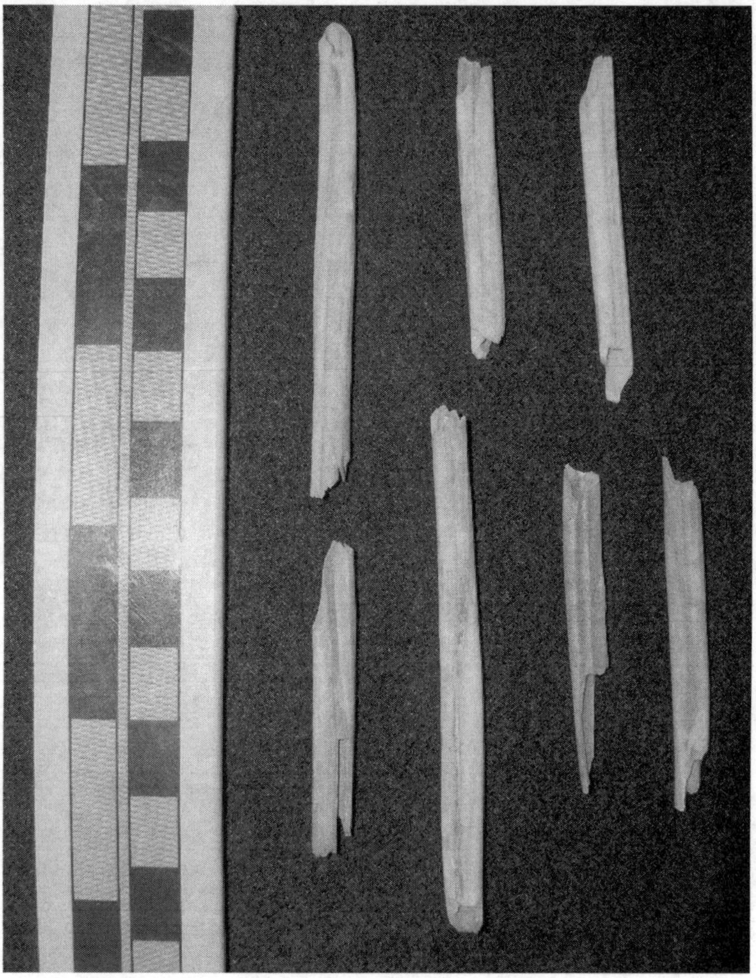

FIGURE 11.8. Late Pleistocene/early Holocene hare radii midsections recovered from Bonneville Estates Rockshelter.

ends to extract marrow (e.g., Schmitt 1986; Hockett 1991, 1994; Hockett and Haws 2002). These cylinders were previously known to occur in Great Basin assemblages dated as early as 7500 to 8500 ^{14}C BP (8400 to 9500 cal BP) (Hockett 1994), and with this analysis they are now known to occur as early as 10,500 to 10,800 ^{14}C BP (12,600 to 12,800 cal BP). The front limbs were also separated from the remainder of the carcasses by snapping the joints just below the proximal ulna and radius and again above the distal ends of these bones, leaving almost all radius-ulna units as midsections (Figure 11.8). These front limb bones look precisely like those recently illustrated in the middle Holocene levels from Camels Back Cave (Schmitt et al. 2004:91, Figure 4). Thus, this method of processing hare carcasses also dates to ca. 10,800 ^{14}C BP (12,800 cal BP) in the Great Basin. In addition, seven hare bones were burned, as were almost 200 leporid-sized bone fragments. Overall, a total of 13 of the 169 hare bones recovered (8 percent) were either cut or burned.

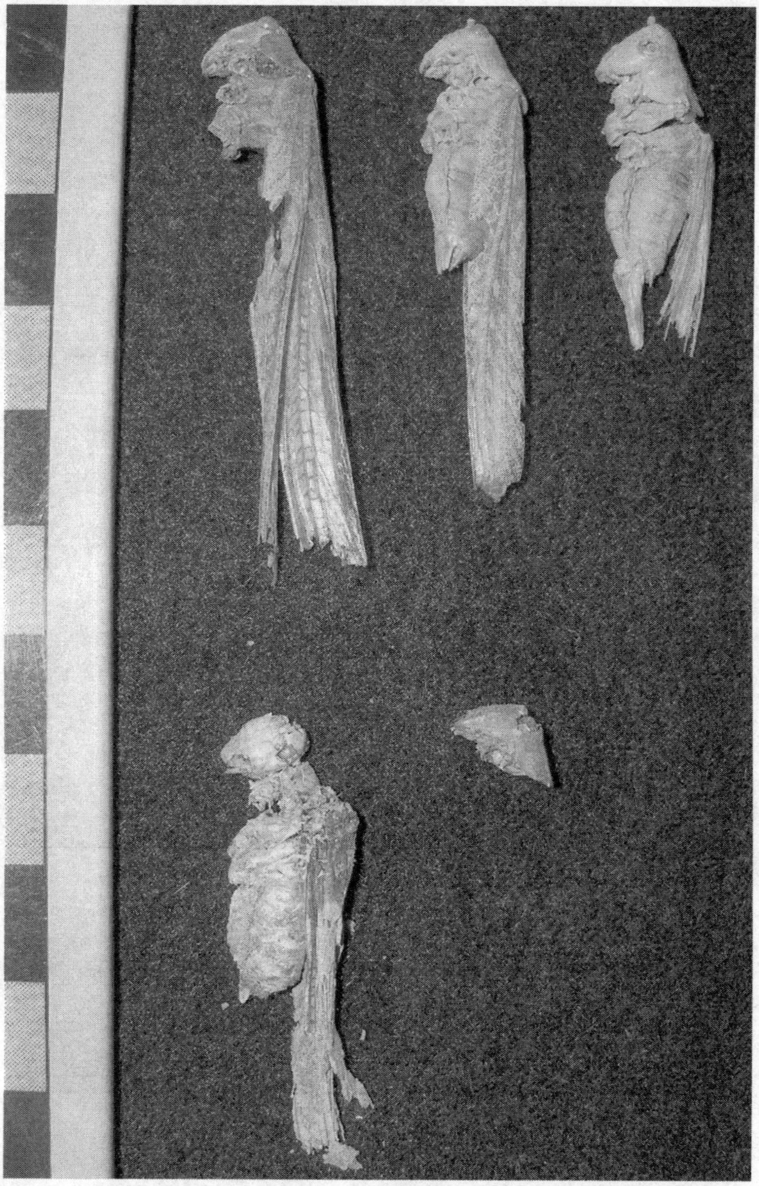

FIGURE 11.9. Late Pleistocene/early Holocene grasshoppers recovered from Bonneville Estates Rockshelter.

One central phalanx of a black bear (*Ursus americanus*) was recovered from Stratum 18a. The bone was thoroughly charred, although no cut marks were visible. It therefore appears likely that late Pleistocene and early Holocene foragers occasionally hunted large carnivores in the west Bonneville Basin, although scavenging of bear parts cannot be discounted.

Finally, Stratum 18a produced 18 well-preserved grasshoppers surrounding several hearths (Figure 11.9). These grasshoppers were typically complete or near-complete specimens, and most had their hind legs removed. Although grasshopper consumption is known from late Holocene–aged contexts in the

Bonneville Basin (Madsen and Kirkman 1988), the grasshoppers from BER may represent the earliest evidence for insect consumption in the Great Basin.

Middle Holocene Animal Procurement

The initial response of human foragers to the warm and arid conditions of the early-middle Holocene was to abandon the area surrounding BER. This may have occurred as early as 8800 ^{14}C BP (9900 cal BP). During the first middle Holocene–aged reoccupations of the shelter that correspond closely to the 8k climate event, the primary prey was artiodactyls (Table 11.2). Pronghorn, mountain sheep, deer, and bison also were taken. Hares were taken in smaller proportion compared to the late Pleistocene/early Holocene levels. No sage grouse or grasshoppers were consumed inside the shelter. Overall, species diversity and, perhaps, mammalian and avian biomass available near the shelter witnessed dramatic reductions. Near BER, human foragers responded in their hunting guild first by changing their settlement patterns (abandonment) and then by raising their trophic level, focusing their hunting efforts primarily on artiodactyls during the 8k reoccupation of the region.

Following the brief 8k reoccupations of the shelter, human foragers again abandoned BER for approximately 1,000 years. The 7k reoccupation that corresponds to another wet spike within the overall warm and arid middle Holocene was more intensive than the previous 8k reoccupation. The diversity of animals taken increased slightly, although artiodactyls continued to be the prey of choice (Table 11.2). Hares and sage grouse (the latter representing a single individual bird) were both consumed inside the shelter.

Following the 7k reoccupation, the shelter was again abandoned for approximately 1,000 years. Similar to the 8k reoccupation, the hunting guild associated with the 6k reoccupation saw a near complete reliance on artiodactyls (Table 11.2). Unlike the 8k reoccupation, however, which saw four different species of artiodactyls taken, human foragers during the 6k reoccupation took primarily pronghorn. Few hares and no sage grouse were taken. There is no evidence for grasshopper consumption.

THE NUTRITIONAL ECOLOGY OF LATE PLEISTOCENE/ EARLY HOLOCENE AND MIDDLE HOLOCENE FORAGERS IN THE WEST BONNEVILLE BASIN

General Patterns

What might these data mean for our understanding of prehistoric subsistence and demographic patterns in the west Bonneville Basin? The first relatively substantial demographic increase throughout the Great Basin, as judged by overall site density, is associated with populations that manufactured Western Stemmed projectile points (e.g., Jones and Beck 1999). These populations are considered here to postdate the Clovis phenomenon (or to postdate ca. 10,800 ^{14}C BP [12,800 cal BP]), although the Cooper's Ferry site in Idaho suggests that stemmed points may be as old as Clovis in some regions (Davis and Schweger 2004).

After two decades of survey in the region surrounding BER in northeastern Nevada, a total of seven isolated fluted points have been recorded. This suggests extremely low population densities for the region's earliest inhabitants. In fact, with only seven isolated points known, it cannot be argued that Clovis/fluted populations were in fact residents. Instead, they likely were only occasional visitors (after Marks 2000). If compared to the Younger Dryas to early Holocene subsistence/settlement patterning described above, this suggests that Clovis populations had very different subsistence and mobility patterns than did the latest Pleistocene populations that inhabited the west Bonneville Basin near BER. Presumably, all of the resources taken by Western Stemmed populations within a context of "seasonal rounds"—including sage grouse, medium-sized artiodactyls, insects, and probably waterfowl and fish (see discussions below)—would have been available to Clovis foragers as well. And yet we do not find an occupational pattern coupled with a consistent, diverse diet in BER for Clovis/fluted populations. Support for the interpretation of reduced mobility within a context of subsistence-related activities targeting a diverse resource base, possibly including marsh resources, has recently been realized at the Old River Bed paleodelta sites, where as many as 500 Western Stemmed sites may be present (Oviatt et al. 2003; Rhode et al. 2005). This is not meant to imply that Western Stemmed populations were tethered year-round to lowland marsh resources. On the contrary, stemmed points have been found in the uplands of the Goshute Mountains west of BER, and two Paleoarchaic points were found slightly farther to the west at an elevation of 2,425 m (8,000 ft) in the Cherry Creek Range. Also, BER itself contains evidence for terrestrial resource use only. Nevertheless, while upland resources were not eschewed by Western Stemmed populations, sites dating to this time period appear to be much more common in lowland and midslope settings.

These data corroborate the suggestion by Haynes (2002) that Clovis populations had fundamentally different subsistence and mobility patterns than did Younger Dryas and early Holocene populations. Not only do fluted sites in the region consist of isolated artifacts only, but none of them have been found in upland settings, suggesting that when the pre–Western Stemmed populations were moving through the region, they were not bothering to explore the uplands as the Western Stemmed groups clearly did. From a nutritional ecology perspective, if Clovis populations chose to restrict their diet by focusing on large terrestrial herbivores within a context of high mobility, then they likely suffered from higher incidences of fetal (and perhaps maternal) mortality compared to Western Stemmed populations. Such behaviors would not spur significant population growth in the region.

In contrast, a major change in occupational intensity occurred in the Great Basin after ca. 11,000 ^{14}C BP (12,900 cal BP). According to the nutritional ecology model, such a population pulse should correlate with an increase in dietary diversity, and perhaps lower mobility, the latter of which would have reduced stress on pregnant females and mothers. The Western Stemmed populations that inhabited BER between ca. 11,000 and 9500 ^{14}C BP (12,900 and 10,700 cal BP) hunted sage grouse, a variety of large and small terrestrial mammals, and grasshoppers. This represents a relatively broad-based diet as seen from a single site. In addition, unlike Clovis, Western Stemmed mobility patterns probably were reduced in the west Bonneville Basin as the latter groups occupied sites such as BER for much longer periods of time, on a more frequent basis, or both. Thus, these data are in conformance with the predictions of the nutritional ecology model.

Support for the link between dietary diversity and population increases in the latest Pleistocene and early Holocene of the Great Basin, however, extends far beyond the data from BER. Coprolites from Spirit Cave suggest that fish such as chub (*Sipateles bicolor*) and the seeds of bulrush (*Scirpus* spp.) were important dietary components for early Holocene populations in the western Great Basin (Napton 1997). Faunal remains from the Weed Lake Ditch, Nials, and Paulina Lake sites in Oregon suggest that waterfowl, small terrestrial game such as rabbits, and plant foods were widely consumed there (Wriston 2003). Numerous specimens of mollusk were recovered from the early Holocene layers in Last Supper Cave in northwestern Nevada (Grayson 1988). Sage grouse, waterfowl, fish, artiodactyls, rabbits, mollusks, grasshoppers and plant foods: it simply doesn't get more diverse than that in the Great Basin. As a result, these groups probably enjoyed relatively higher rates of successful fetal development, lower rates of maternal and infant mortality, and perhaps longer life expectancies than did earlier foragers who occasionally visited the region carrying fluted points. Population growth was one result for Western Stemmed foragers.

The subsistence data from the middle Holocene occupations at BER tell a very different tale. In contrast to the repeated occupations of the shelter during the late Pleistocene and early Holocene over many centuries, the middle Holocene use of BER was characterized by relatively short-term occupations of the immediate region (for perhaps no more than a couple centuries) followed by long periods of abandonment. These latter groups may have actually stayed inside the shelter for longer periods of time during each occupation compared to earlier occupations, but paradoxically, the diversity of animals taken dropped dramatically. Artiodactyl hunting was the primary focus of their animal procurement activities. Preliminary evidence suggests that the overall diversity of plant use did not increase significantly during the middle Holocene compared to earlier occupations, with both groups consuming primarily terrestrial-based foods such as seeds and cacti (Rhode and Louderback, this volume). However, the manner in which these plant resources were processed, as well as perhaps the gross amount consumed, appears to have differed with the advent of grinding stone technology in the middle Holocene (Rhode and Louderback, this volume). Overall, then, there appears to be another correlation between population and the diversity of resources eaten during the middle Holocene, particularly in the animal fraction. In this case, as overall food diversity dropped, so too did the longer-term support of human populations in the region. This is precisely what the nutritional ecology model would predict. These general interpretations are elaborated upon in further detail below.

The Nutritional Well-Being of Late Pleistocene/Early Holocene and Middle Holocene Foragers Near BER: A First Approximation

Tables 11.6 and 11.7 display the nutrient values per 100 g of various animal and plant food types that are known to have been consumed by late Pleistocene/early Holocene or middle Holocene foragers in the Great Basin. As a general rule, foragers would have increased the dietary diversity of essential nutrients they consumed by including different food types in their diet (across the table columns), not by including different forms of the same food type (up or down a single column). There is considerable variability in the amount of nutrient diversity found within certain food types, however. In general, plant foods such as green leafy vegetables (GLVs) can be highly variable in nutrient content between plant genera, but the same cannot be said for many of the genera and species of animals, particularly terrestrial herbivores. Thus, a forager is more apt to significantly diversify his or her essential nutrient intake by adding birds to a diet previously based on pronghorn, rabbits, and seeds, rather than by adding mountain sheep to that same base diet. This principle is discussed in further detail below.

If we compare the nutrient values listed in Tables 11.6 and 11.7 between food types, a number of significant points may be highlighted. Within the animal fraction, bird meat is significantly fatter than terrestrial herbivore meat. Foragers could "make up" for the leanness of terrestrial herbivore meat, however, by consuming bone marrow or rendering grease from bones through boiling (e.g., Lupo and Schmitt 1997). Grasshoppers contain as much as four times the amount of fat as terrestrial herbivore meat, so foragers could also "make up" for a loss of terrestrial herbivore fat by consuming foods such as birds, fish, insects, and certain seeds. The point is that there are various ways that Great Basin foragers could have obtained adequate amounts of fat that either included or excluded terrestrial herbivores.

While terrestrial herbivores are high in protein, so too are birds, fish, insects, mussels, and many seeds. Carbohydrates are found mainly in plant resources but also may be obtained by eating animal resources such as grasshoppers and mussels.

Diversity in micronutrient intake can also be achieved through a number of foraging strategies, but again it is the diversity of food types consumed that is the key. Minerals such as calcium are significantly more abundant in nonterrestrial animal resources such as bird eggs, fish, mussels, and many plant foods. Bird organs are also significantly higher in iron and zinc compared to terrestrial

TABLE 11.6. Macro- and Micronutrient Values of Animal Foods Potentially Consumed by Great Basin Foragers

	Artiodactyls				Birds					
	Meat[a]	Marrow[b]	Organs[c] and Skin[d]	Hare Meat[e]	Meat	Organs[f]	Eggs[g]	Fish[h]	Grass-hoppers[i]	Mussels[j]
Calories	114	408	183	105	244	180	185	92	380	86
Fat	2.0	41.0	11.7	1.3	11.1	6.7	13.8	2.3	8.9	2.2
Protein	22.4	8.9	20.0	22.0	23.4	25.6	12.8	16.8	66.9	11.9
Carbohydrates	0.0	0.0	0.0	0.0	0.0	2.6	1.5	0.0	19.0	3.7
Micronutrients										
Calcium	3.0	6.1	4.1	5.0	5.9	6.6	64.0	70.0	—	26.0
Iron	3.2	6.9	19.0	4.0	4.6	26.0	3.9	1.3	—	4.0
Zinc	1.3	.98	1.8	1.6	1.9	4.6	1.4	.75	—	1.6
Magnesium	trace	34.0	21.0	32.0	25.0	24.0	17.0	30.0	—	34.0
Copper	.18	.50	1.4	.30	.26	1.1	0.6	.20	—	.09
Phosphorus	188	34.0	194	212	179	261	220	210	—	197
Potassium	353	36.0	306	449	288	230	222	380	—	320
Selenium	9.7	0.0	trace	.01	6.4	68.1	36.4	12.6	—	44.8
Vitamin A (IU)	0.0	240	9,196	0.0	88.0	31,000	674	170	—	160
Vitamin B-6	.37	0.0	.50	.50	.39	.76	.25	.20	—	.05
Vitamin B-12	6.3	0.0	27.6	7.2	.65	54.0	5.4	2.0	—	12.0
Vitamin C	0.0	0.0	12.7	0.0	5.2	4.5	0.0	0.0	—	8.0
Vitamin D	0.0	0.0	.75	0.0	0.0	0.0	.60	0.0	—	0.0
Vitamin E	.20	0.0	1.2	0.0	.70	0.0	1.3	0.0	—	.55
Vitamin K	1.1	0.0	0.0	0.0	0.0	0.0	.40	0.0	—	.10
Thiamin	.22	.04	.16	.03	.35	.56	.16	.01	—	.16
Riboflavin	.48	0.0	2.4	.06	.27	.89	.40	.07	—	.21
Niacin	6.4	.20	9.4	6.5	3.3	6.5	.20	1.2	—	1.6
Folate	4.0	0.0	105	8.0	21.0	738	80.0	15.0	—	42.0

Note: All values are per 100 g of food.

[a] Macronutrient and mineral values based on the pronghorn (*Antilocapra americana*); vitamin values based on deer (*Odocoileus* spp.) (USDA 2005).

[b] Values based on the cervid caribou/reindeer (*Rangifer tarandus*) (Kuhnlein et al. 2002; USDA 2005).

[c] Macronutrient and mineral values based on an average of caribou/reindeer heart, liver, and tongue (Kuhnlein et al. 2002); vitamin values based on beef liver, brains, and kidney (USDA 2005).

[d] Macronutrient and mineral values based on spruce grouse meat (*Dendragus canadensis*), fire-roasted leg with skin of Canada goose (*Branta canadensis*), and duck (*Anas* spp.) meat with skin (Belinsky and Kuhnlein 2000; Kuhnlein et al. 2002; USDA 2005).

[e] Macronutrient and mineral values based on the hare (*Lepus americanus*) (Kuhnlein et al. 2002); vitamin values based on the wild rabbit (*Oryctolagus cuniculus* and *Sylvilagus* spp., except for folate and the B vitamins, which are based on domesticated rabbit (USDA 2005).

[f] Macronutrient and iron, zinc, calcium, and copper values based on Canada goose liver, gizzard, and heart (Belinsky and Kuhnlein 2000); all other values based on goose liver (USDA 2005).

[g] Values based on duck egg (USDA 2005).

[h] Values based on the sucker (*Catostomus commersoni*) (USDA 2005).

[i] Macronutrient values based on averages of seven species of grasshoppers and locusts.
representing the genera *Sphenarium*, *Plectotettia*, *Melanoplus*, and *Schistocerca* (Ramos-Elorduy et al. 1997).

[j] Values based on the blue mussel (*Mytilus edulis*) (USDA 2005).

TABLE 11.7. Macro- and Micronutrient Values of Plant Foods Potentially Consumed by Great Basin Foragers

	GLVs[a]	Seeds[b]	Fruits[c]	Pine "Nuts" (Seeds)[d]	Bulbs[e]
Calories	82	229	56	629	32
Fat	2.0	1.5	.20	61.0	.19
Protein	2.8	8.5	1.4	11.6	1.8
Carbohydrates	35.1	49.0	13.8	19.3	7.3
Micronutrients					
Calcium	129	22.0	33.0	8.0	72.0
Iron	2.2	2.0	1.0	3.1	1.5
Zinc	.54	2.3	.23	4.3	.39
Magnesium	91.1	71.5	13.0	234	20.0
Copper	.10	.28	.11	1.0	.08
Phosphorus	31.0	214	44.0	35.0	37.0
Potassium	460	179	275	628	276
Selenium	—	22.0	.60	0.0	.60
Vitamin A (IU)	2918	9.0	42.0	29.0	997
Vitamin B-6	.09	.20	.07	.11	.06
Vitamin B-12	0.0	0.0	0.0	0.0	0.0
Vitamin C	18.5	0.0	41.0	2.0	18.8
Vitamin D	0.0	0.0	0.0	0.0	0.0
Vitamin E	0.0	.65	.10	0.0	.55
Vitamin K	174	3.4	11.0	0.0	207
Thiamin	30.0	.20	.04	1.2	.06
Riboflavin	8.3	.16	.05	.22	.08
Niacin	.38	3.2	.10	4.4	.53
Folate	38.0	38.0	8.0	58.0	64.0

Note: Values based on 100 g of food.

[a] GLVs = green leafy vegetables. Macronutrient and mineral values based on averages of numerous GLVs reported in Guerrero et al. 1998; Agte et al. 2000; and Lockett et al. 2000. Most vitamin values based on averages of numerous GLVs reported in Anderson and Deskins 1995 and Agte et al. 2000; other vitamin values are based on green and red lettuce (*Lactuca sativa*) (USDA 2005).

[b] Nutrient values of seeds commonly consumed in the Great Basin are not readily available; these values are based on an average of a relatively low nutrient-rich seed (barley) combined with a relatively high nutrient-rich seed (rye) (USDA 2005).

[c] Values based on the currant (*Ribes rubrum*) (USDA 2005).

[d] Values based on the Colorado pinyon pine (*Pinus edulis*) nut or seed (USDA 2005).

[e] Values based on the tops and bulb of the onions *Allium cepa* and *Allium fistulosum* (USDA 2005).

herbivores and plant foods. For the vitamins, selenium is significantly more abundant in nonterrestrial herbivore resources. Bird organs can substitute for terrestrial herbivore organs for adequate consumption of vitamin B-12. On average, plant resources would provide a more consistent consumption of vitamin C, vitamin K, and folate. Nonterrestrial herbivore animals such as birds and mussels would also provide a more consistent consumption of folate, and

TABLE 11.8. Hypothetical Average Diets of Late Pleistocene/Early Holocene vs. Middle Holocene Foragers in the Great Basin

FOODS CONSUMED	LATE PLEISTOCENE/ EARLY HOLOCENE DIET	MIDDLE HOLOCENE DIET
Rabbit	1 leg (leg and thigh)	1 leg (leg and thigh)
Pronghorn	8 ounces boneless	16 ounces boneless
Liver	1 ounce	2 ounces
Brains	1 ounce	2 ounces
Tongue	1 ounce	2 ounces
GLVs[a]	1 large leaf	2 large leaves
Seeds[b]	1 cup	6 cups
Duck	1 leg (drumstick and thigh)	—
Grouse/quail	4 ounces	—
Duck egg	1 egg	—
Fruit	1 cup currants	—
Fish	½ sucker	—

[a] Lettuce was used here as an approximation of nutrients consumed by eating GLVs.
[b] Barley was used here as an approximation of nutrients consumed by eating seeds.

birds and GLVs would provide a more consistent consumption of thiamin and riboflavin. These data are particularly significant because the underconsumption of adequate micronutrients such as vitamins A, C and E—as well as folic acid, zinc, copper, and iron—has been associated with higher incidences of failed human births (e.g., Ashworth and Antipatis 2001; Bendich 2001).

The next step in assessing whether there was a difference in overall forager health between the late Pleistocene/early Holocene and the middle Holocene is to begin modeling average dietary consumption between the two groups and analyzing the nutrients consumed as a result of those diets. Table 11.8 represents a first approximation toward this goal.

It is certainly true that we cannot know the exact proportions of foods eaten on each and every plate each and every day in the past, but we do not necessarily need access to this kind of detailed information in order to model average prehistoric diets. If we know the types of foods eaten on a consistent basis, and if we know to some degree the relative proportion that each food type contributed to the overall diet, then we can obtain an approximation of essential nutrient intake in prehistoric populations because the sum calories of the proportions of each food type consumed will be moderated by human needs and tolerance levels. In other words, if we grossly overestimate proportions of each food type known to have been consumed, then any final tally would show a calorie consumption beyond human tolerance and gut size. If we grossly underestimate these proportions, then it would show a level of chronic energy deficiency that would lead to extinction. Absent evidence for either scenario, the proportions listed in Table 11.8 are within the boundaries of human caloric requirements and conform to the proportions listed by the USDA (2005) so that accurate nutrient content can be calculated.

Based on the faunal and plant resources retrieved from Great Basin sites that date between the late Pleistocene and middle Holocene, Table 11.8 outlines an average diet of the two groups. The emphasis placed on the late Pleistocene/early Holocene group is on diversity, with large and small terrestrial herbivores, birds, fish, and some plant resources—including moderate amounts of fruits, seeds and GLVs—included in the diet. The rationale for placing fruits as an important part of the diet during this time is the fact that a cooler and wetter climate would have fostered more widespread and dense populations of a variety of Great Basin fruits, including currant, chokecherry, and rose (see also Schmitt and Madsen 2005). Unfortunately, grasshoppers that were eaten by late Pleistocene/early Holocene foragers at BER could not be included in Table 11.8 because the U.S. Department of Agriculture (USDA) Web site that calculates nutrient values does not currently include insects. Consequently, the macronutrient values (discussed in detail below) for this group will probably underestimate the consumption of fat, protein, and carbohydrates, as grasshoppers provide an abundance of all three.

The middle Holocene diet was less diverse overall but may have become more "diverse" where it counted less (up and down a column rather than across columns) in terms of diversity of essential nutrient intake. The middle Holocene diet seemed to represent a "splitting apart" of the diverse diet seen in the late Pleistocene/early Holocene, with greater emphasis placed on either extremes of the foraging spectrum: large terrestrial herbivores and, with an increasing abundance of milling stones seen in the archaeological record at this time, seeds. Thus, for middle Holocene foragers I doubled the intake of terrestrial herbivores and GLVs and placed a much greater emphasis on seeds (Table 11.8). Lost during this time was a more consistent consumption of resources such as birds, fruit, and fish. Certain other resources, such as bulbs, were not included in either diet for this exercise, although further research may suggest that these potentially important resources should be included in the diets of either one or both of these groups.

Following this exercise, I then placed these values into the USDA's (2005) newly created "Diet Analysis" Web site, which calculated the individual nutrients consumed based on these diets, as well as the current recommended daily allowance (RDA) for daily nutrient consumption (Table 11.9). RDA values vary based on the sex, weight, and age of the individual. They also vary for women who are pregnant, as expecting mothers need to consume additional macro- and micronutrients to ensure a successful pregnancy. Because dietary diversity most influences human demographic trends at the level of the mother and child, I based the diet on a female that was 5'3" tall, weighing 125 lbs. For this modeling procedure, the RDA values are based on a nonpregnant female. I will specifically discuss the dietary implications of Great Basin diets through time on pregnant females in an upcoming essay.

TABLE 11.9. Diet Analysis Comparison of Two Hypothetical Prehistoric Females Living in the West Bonneville Basin during the Late Pleistocene/Early Holocene and the Middle Holocene

NUTRIENT	LATE PLEISTOCENE/ EARLY HOLOCENE FORAGER	MIDDLE HOLOCENE FOAGER	CURRENT USDA RECOMMENDATION OR ACCEPTABLE RANGE
Food energy (kcals)	1,838	2,448	1,912
Protein (gm)	165	230	46
Carbohydrate (gm)	165	277	130
Total fiber (gm)	16	37	25
Total fat (gm)	60	44.6	40.9–71.5
Saturated fat (gm)	20	17	< 27.2
Monounsaturated fat (gm)	22	12	not established
Polyunsaturated fat (gm)	8	4	not established
Linoleic (omega-6) (gm)	6.1	3.2	12
Alpha linoleic (omega-3) (gm)	.90	.50	1.1
Vitamin A (mcg RAE)	2,112	3,744	700
Vitamin C (mg)	15	12	75
Vitamin E (mg)	4.7	5.1	15
Thiamin (mg)	1.2	1.6	1.1
Riboflavin (mg)	3.5	5.6	1.1
Niacin (mg)	41	69	14
Folate (mcg)	227	424	400
Vitamin B-6 (mg)	3	4.2	1.3
Vitamin B-12 (mcg)	35	60	2.4
Calcium (mg)	236	159	1,000
Phosphorus (mg)	1,708	2,326	700
Magnesium (mg)	226	383	310
Iron (mg)	32	44	18
Zinc (mg)	41	74	8
Selenium (mcg)	140	214	55
Potassium (mg)	2,968	2,910	4,700
Sodium (mg)	1,875	3,795	1,500–2,300

Note: This analysis is based on ecofacts recovered from BER and other similar-aged sites across the Great Basin, and on the serving sizes and foods consumed listed in Table 11.8. In both cases, the nutrient composition analyses were based on a 21-year-old female with a stature of 5'3" and a weight of 125 lbs. (USDA 2005).

The results of the dietary analyses, displayed in Table 11.9, are interesting. Both groups show sufficient calorie consumption, with the middle Holocene group consuming approximately 25 percent additional calories than the late Pleistocene/early Holocene group. This may well have been necessary for the middle Holocene group. Evidence suggests that many late Pleistocene/early

Holocene foragers were able to tether themselves more closely to marshland and adjacent midslope resources, so these populations may have been less mobile on a day-to-day basis and, as a result, may have required fewer calories than the middle Holocene group.

Both groups show abundant consumption of proteins and carbohydrates. The middle Holocene group, however, shows a rather severe overconsumption of both types of macronutrients, but the late Pleistocene/early Holocene group less so. These data conform to the results of modern nutrition science research indicating that as dietary diversity drops, humans tend to underconsume certain nutrients and overconsume others. Both conditions can be potentially devastating to human health, particularly for a pregnant female and her developing fetus (Ashworth and Antipatis 2001).

Another interesting pattern within the macronutrient fraction is seen in the two groups' fat consumption. While both groups show adequate consumption of fat, the middle Holocene group is at the low end of the RDA scale, while the late Pleistocene/early Holocene group is at the upper end. Additionally, the late Pleistocene/early Holocene group consumed twice as much omega-6 and omega-3 fatty acids, which are critical to many aspects of human health, including brain development.

The micronutrients also tell an interesting tale. Both groups show adequate consumption of certain vitamins and minerals such as thiamin and magnesium. And while both groups show an overconsumption of many micronutrients based on the current RDA standard, the middle Holocene group appears to have exacerbated this situation to a much greater degree. Thus, for example, while the late Pleistocene/early Holocene group may have overconsumed riboflavin by three times the RDA standard, the middle Holocene group may have overconsumed this vitamin by five times the standard. A similar situation applies to vitamin A, niacin, vitamin B-6, vitamin B-12, phosphorous, iron, zinc, selenium, and sodium.

In contrast, for some of the micronutrients that were underconsumed by both groups, including vitamin C and calcium, the late Pleistocene/early Holocene group probably consumed more than the middle Holocene group. For either human group, however, the underconsumption of calcium would be especially troubling for long-term health in later years, particularly for groups that had higher rates of mobility. For the latter group, we should expect to see, for example, earlier incidences of, and greater degrees of, bone mass loss and arthritic conditions. Overall, then, the intake of micronutrients parallels the nutrition principle applied above for the macronutrients: the less diverse diet seen in middle Holocene populations likely led to more persistent hyper-overconsumption of some micronutrients and hyper-underconsumption of others. This situation likely created additional health problems for these individuals, and likely led to decreases in the percentage of successful pregnancies compared to the late Pleistocene/early Holocene group.

CONCLUDING REMARK

Faunal evidence from BER and other Western Stemmed assemblages suggests that during the initial phases of population growth in the Great Basin, groups exploited a quite diverse resource base. This diversity is linked to the consumption of a variety of artiodactyls, small mammals such as hares, birds such as sage grouse and waterfowl, fish, and grasshoppers, as well as plant foods. From a broader perspective, Western Stemmed populations set the nutritional framework under which their populations could expand by consuming a broad-based diet. And expand they did.

12

Dietary Plant Use in the Bonneville Basin During the Terminal Pleistocene/ Early Holocene Transition

David Rhode and Lisbeth A. Louderback

The antiquity of Archaic subsistence practices has long been an important issue in Great Basin prehistory. When did people adopt a broad-spectrum diet emphasizing plant foods such as small seeds, roots and tubers, and small animals? Did human groups in the terminal Pleistocene/ early Holocene transition (TP/EH) subsist in much the same manner as later Archaic foraging societies? Were plant resources important in the late Pleistocene and early Holocene diet, and if so, which plants?

Questions of plant use in the TP/EH and after ties directly to one of the central distinctions drawn between the so-called Pre-Archaic and Paleoarchaic conceptions of foraging strategies practiced by early occupants of the Great Basin. Opinions vary about the nature of subsistence during this period. These range from notions of a relatively narrow diet breadth focusing on high-quality resources from primarily wetland patches, a view associated with the Pre-Archaic concept (Elston 1982, 1986; Elston and Zeanah 2002), to notions of a broader diet breadth quite akin to subsequent Archaic foragers but set apart from the Archaic by the lack of use of low-ranked small hard seeds and associated milling stone technology (Madsen, this volume), a view most closely associated with the Paleoarchaic concept (Beck and Jones 1997; Jones and Beck 1999; Willig and Aikens 1988).

Robert Elston (1982), drawing on the Western Pluvial Lakes Tradition concept (Bedwell 1973), originally used the term *Pre-Archaic* to contrast the earliest cultures inhabiting the western Great Basin with later Archaic groups: "site content and distribution of the Pre-Archaic cultures is nothing like those of the Archaic" (Elston 1982:189). Pre-Archaic "'free-roaming' nomads" traveled among rich resource patches, mainly wetlands, that were "high on the food chain," using a lithic toolkit centered on large bifaces, stemmed or concave-based projectile points, and an impressive array of other formal tools, including crescents, beaked gravers, side and end scrapers, scraper planes, and choppers.

Basketry and other textiles, known from several cave sites in the western basin, attest to a sophisticated perishable subsistence technology, but ground stone implements, indicative of grinding and processing of seeds, are conspicuously absent or very rare. In this view, Pre-Archaic foragers maintained a narrow diet breadth in which high-quality resource areas were the focus of settlement and the use of low-ranking plant and animal foods was limited.

Elston has since reconsidered this view of the Pre-Archaic in light of more recent research (Elston and Zeanah 2002), finding that the concept applies more generally to TP/EH groups throughout the Great Basin and that the differences recognized between these populations and later Archaic groups are still valid: these include "dramatically different technological organization and site distribution," an emphasis on hunting of large game by men, and the collection of smaller game and occasionally plant resources by women. Elston and Zeanah (2002:119–120) acknowledge that "broad-spectrum foraging and sexual division of labor were always components of Great Basin foraging strategies," but they argue that the abundance of large game and lack of competition from other foragers "allowed Prearchaic foragers high mobility that maximized men's encounters with large game without sacrificing women's foraging interests" or requiring heavy investment in plant processing technology.

The Paleoarchaic concept was initially articulated by Beck and Jones to avoid distinguishing between "fluted and stemmed [projectile point] occupations as 'Paleoindian' and 'Archaic' as they are often termed."

> Use of these terms implies knowledge that there was a distinct temporal and adaptational boundary between the two, where such knowledge does not exist (Simms, 1988; Beck, 1996). Thus, we use the term "Paleoarchaic" to refer to the populations who lived in the Great Basin during the TP/EH interval. (Beck and Jones 1997:163)

The ensuing discussion of the archaeological records of such populations accords well with Elston's (1982) characterization of the same interval in the western Great Basin, so it seems that the concepts of Pre-Archaic and Paleoarchaic may not be so far apart, at least as Elston (1982) and Beck and Jones (1997) used them. Use of the term Paleoarchaic in this context served mainly to distinguish the occupations in the Great Basin from Paleoindian occupations outside of the Great Basin, with its implied emphasis on fluted points and megafaunal hunting (but cf. Cannon and Meltzer 2004 for a current assessment of the lack of megafaunal specialization among Paleoindian populations).

The term *Paleoarchaic*, however, certainly does imply a strong connection and likely continuity with Archaic populations, along with attendant notions of broad-spectrum subsistence and settlement strategies. Jones and Beck (1999:94) subsequently made this aspect of the term *Paleoarchaic* clear: "As the TP/EH cultural record presently stands, there is no clear basis on which to distinguish 'Paleoindian' and 'Archaic' occupations. For the entire interval, as far as we know from current evidence, Great Basin peoples were generalized mobile foragers, and thus exhibited the essential features of an Archaic lifeway." On the other hand, they also note that "at the risk of over-generalization, it seems clear that this [Paleoarchaic] adaptive pattern shares elements of later Archaic practices, but that Paleoarchaic adaptation was organized in fundamentally different ways than that in mid- and late Holocene" (Jones and Beck 1999:94).

Other archaeologists also consider the Paleoarchaic to be similar to the succeeding Archaic in most particulars:

> In the Great Basin…what little subsistence data are available suggest that these peoples were essentially broad-spectrum foragers who did not rely on Pleistocene megafauna and who were more similar to foragers of the following Archaic period (e.g., Beck and Jones 1997; Elston and Zeanah 2002; Jones et al. 2003). As a result, terms such as *Initial Archaic*, *Prearchaic*, and *Paleoarchaic* are often substituted [for the term *Paleoindian*] to emphasize this broad adaptive focus while maintaining a chronological separation from the period that followed.… But if these subsistence data suggest a broad-spectrum adaptation tethered to marsh ecosystems, then what separates Paleoarchaic from Archaic foragers, if indeed they can be separated at all? Basically, there is but a single factor: the appearance of ground stone by about 8500 ^{14}C BP (9500 cal BP)(Rhode et al. 2006; see also Rhode and Louderback, this volume) and the processing of hard, small seeds that, because of their high processing costs, have relatively low return rates. These return rates are lower than those associated with small mammals, fish, rhizomes, and other taxa associated with the Paleoarchaic subsistence focus (e.g., Simms 1987), suggesting that the diet breadth of Archaic foragers was even broader than that of the Paleoarchaic. That is, the Paleoarchaic to Archaic transition in the Great Basin consisted essentially of a shift from broad-spectrum foraging to very broad-spectrum foraging, with the addition of movement to previously unoccupied areas where these lower-ranked resources were found. (Madsen, this volume, 14–15)

It seems somewhat incongruous to us to argue that Paleoarchaic peoples were essentially Archaic in character, only lacking small-seed consumption and milling stone technology, when a small-seed subsistence emphasis and ground stone milling technology are rightly regarded as major "hallmarks" of the Desert Archaic (Jennings 1957; D. Fowler 1986). Be that as it may, these various quotations highlight a potential distinction between a view that people in the TP/EH boasted a relatively narrow diet focused on high-quality resources taken in resource-rich wetland patches (i.e., the original Pre-Archaic concept) vs. a view that these people had a broad diet but not one so broad as to emphasize small, hard seeds and seed-grinding technology. As such, they raise questions about the nature of dietary plant use during the TP/EH—questions that deserve empirical testing. Our purpose here is to examine the TP/EH archaeological

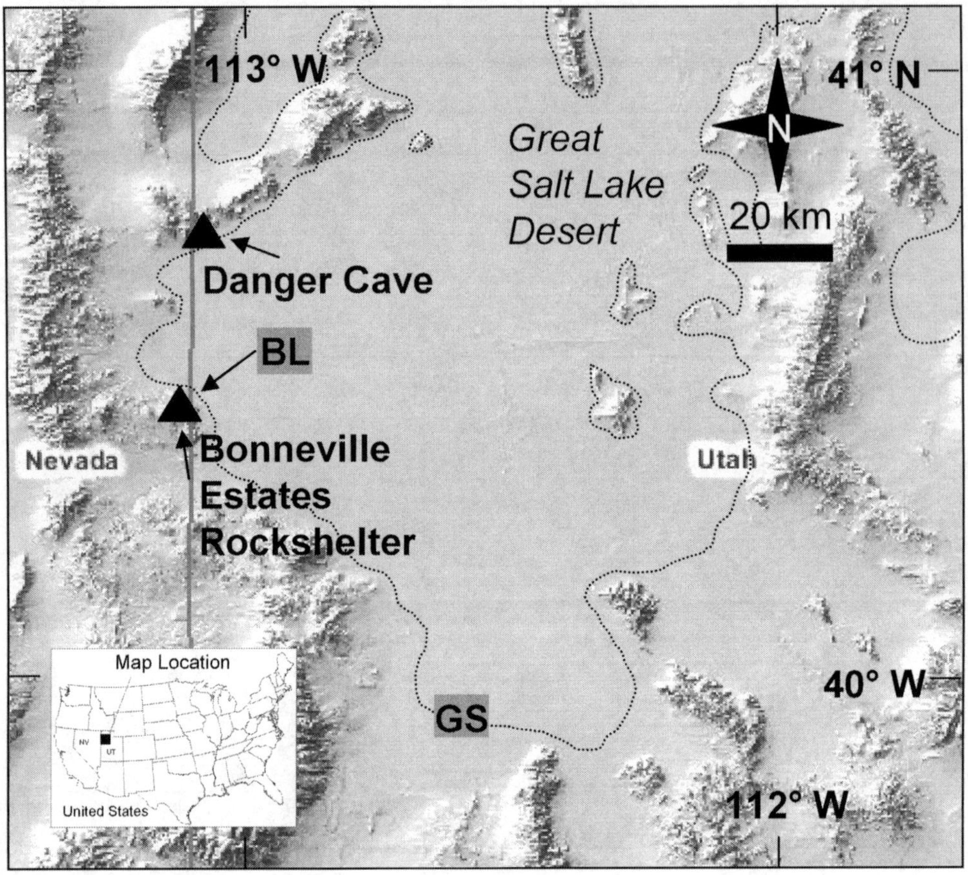

FIGURE 12.1. Map showing locations of Danger Cave and Bonneville Estates Rockshelter in the Great Salt Lake Basin, western Utah/eastern Nevada. *GS* refers to the approximate location of the Gilbert Shoreline, the last significant stand of Pleistocene Lake Bonneville; *BL* refers to the approximate location of Blue Lake and Mosquito Willy's wetland areas.

records from two sites, Bonneville Estates Rockshelter and Danger Cave, to learn about the possible use of plants in the diets of the Pre-Archaic and/or Paleoarchaic foragers in the Bonneville Basin.

BONNEVILLE ESTATES ROCKSHELTER

Bonneville Estates Rockshelter (BER) is located on the uppermost shoreline of Pleistocene Lake Bonneville, overlooking the Great Salt Lake Desert about 40 km south of the town of Wendover on the Nevada/Utah state line (Figure 12.1) (Goebel et al. 2003; Rhode et al. 2005). This large, open rockshelter (Figure 12.2) contains extensive, well-stratified archaeological deposits as old as 13,000 cal BP. Details of this site have been discussed previously in other chapters (Graf; Goebel; and Hockett; this volume).

To examine plant use at BER during the TP/EH, we rely on samples taken from several pit, hearth, and hearth-like features in both the West and East blocks (Figure 12.2). The features in question range from approximately 13,100 cal BP to approximately 10,600 cal BP and correspond to Cultural Component 6, the Pre- or Paleoarchaic occupation that ranges in age from ~13,100 to 9600

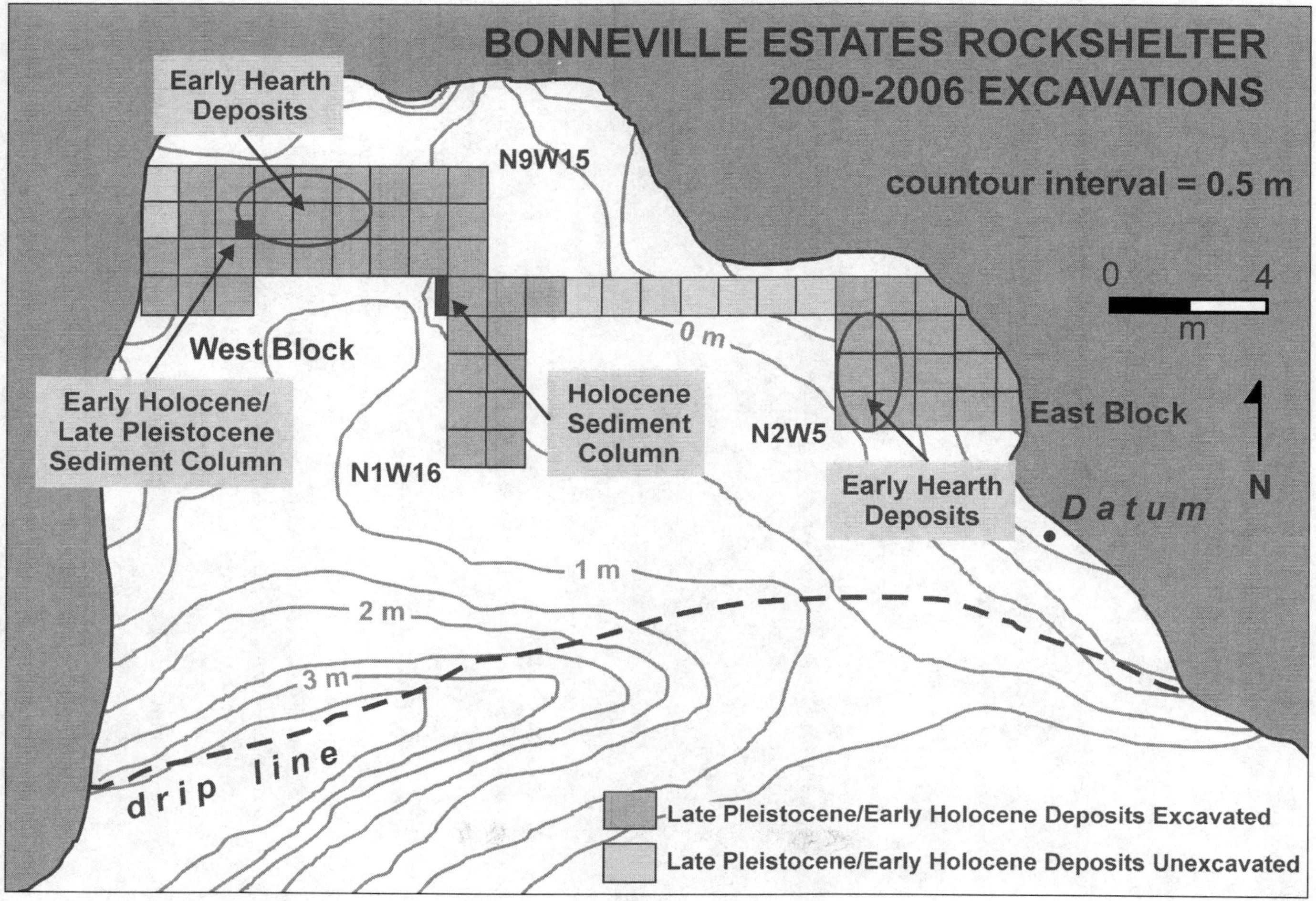

FIGURE 12.2. Plan view of Bonneville Estates Rockshelter showing extent of excavations through 2004. Note the locations of sediment column samples and hearth samples discussed in text.

cal BP in BER (see Graf, this volume). In the West Block, five features from Stratum 18b and one from mixed 18b/19 were sampled. The overlying strata were sampled by F01.01 in Stratum 18a, F05.02 at the Stratum 17b'/18a contact, and F03.13 in Stratum 17b'. In the East Block, two features from Stratum 10 were sampled. Samples from a stratigraphic sediment column in the West Block area provide a comparison of materials from nonfeature contexts.

Sediment samples, collected in bulk in the field, were first dry-sieved into > 6.2, > 3.1, > 1.6 mm, and < 1.6 mm fractions in the lab. The > 1.6 and < 1.6 mm fractions were floated in a water bath to remove heavy sands and silts, and the light fraction was air-dried for sorting; the > 3.1 and > 6.2 mm fractions needed no flotation for sorting. Potentially dietary plant remains were picked from each fraction under 10–60× magnification and identified using comparative collections housed at the Desert Research Institute.

Results

Table 12.1 gives counts of identified potential dietary plant remains from features. Table 12.2 presents counts of dietary plant materials from different strata samples (Strata 17b–19) from the stratigraphic column and one packrat midden sample (PM2, associated with culturally sterile Stratum 17b), located in the West Block. Available radiocarbon dates and calendar age estimates are given in Graf, this volume.

The assemblages of dietary plant materials are heavily small-seed oriented. Among potential dietary taxa represented in features by seeds are saltbush (*Atriplex;* mostly the silver saltbush, cf. *A. argentea*) and other members of the goosefoot family (Chenopodiaceae); grasses, including ricegrass (*Achnatherum hymenoides*), dropseed sandgrass (*Sporobolus* sp.), and wild rye (*Leymus* cf. *cinereus*); bulrush (*Scirpus;* probably alkali bulrush, *S. maritima*); sedge (*Carex*); cattail (*Typha*); sunflower family (Asteraceae); mustard family (Brassicaceae); cacti (Cactaceae); greasebush (*Forsellesia spinescens*); and snowberry (*Symphoricarpos* cf. *longiflorus*). Cacti are heavily represented by burned spines, spine clusters, and epidermis, indicating roasting and use of the fleshy stems. Several potentially important dietary items are absent or rare, such as roots and tubers, berries, and greens. Some of these would not be expected to show up simply for reasons of differential preservation, but some—particularly rhizomes and seeds from edible berries—might be expected if they were used in significant quantities.

A number of complicating factors hamper our ability to identify which components of the BER seed assemblage can be reliably attributed to human agency and human diet. Various taphonomic agents besides people (especially rodents such as packrats) could have introduced seeds and other plant remains into these features. Many of the same seeds and fruits favored by people for food were also collected, stored, and eaten by rodents or consumed by other animals and potentially deposited in excreta. For example, the culturally sterile Stratum 17b and the packrat midden PM2 contain abundant saltbush fruits and

TABLE 12.1. Potentially Economic Plant Remains from Cultural Component 6, Bonneville Estates Rockshelter

	FEATURE	E4-10-C3	E6-10-C10	F3.13	F5.02	F1.01	F4.14	F3.15	F5.06	F4.15	F5.03	F3.16/F4.13
	STRATUM	10[a]	10[a]	17B'[b]	17B'/18A[b]	18A[b]	18B[b]	18B[b]	18B[b]	18B[b]	18B[b]	18B/19[b]
TAXON	CONDITION											
Saltbush fruit	burnt	—	—	—	1	—	12	1	—	18	—	20
Saltbush fruit	unburnt	—	1	—	5	—	9	—	—	6	—	9
Saltbush seed	burnt	—	—	—	17	—	30	—	—	20	2	7
Saltbush seed	unburnt	—	—	—	1	—	—	—	—	—	1	—
Goosefoot seed	burnt	2	—	—	—	17	—	—	—	3	3	—
Goosefoot seed	unburnt	—	—	—	—	4	1	—	—	—	—	2
Goosefoot family seed	burnt	—	—	4	25	—	22	2	—	6	—	18
Goosefoot family seed/embryo	unburnt	—	—	5	—	—	—	—	—	—	—	2
Ricegrass seed	burnt	—	—	1	4	—	7	—	—	32	—	1
Ricegrass seed	unburnt	—	—	10	3	—	9	—	—	—	—	9
Dropseed sandgrass seed	burnt	—	—	—	10	1	8	1	—	—	—	—
Wild rye seed	burnt	—	—	—	—	—	—	—	—	—	—	1
Grass family seed	burnt	—	3	—	—	—	1	1	—	1	—	—
Sedge seed	burnt	—	—	—	—	1	—	—	—	—	—	—
Bulrush seed	burnt	—	—	—	—	1	—	—	—	—	—	—
Bulrush seed	unburnt	—	—	—	—	4	—	1	—	—	—	—
Cattail seed	burnt	—	—	—	—	3	—	—	—	—	—	2
cf. Sunflower (*Helianthus*) seed	burnt	—	—	—	—	—	—	—	—	1	—	—
Sagebrush seed	burnt	—	—	—	—	—	5	—	—	—	—	—
Sunflower family seed	burnt	—	—	—	1	—	1	—	—	—	—	—
Mustard family seed	burnt	5	1	1	1	—	1	—	—	—	—	1
Mustard family seed	unburnt	—	—	—	—	—	—	—	—	—	—	—
Cactus family seed	unburnt	—	—	—	—	—	1	—	—	—	—	—
Cactus family spine cluster	burnt/pt burnt	—	—	1	4	—	2	—	—	—	2	1
Cactus family spine cluster	unburnt	—	—	—	1	—	1	—	—	—	—	—
Cactus family spines	burnt	—	—	—	1	—	1	—	—	—	—	—
Greasebush seed	burnt	—	—	1	4	—	3	—	—	—	3	5
Greasebush seed	unburnt	—	—	—	2	—	6	—	—	—	—	—
Snowberry seed	burnt	—	—	—	1	—	—	—	—	1	—	—
Snowberry seed	unburnt	—	—	—	—	—	3	—	—	—	1	—
Pea family seed	burnt	—	—	—	—	—	1	—	—	—	—	—
Lily family, bulb?	burnt	—	—	—	—	1	—	—	—	—	—	—

[a] From the East Block excavation.

[b] From the West Block excavation.

TABLE 12.2. Potentially Economic Plant Remains from Early Holocene and Latest Pleistocene Stratigraphic Column Samples, Cultural Component 6, Bonneville Estates Rockshelter

		STRATUM					
TAXON	CONDITION	17B	PM2	17B'	18A	18B	19
Saltbush seed	burnt	—	—	—	—	4	—
	unburnt	—	—	1	—	—	4
Saltbush fruit	burnt	—	—	3	—	—	11
	unburnt	572	30	150	60+136	103	28
Goosefoot seed	burnt	—	—	1	—	4	—
	unburnt	6	—	6	0+1	—	—
Goosefoot family seed	burnt	—	—	—	0+1	—	—
	unburnt	108	15	—	0+24	7	—
Bulrush seed	unburnt	1	—	—	—	—	—
Sedge seed	unburnt	—	—	—	0+1	—	—
Ricegrass seed	unburnt	4	—	7	—	1	1
Grass family seed	burnt	—	—	—	0+1	—	—
Sagebrush seed	burnt	—	—	2	—	1	1
	unburnt	—	1	—	0+3	—	3
Cactus family spines	burnt	—	—	1	—	—	—
	unburnt	8	—	4	—	—	—
Cactus family seed	unburnt	—	—	2	0+2	5	—
Blazingstar seed	unburnt	1	2	2	—	—	1
Beeweed seed	unburnt	—	—	14	0+4	—	—
Greasebush seed	burnt	—	—	—	—	2	—
	unburnt	4	5	1	5+9	7	9
Wild rose seed	unburnt	1	—	—	0+1	1	—
Snowberry seed	unburnt	—	1	—	1+6	3	—

Note: Stratum 18a contains two column samples, the counts of which are separated by a + sign.

goosefoot family seeds, cactus spines, and lesser numbers of seeds of ricegrass, sagebrush (*Artemisia*), greasebush, blazingstar (*Mentzelia* cf. *albicaulis* and *M. laevigata*), snowberry, wild rose (*Rosa* cf. *woodsii*), and even a bulrush seed (Table 12.3). All of the specimens in the sterile deposits are unburned. We found unmodified and unburned fruits and seeds from saltbush and several other taxa in all the stratigraphic fill samples and in several of the features, including hearths. Unburned seeds and fruits in hearths may have been deposited long after the hearth ceased to function; some may be modern contaminants. Seeds and fruits deposited on the ground surface could become burned if a hearth was built on that surface, so some of the burned fruits and seeds found in hearth fills may not be human dietary residues but instead may have been incidentally incorporated into the hearth as part of fuelwood.

With these caveats in mind, we now describe the plant remains in Cultural

TABLE 12.3. Identified Plant Remains from Early Deposits, Danger Cave

TAXON	CONDITION	DI HEARTHS (F111/112)	SAND I SURFACE (F113)	DII LOWEST LAYER (F115)
Greasewood fruit	unburnt	12	6	—
Saltbush fruit	unburnt	12	65	2
Saltbush fruit	burnt	—	—	1
Ricegrass seed	unburnt	3	1	—
Ricegrass seed	burnt	—	—	10
Cactus seed	unburnt	2	—	—
Fiddleneck seed	unburnt	2	—	—
Goosefoot family seed	unburnt	1	—	—
Dock (Rumex) seed	unburnt	—	1	—
Bulrush seed	unburnt	—	1	—

Component 6. Most abundant are saltbush fruits and seeds (mostly burned), Chenopodiaceae seeds (many of which are likely to be saltbush, but goosefoot [*Chenopodium*] is represented as well), ricegrass seeds, dropseed sandgrass seeds, greasebush seeds, and burned cactus spines and spine clusters (often with the spines twisted off) (Table 12.1). Minor quantities of other plants are scattered through the hearths, including mustard, sunflower (*Helianthus*), snowberry, bulrush, cattail, and one possible unidentified bulb of the lily family.

Feature samples and strata samples appear generally similar in taxonomic content, though several important differences emerge on close comparison. The vast majority of plant specimens found in the strata samples are unburned; by contrast, more than 75 percent of the plant remains found in the features are burned. (The unburned minority of specimens include ricegrass seed coats, saltbush fruits, and Chenopodiaceae and greasebush seeds, most likely noneconomic contaminants.) Saltbush and Chenopodiaceae are common in both strata and feature samples, but the feature samples are dominated by seeds, while the strata samples are dominated by unburned saltbush fruits. Feature samples also contain relatively greater numbers of ricegrass seeds, dropseed sandgrass seeds, and burned cactus spine clusters than do the strata samples. Finally, a number of potentially economically important seeds such as sunflower, mustard, and bulrush are present in several of the features, suggesting a somewhat common pattern of economic seed deposition in features that is not similarly represented in the strata samples (though, it must be noted, the strata samples are fewer in number and smaller in size than the feature samples).

The bulrush and cattail seeds must have been brought into the shelter from the Blue Lake or nearby Mosquito Willy's wetlands on the margin of the Great Salt Lake Desert about 4 km east of BER (Figure 12.1). The tiny cattail seeds may indicate that the cattail fluff was used as fire-starter rather than as food. The single cactus seed is from a small hedgehog cactus, probably *Pediocactus simpsonii*, a cold-hardy mountain ball cactus that blooms early in spring and produces small dry fruits in summer. Abundant cactus spines are found in these deposits, including spine clusters that have the spines burned or twisted off, indicating people's use of the cactus flesh for food. The types of cacti represented by the spine clusters and spines are not yet known.

These results are compelling, but not yet conclusive, evidence of human plant use during Cultural Component 6. The fact that many of the plant remains from economically important plants are found burned or modified in cultural features, and yet they are rare or absent from strata samples, suggests deliberate plant use by the makers of those cultural features. In particular, the use of certain grass seeds (such as ricegrass, dropseed sandgrass, and possibly wild rye), seeds of members of the goosefoot family (particularly goosefoot, possibly saltbush), seeds of the mustard family, sunflower seeds, and cactus pads and stems are most suggestive. The bulrush and cattail seeds must have been transported from wetland environments not existing around BER, and the cacti spine clusters were deliberately modified by burning and twisting off spines.

It is also notable that cacti, ricegrass, and seeds of saltbush and other chenopods were common dietary plants found in abundance in later strata of BER (Rhode 2006) and were important plant foods among historic Shoshoni and Gosiute (Chamberlin 1911; C. Fowler 1986; Steward 1938). The evidence for use of these taxa in these oldest cultural features appears to be as strong as in features of younger age in the shelter.

As yet we lack the substantive confirmation of dietary use that paleofecal specimens (coprolites) could provide. Moreover, certain potential dietary plant remains found in cultural feature samples (saltbush fruits, unburned seeds, and unburned cactus spines) are likely to have been deposited by rodents or other nonhuman agents, and accidental burning of various plant parts cannot be ruled out. Nevertheless, the patterning and relative abundance of economic plants in these features from Cultural Component 6 make a persuasive case for the use of small seeds of certain taxa, both wetland and terrestrial, as well as cacti in the diet of the people who occupied BER during the TP/EH.

What differs between these remains from Cultural Component 6 and the later cultural components at BER is the presence of grinding stone technology. Ground stone artifacts are absent in Cultural Component 6 and first appear during the Early Archaic in overlying Cultural Component 5. Whatever people were doing with small seeds and cacti during Cultural Component 6, they apparently were doing it without the benefit of grinding stone processing technology. Grinding stone technology was presumably adopted to make flour or meal from small seeds, which is likely to have increased digestibility, but it also presumably increased processing time, an indicator of subsistence intensification. Use of small seeds was therefore apparently not intensive until the Early Archaic, after ~8400 cal BP, at BER. This finding is consistent with the advent of grinding stone technology at nearby Danger Cave, to which we now turn.

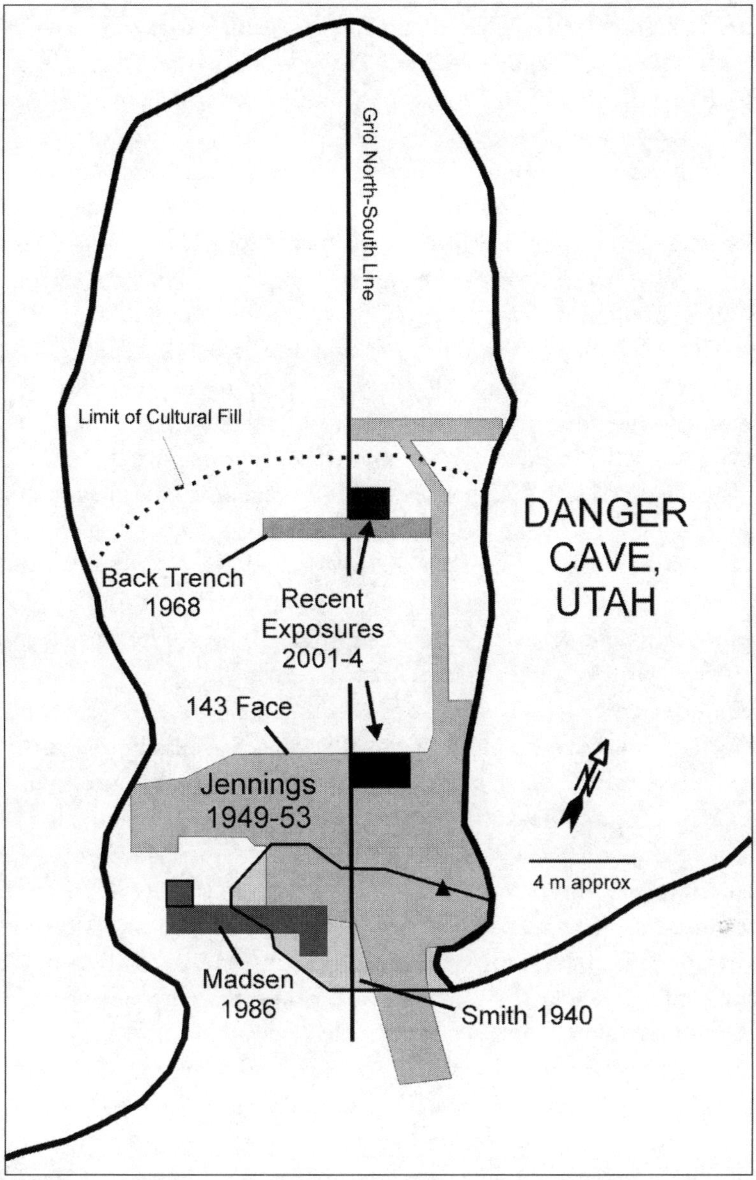

FIGURE 12.3.
Plan view of Danger Cave showing locations of previous archaeological excavations. Recent exposures conducted in 2001–2004 and discussed here are shown in black.

DANGER CAVE

Located on the edge of the Bonneville Salt Flats just outside of the town of Wendover at the Utah/Nevada state line (Figure 12.1), Danger Cave is a large, oval chamber formed in Paleozoic limestone by solution weathering in fractures and subsequently enlarged by spalling of the walls and wave action by Pleistocene Lake Bonneville (Figure 12.3). It contained a remarkable set of deep, well-stratified cultural deposits with excellent preservation, left during repeated occupations by people throughout the Holocene. The site has been the subject of numerous archaeological campaigns (Figure 12.3), most extensively by Jesse

Jennings a half century ago (Jennings 1957). Subsequent excavations were conducted in 1968 by Fry and colleagues (Fry 1976; Harper and Alder 1972) and in 1986 by David Madsen and his group (Madsen and Rhode 1990; Rhode and Madsen 1998; Rhode et al. 2005).

The rich deposits at Danger Cave have long been used to argue for intensive small-seed use beginning during the earliest Holocene (Aikens and Madsen 1986; Beck and Jones 1997; Fry 1976; Grayson 1988, 1993; Harper and Alder 1972; Jennings 1957, 1978; O'Connell et al. 1982; Willig and Aikens 1988). Seed-processing residue, human paleofecal samples, and abundant ground stone artifacts associated with small-seed processing were all found in deposits that were thought to date to ~12,000 cal BP. However, recent reanalysis and redating of these various lines of evidence now demonstrate that intensive small seed use began at Danger Cave after ~9700 cal BP (Rhode et al. 2006).

To better characterize the pre-9700 cal BP occupations at Danger Cave and the possible plant use in the diet of these earliest inhabitants, the Utah Division of State History and the Division of State Parks authorized a reconnaissance of previously excavated areas, allowing the removal of backfill to assess the extent and research potential of remaining intact deposits. These efforts began in 2001 at Jennings's 143 Face and Fry's Back Trench.

Jennings terminated his excavations at an east-west sidewall running perpendicular to the 143 ft point on the main grid north-south line (Figure 12.4). In our investigations, we exposed intact deposits in a portion of Jennings's original 143 Face, protected beneath large rocks and an impenetrable plate of calcium-cemented ash. These lower strata contain a remarkably well-preserved set of cultural deposits dating from ~9200 to > 12,100 cal BP (Figure 12.4). We also exposed the sidewalls of a trench near the back of the cave, excavated in 1968 by a party from the University of Utah led by Gary Fry (Harper and Alder 1972), revealing a very well-stratified set of pickleweed chaff layers, coarser vegetal deposits, and intervening bands of rockfall (Figure 12.5).

Together, the two exposures nicely reveal the two lowermost stratigraphic divisions originally given by Jennings (1957), encompassing the TP/EH. Jennings divided the occupational deposits of Danger Cave into five gross levels, DI through DV, separated by layers of roof spall that imply long hiatuses of occupation. The upper three (DIII–DV) were composed of thick layers of well-preserved vegetal debris, dust, ash, and artifacts, dating from ~9700 cal BP to historic times. They are largely destroyed in the vicinity of the 143 Face, but a fairly thick remnant of DIII deposits still exists at the top of the Back Trench, including layers of nearly pure pickleweed chaff, the byproduct of processing pickleweed for its edible seeds. The age of the lowest pickleweed processing layer, found at the base of Stratum 04-11 in Figure 12.5, is ~9600 cal BP (8570 ± 40 ^{14}C BP, Beta-193123), and effectively marks the beginning of intensive small seed use with grinding stone technology at Danger Cave, the inception of a truly Archaic way of life (Rhode et al. 2006).

Of greatest significance for our purposes here, however, are the lowest two

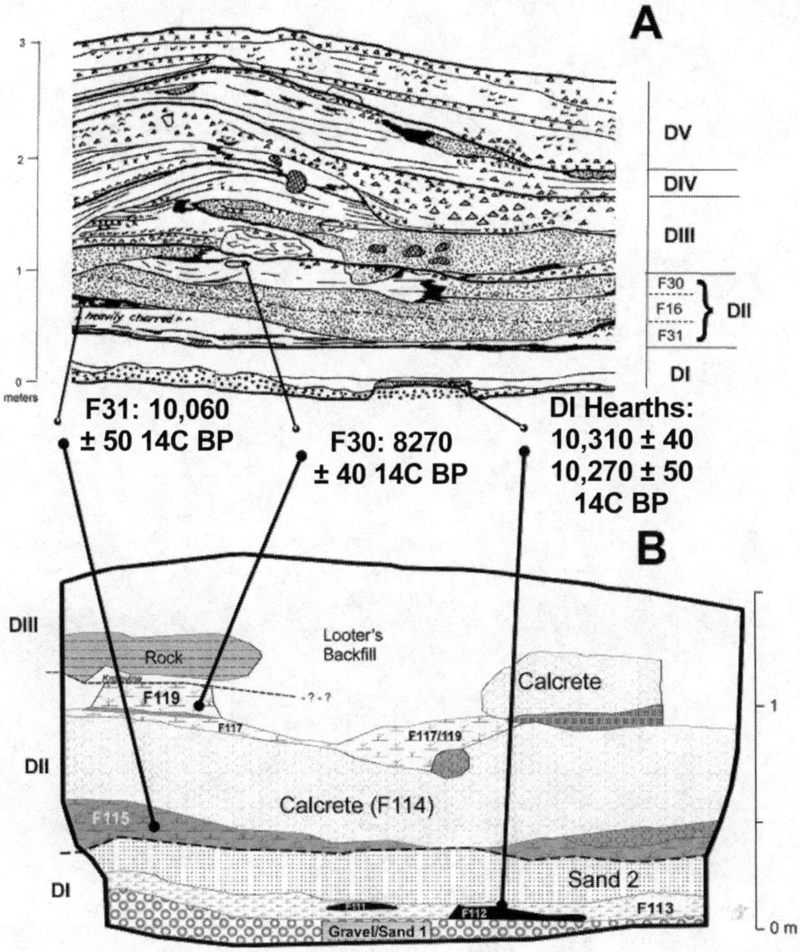

FIGURE 12.4. Stratigraphic profiles of the 143 Face in Danger Cave: *A*, as mapped in 1953 by Jennings (1957: Figure 54); *B*, as it appears today. Radiocarbon dates were obtained from charcoal/ash stains (F111 and F112), charred vegetation from the lowest part of D2 (F115), and pickleweed chaff from the uppermost part of D2 (F119).

of Jennings's levels, which he called DI and DII. What is the archaeobotanical record of the earlier deposits, and what does it suggest for plant use before ~9700 cal BP, before pickleweed began to be exploited in significant quantities?

The DI Occupation

The DI level holds evidence of the earliest human occupation at the site. Several small fire hearths accompanied by a sparse scatter of artifacts and ecofacts lay on a bed of beach sand ("Sand 1"), capped by a variably thick layer of wind-deposited sand containing abundant artiodactyl pellets ("Sand 2") (Figure 12.4). Jennings's original radiocarbon dating and recent radiocarbon dates we obtained from ash lenses exposed in the 143 Face confirm that this earliest occupation took place at ~12,100 cal BP (Rhode et al. 2006).

A small but interesting collection of artifacts was obtained from the DI level (Jennings 1957). These include one lanceolate projectile point of the Agate Basin style, several unifacial scrapers, a few pieces of possible ground stone artifacts

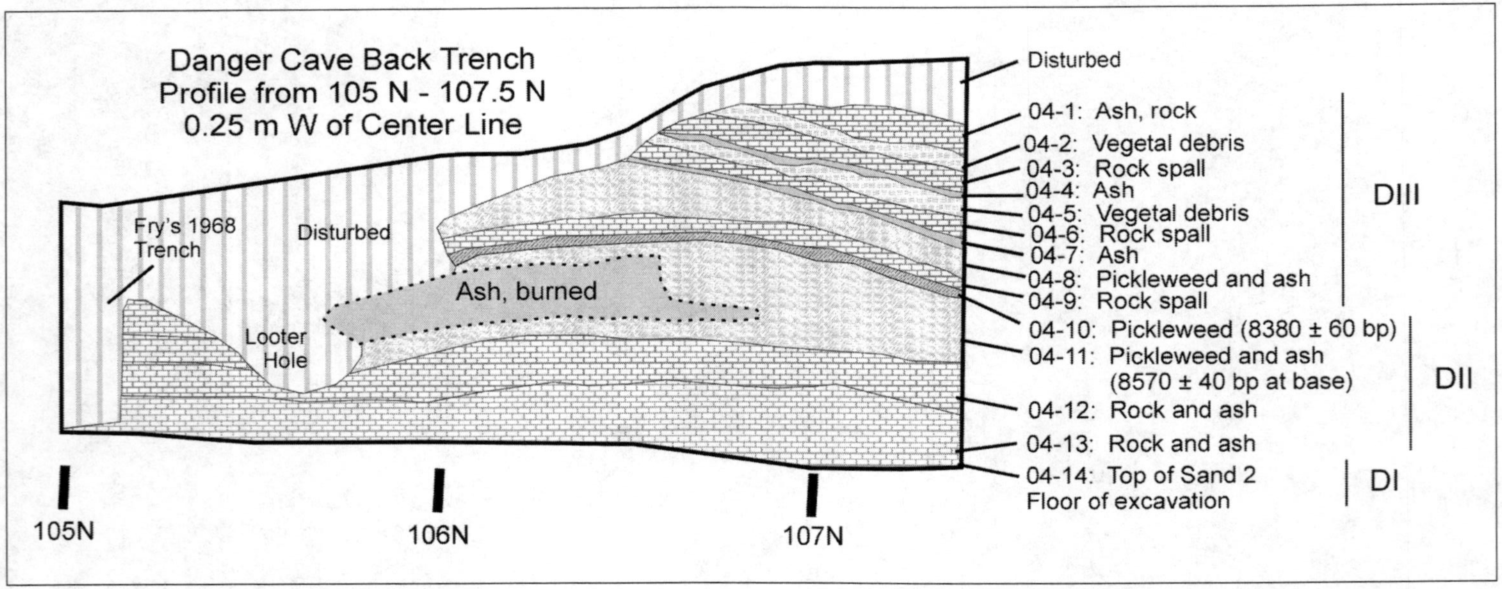

FIGURE 12.5. Stratigraphic profile of intact sediments exposed in the Back Trench area in 2004. Strata 04-1 through 04-9 correspond to Jennings's DIII cultural level; Strata 04-10 through 04-13 correspond to the DII level; and Stratum 04-14 (base of profile) is the DI level. Height of profile at 107.5 N is 115 cm. Radiocarbon dates were obtained from Stratum 04-10 and Stratum 04-11.

(including stones for grinding ochre), numerous chert and obsidian flakes, several knotted pieces of twine of unknown function, modified bone, and assorted fragments of "food bones," as Jennings called them. Six human coprolites were found on this level (Fry 1976), but they date to much later times (Rhode et al. 2006). We took several small samples of sediments from the 143 Face, and these contained an abundance of chipped stone waste flakes but no tools, as well as knotted string fragments and small wood whittling curls. Given the small size of our samples, it is likely that the remaining DI occupation deposit still contains an abundance of artifacts.

Our evidence for plant use from the earliest deposits is limited to small sediment samples taken from the hearths and exposed strata in the 143 Face (Table 12.3). These samples are small but provide some interesting information regarding the plant remains and other artifacts from these lowermost occupations.

From the early hearths, we find a variety of fruits and seeds, including greasewood (*Sarcobatus vermiculatus*), saltbush, ricegrass, cactus, and fiddleneck (*Amsinckia*). Virtually all of them are unburned, and all are probably locally derived from the cave's immediate surroundings. They suggest a scatter of plant debris mainly after the deposition of the ash and charcoal, which probably represents deposition by rodents or other animals rather than by people. There is no evidence of anthropogenic seed deposition in the hearths, though abundant flakes and bone fragments were probably derived from human activities. Charcoal used for fuel was obtained from local greasewood and sagebrush. The sand floor on which the hearths were built contains a very similar record of unburned saltbush and greasewood fruits, greasewood and sagebrush charcoal, chipped stone flakes, burned and unburned bone fragments, a few whittling curls, and some knotted milkweed (*Asclepias*) fibers, but very little demonstrable evidence pointing to dietary use of any particular plants.

The DII Deposits

The overlying DII level is a thick deposit of organic debris, cemented ash, rock spall, bat guano, and artifacts that combines three distinct stratigraphic layers: an upper layer (called F30 in Jennings's field notes) dating ~9200–8300 cal BP, a middle layer (F16) estimated to date ~9700–9450 cal BP, and a lower layer (F31) resting on DI sands dated ~11,500–11,200 cal BP. Most of the DII artifacts, including the vast bulk of the large ground stone assemblage, come from the upper two layers, postdating ~9700 cal BP (Rhode et al. 2006).

The artifact inventory and character of the lowermost part of DII, exposed in the 143 Face (Figure 12.4) and dating ~11,500 cal BP, is poorly known at present; there may have been a substantial pause in occupation between ~11,400–9700 cal BP. The layer does contain plant remains, but taphonomic problems intrude on their interpretation as representing human dietary items. This stratum was almost completely burned after its deposition, and most of the remains from this layer throughout the cave are variously singed, toasted, or charred. The small sample that we took from this stratum contains a few burned ricegrass seeds, saltbush fruits, and snowberry seeds amid abundant singed bones, ungulate and rodent dung, and burned twigs (Table 12.3). No artifacts were encountered. Very little information about dietary use of plants can currently be inferred from this stratum.

In sum, the samples from these lower units do not indicate significant use of plants in the diets of Danger Cave's earliest inhabitants. While plants may have played a role in the inhabitants' diets, our present evidence does not show it. Of course, the samples we have are very small, and we lack horizontal spatial extent to examine patterning across the occupation surfaces. Given the rather rich artifact content of the small samples we obtained, the lower sand layer appears to hold the greatest potential for well-preserved information about occupation of the cave ~12,100 cal BP, and it deserves to be explored more fully in the future. For now, however, the evidence for human dietary plant use prior to that time at Danger Cave is negligible, though still inconclusive.

SUMMARY AND DISCUSSION

The early cultural deposits at BER tentatively support claims for the dietary use of plant foods during the TP/EH transition. Burned seeds of economically important plants such as ricegrass, dropseed sandgrass, goosefoot, and possibly other taxa, as well as stems of cactus, appear in hearths dating as early as 12,000–13,100 cal BP. Many of the same plant taxa that were apparently used in the earliest cultural component at BER were also utilized during later times.

At Danger Cave, occupations dating older than ~11,500 cal BP do contain seeds and fruits—primarily of local plants such as shadscale, sagebrush, ricegrass, cacti, and others—but evidence that these represent human use is inconclusive.

The evidence of plant use at BER, the earliest currently available in the Great Basin, suggests that TP/EH human occupation and foraging strategies in the Bonneville Basin are most accurately described as Paleoarchaic. The earliest inhabitants of the Bonneville Basin, so far as is known, apparently had a broadly generalized diet that included small seeds and cacti.

They did not, however, use grinding stone technology to process these plant foods, a technology that was introduced during the Early Archaic. At Danger Cave, this introduction occurred after ~9700 cal BP, when layers of pickleweed processing residue first appear, milling stones become abundant, and human coprolites commonly contain pickleweed seeds (Rhode et al. 2006). At BER, this use commenced after ~8400 cal BP, following a long hiatus in occupation (Graf, this volume). The timing of intensive small-seed processing and use at Danger Cave coincides with an interval of significant aridification of the region (Grayson 1993, 2000b; Madsen et al. 2001; Oviatt et al. 2003; Rhode 2000; Rhode et al. 2005; Schmitt et al. 2002b; Wigand and Rhode 2002). People may have intensified their processing and use of small seeds in the diet in response to broad-scale environmental change in the Bonneville Basin. In this regard it is of interest to note that other cave sites in the Bonneville Basin began to be occupied at about this time or somewhat later, including Hogup Cave (Aikens 1970) and Camels Back Cave (Schmitt and Madsen 2005), providing additional evidence for a broad-scale adaptation to desert resources.

What the evidence of plant remains from Paleoarchaic Cultural Component 6 at BER suggests is that a broad-scale adaptation to resources such as desert grass seeds and cacti may already have had quite a long history.

ACKNOWLEDGMENTS

This work was supported by the National Science Foundation (BCS-0312252 and BCS-0313813); the Sundance Foundation at the Department of Anthropology, University of Nevada, Reno; and the Lander Fund at the Desert Research Institute. We thank our many associates in the field, especially John Varnon and Jonathan Grant, whose dedicated work in the archaeobotany lab enabled us to complete this study.

PART V

Afterword

13

Paleoindian or Paleoarchaic?

Gary Haynes

THIS VOLUME IS A WATERSHED IN GREAT BASIN PREHISTORY. ITS CHAPters show how the first thin waves of people entered, explored, and began settling into an enormous region that was evolving from a unique late Glacial ecosystem into the more severe landscapes of the Holocene. Some chapters provide new perspectives on familiar Great Basin issues such as spatial patterning in toolstone provisioning and its effects on mobility, and the long-standing ambiguity about stemmed-point makers' subsistence practices. But the importance of this volume goes beyond the resolution of regional problems. It also tells us something of broader significance about the continent-wide first phases of human dispersal and colonization. An important shift in thinking to follow upon this volume will be the wider acceptance of the now certain fact that the Great Basin's earliest human settlement was multiphase and no less complex than what must have occurred in the rest of North America's Paleoindian stage. Adding to the complexity is the discovery of an archaeologically cryptic earliest human presence that preceded an actual settling-in.

The nearly invisible first phase has begun emerging from a time interval for which the human presence was impossible to prove for a long time. This phase may have lasted a thousand years, but it is still wispy and vague and has provided so few solid clues about behavior, technology, and mobility that little can yet be said.

The succeeding phases of exploration and settling-in have been given several different names. A resistance to the term *Paleoindian* seems pervasive in the Great Basin. *Paleoindian* is understood to be Clovis-specific, whereas the term *Paleoarchaic* is stemmed-point-specific and tied to pluvial lake basins. Yet the peoples who made each kind of projectile point were frequently selective for the highest quality toolstone, curated their toolkits, created portable and flexible tools, made generalized tools for more than one task, and carried tools made of distant raw materials (although Beck and Jones [this volume] see different technological trajectories for the bifacial tools, especially points).

Various alternatives to the term *Paleoindian*—such as *Pre-Archaic, Initial Archaic, or Paleoarchaic*—are awkward, misleading, and unnecessary. Several chapters in this volume (such as Duke and Young; Goebel; and Hockett) make it clear that the first stage in human dispersal was not simply the beginning episode of the Archaic (as implied by the term *Initial Archaic*), and it certainly was not an old-fashioned or outmoded form of the Archaic, as implied by the word *Paleoarchaic*.[1] The Archaic itself anywhere in North America has long been understood to mean the stage when people had wide diet breadths and made use of food resources that demanded high expense in time and energy, such as small, hard seeds that required milling technology. The Archaic is also marked by the first appearance of notched projectile points in most of North America. Both of these cultural features—intensive milling stone use and the beginning of notched-point manufacturing—are not to be found in the phase called Paleoindian anywhere in the continent. Along with other dramatic differences in mobility, paleoenvironments, and technological organization, the already well-known distinctions between Archaic and Paleoindian justify what Goebel suggests in his chapter (this volume)—that the earliest stage deserves to be called Paleoindian in the Great Basin as it is almost everywhere else.

The Paisley Cave coprolites (Jenkins, this volume) show that one early human subpopulation in the Great Basin must have eaten at least some megafauna (bison has been identified) about a thousand years before many species went extinct. The later Paleoindian occupants of the Great Basin may or may not have continued targeting the largest mammals, but they left virtually no record of megafaunal hunting. Jennings (1986:115) pointed out 20 years ago that no extinct megafaunal remains have been found in behavioral association with fluted or stemmed points in the Great Basin, in spite of a search by archaeologists and paleontologists that continues to this day (for examples in this volume, see Jenkins; and Henriksen and Long). The argument has been advanced by some archaeologists (see Madsen, this volume; and Beck and Jones, this volume) that a narrow prey-choice strategy of hunting big game animals was not used by the highly mobile people making fluted and stemmed points in the Great Basin, and therefore the label *Paleoindian* cannot be applied. But such reasoning is out of step with current thinking about the tactics used by the earliest human settlers in most of North America. What archaeologists have decided to call Paleoindian in other regions of North America such as the eastern United States (see, for example, Anderson and Faught 2006; Haynes 2002; Meltzer 1984) is rarely or never associated with a narrow diet breadth or with the hyperspecialized hunting of now-extinct megafauna. The evidence suggesting a wider and more diverse diet among Paleoindians does not by itself make this first phase congruent with the stage called Archaic because so much else differs.

Several chapters in this volume well describe the very characteristics of this earliest Paleoindian stage that set it apart from the succeeding Archaic in the Great Basin and elsewhere in North America. One important distinguishing feature is forager mobility as it may relate to lithic technological organization.

MOBILITY AND LITHIC TECHNOLOGICAL ORGANIZATION

Mobility remained high throughout the Paleoindian and Archaic stages in the Great Basin, but it was very different in each period. Madsen (this volume) thinks that the decline and loss of marshes at the Pleistocene/Holocene transition was the deciding factor behind shifts in mobility. In Madsen's view, late Pleistocene men might have moved long distances to forage and visit preferred toolstone sources while women stayed in smaller lowland ranges and made marsh-side camps. When the marshes dried in the Holocene, a different kind of foraging was developed, perhaps with less-gendered residential and logistical mobility.

Duke and Young (this volume) tell us that the manufacture and use of crude tools of basalt in the earliest period are not examples of stop-gap tactics tiding people over until they could reach preferred obsidian sources, but instead reflect a unique technological strategy for spending longer times within an obsidian-poor patch. The people frequently made tools from basalt, provisioning their long-use patches, whereas they were much more economical with obsidian, which was made into tools for use at many places that could be very distant from each other. It would appear that the occupants of the Bonneville Basin locality were wetland hunter-gatherers whose localized foraging was a serial activity within larger-scale mobility systems involving perhaps cyclic movements between basins and sometimes distant toolstone sources. This pattern differs from Archaic strategies because the later patches were smaller, less productive, and did not allow such lengthy stays.

Schmitt et al. (this volume) provide findings that are similar to those of Duke and Young (this volume) from the same subregion: stemmed-point-making people managed long stays within a large wetland patch but frequently moved to other patches, too. This should remind us of eastern Paleoindian mobility patterns, in which there was a cycling attachment to certain separate parts of the landscape, and we can imagine that a similar process of settling-in was evolving in both the East and the Great Basin.

Smith (this volume) concludes that people brought worn tools made of nonlocal materials to the Parman Lake sites, made new tools from nearby toolstone material, and presumably took away some of the new tools to other sites where they repeated this sort of provisioning activity. The pattern emerged because people planned the sequence of tool replacement. Whatever other insights might come from future research, as Smith clearly knows, to continue shoehorning such patterns into as yet poorly defined Paleoindian and Paleoarchaic monikers is not the goal of the best studies. Of course it would be interesting to know if the Lake Parman people who made stemmed points considered themselves to be "limnosedentary" due to their attachment to the edges of pluvial Lake Parman, but Smith's study reveals far more than the accuracy or inaccuracy of this particular label. The people's wide travels to lithic sources are similar to what can be seen in other parts of the Great Basin in the early period, and now we want to know if their mobility is closely comparable to that of the earlier fluted-point makers. Were they more or less wide-ranging, more or less mobile? Thanks to studies like Smith's, we are a step closer to finding out.

Goebel (this volume) demonstrates for Bonneville Estates Rockshelter that in the earliest occupations (before 10,000 ^{14}C BP or ~11,450 cal BP), the people did not use local raw materials as much as did later peoples (after 7500 ^{14}C BP or ~8350 cal BP), and they also did not stay as long in the site or its locality. The oldest assemblages indicate that resharpening of bifaces and tools dominated activities at the site, and that most tools were made of raw materials brought from some distance away. The later assemblages (Early Archaic) also reflect biface thinning, but there is more local toolstone represented and more early-stage biface reduction, plus seed-grinding technology and side-notched points.

Studies like Goebel's and Smith's not only show us how the lithic technological organization of the Great Basin's earliest people differed from that of the later Archaic foragers, but also suggest to us how they might be similar to contemporary people who were beginning to occupy other parts of the continent. The long-distance movement of toolstone in the Great Basin's earliest period—presumably as direct transport rather than as stops along routes of sequential exchange—reflects the long-distance movement of Paleoindian people without watercraft. There were no large lakes or navigable rivers connecting the outermost limits in each of the Great Basin's huge lithic conveyance zones. Extreme movements were all on foot. Toolstone traveled in bags or pockets, not in canoes. Pedestrian explorers were familiar with the best toolstone sources and frequently traveled long distances between them.

VARIABILITY IN THE EARLY PHASE

Within the separate phases of the earliest period, even in the brief initial fluted-point dispersal, variability is to be expected in material culture region-wide, as Beck and Jones (this volume) describe. Beck and Jones's cladistic analysis unavoidably suggests some sort of fluted point descent with modification in the Great Basin. The message is valuable and interesting, although as with any analysis lacking chronostratigraphic control, the layered assumptions can magnify the already existing problems of uncertainty. For example, the change from multiple to single flute scars supposedly seen in all later fluted forms in the eastern United States and the Plains (Beck and Jones, this volume) is as yet not well enough supported by stratified and in situ discoveries or chronometry; also, as acknowledged, the radiometric dates in Chapter 2, Table 1, are not thought to be behaviorally associated with Great Basin fluted points. Therefore the true ages of all fluted points are unknown, and none can be automatically assumed younger (or older) than Clovis.

Another questionable assumption in some analyses of fluted points is that the fundamental Clovis adaptations began at a single point in space and spread from there in a ripple-like set of clear wave-fronts. This is a naive and impractical model of human dispersal. Other patterns of spread were more likely, such as long-distance leapfrogging (such as Jochim et al. [1999] have suggested for the Magdalenian expansion in postglacial Europe) or temporary explorations-and-returns to staging points (Anderson 1990; Dincauze 1993b) that could have created enormous geographic gaps in spatial ranges later filled by return movements, thus erasing temporal patterns that could suggest the direction of diffusion fronts.

SUBSISTENCE AND DIET

A major theme in this volume is the difference between early-period and Archaic subsistence. Rhode and Louderback (this volume) make the case that the presumably broad-spectrum Pre-Archaic diet in the Great Basin was not as broad as the Archaic diet, which much more strongly emphasized plants (seeds, roots, tubers). At Danger Cave, small-seed use was negligible in the earlier levels until about 8600 ^{14}C BP (~9545 cal BP), after a drying interval. At Bonneville Estates Rockshelter, a diversity of seed taxa is represented in the earlier levels, but the evidence for intensive use of seeds dates later, around 7500 ^{14}C BP (~8350 cal BP). Some seed taxa found in older deposits were likely packrat-deposited, but others were not and might represent a part (but not yet a big part) of the human diet. Pinson's chapter (this volume) proposes that northern Great Basin foragers depended on small game, birds, and fish, while occasionally hunting large artiodactyls, in contrast to the later Archaic pattern with its even broader diet breadth, including more artiodactyls. Once again, the early period is shown to be very different from the Archaic.

Hockett (this volume) argues that stemmed-point makers had a relatively more diverse diet than later people (apparently contra Pinson; and Rhode and Louderback, this volume), which makes sense for a postulated expansion phase of human populations. The stemmed-point makers thoroughly processed grouse, hare, and artiodactyls, and presumably also made use of selected plant foods, whereas later Archaic people ate fewer different types of animals. Hockett emphasizes the nutritional advantages of a diverse diet and makes an effort to contrast this interpretation of prehistoric patterns against what he sees as the overly restrictive predictions of evolutionary ecology, which would attribute the different diet breadths to changes in resource abundance and patchiness, handling costs, and energetic returns. Perhaps both sets of motivational factors were selected for: the reproductive factors that pertain to nutritional ecology, and the somatic factors that pertain to optimal foraging models. Whatever the motivation may have been to change diet breadth—to minimize foraging costs, to avoid risk, or to diversify nutrients—the early-period forager diet in the Great Basin remains inadequately known.

A central theme in Great Basin studies is the apparent disconnect between early-period lithic technology, which seems oriented toward larger game hunting, and the presumed relatively broad diet breadth. Some researchers have wondered if the lithics and the organic remains at early sites are almost at odds with each other in terms of lifeways suggested. Elston and Zeanah (2002) proposed that the reason Pleistocene/Holocene diets were broad even though lithic assemblages seem to be only hunting oriented is because of a division of labor, with men hunting the higher-ranked game animals while women gathered lower-ranked resources such as waterfowl and seeds. Madsen's chapter (this volume) presents a view in line with this sort of possibility as well.

In my opinion, early-period diets are still unsolved mysteries. Any discriminating examination of organic remains from Great Basin early-period sites, such as was included in the broader archaeofaunal study done by Cannon and Meltzer (2004), will show that less than a handful of sites have data that can be considered minimally acceptable as dietary evidence. There is still a long way to go before the foraging practices of the Pleistocene/Holocene interval are understood.

THE EARLIEST HUMAN USE OF ROCKSHELTERS

The picture that has arisen from most of the chapters is one of a discontinuous archaeological record in which technology, mobility, range, and subsistence practices are not a continuum linking oldest to youngest. Instead the record is one of temporary abandonment of landscapes and subsequent reentry during a renewed phase of dispersal. This picture is solidly supported by the chapters that focus on rockshelter excavations.

Graf describes the chronostratigraphic setting at Bonneville Estates Rockshelter, which provides further evidence for the multiphase colonization process. The site was initially occupied while an Allerød-age surface was exposed, and a series of fire features were created with small amounts of debitage around them; then the site appears to have been temporarily abandoned after 9500 ^{14}C BP (~10,740 cal BP), to be reoccupied 2,000 years later and never again to be so abandoned, most likely because of changes in wetland distribution in response to climate shifts. The radiometric dates on unlined fire features in the latest Pleistocene levels are both pre-Clovis and late Clovis. The earliest diagnostic cultural item in the site is one fluted point base recovered far out of expected stratigraphic context. Perhaps a Clovis person or band made a fire in one part of the site during a cool period but never returned. Excavated in situ were also Parman, Haskett, and Windust Stemmed points. The economical interpretation of the site is that the region's "colonizers"—that is, the first occupants who returned to the rockshelter several times—were stemmed-point makers.

Jenkins (this volume), in describing the unvandalized units in Cave 5 of the Paisley Five Mile Point Caves, notes that no clear overprint or underprint of human behavior is to be found in the oldest excavated faunal materials, although the cave did have human occupants. The first people apparently cleared the cave floor by sweeping off the Pleistocene bones that date to the Allerød. More than 150 cultural items were found below the bones of extinct megafauna, and obsidian hydration dating extends the human occupation beyond the earliest available radiometric dates on rope and twisted grass. But the oldest radiocarbon dates on artifacts at the site are younger than the admittedly uncertain obsidian hydration dates, and the presence of possible paper and woven fabric below the early dated materials calls for caution about the mixing effects of burrowing and sediment turbation. However, in spite of the stratigraphic uncertainties, several profoundly important implications from the site may change most of what we thought we knew for sure about the peopling of the Great Basin: people were present before the Younger Dryas (and appear to be pre-Clovis), they were descended from populations that had originated in Siberia, they coexisted (for a short time, anyway) with now-extinct megafauna, and they made lithic tools (as yet undiagnostic) from local materials.

Henrikson and Long (this volume) report on Kelvin's Cave, where the deposition reflects a relatively drawn-out accumulation of bones after the Last Glacial Maximum in a cool and wet period. The radiocarbon dates indicate Pleistocene camel, horse, musk ox, bison, and micromammals disappeared before the Younger Dryas, and bioturbation and sediment deformation during freeze/thaw cycles were potential causes of the vertical displacement downward of 47 lithic flakes from overlying occupations. As in any bioturbated rockshelter sediments, even the existence of burning does not prove the human hand was present in the oldest dated level. The residual bone assemblage is typical of a carnivore's choice of bones to collect and also appears to be density-mediated.[2]

THE EARLIEST HUMAN PRESENCE: INITIAL DISPERSAL AND THE LATER SETTLING-IN PHASE

The evidence from Great Basin lowland ranges and rockshelters along the edges of pluvial lake basins, especially those described in this volume, suggests that many were truly inviting for human occupation in the intra-Allerød cold period, around 12,000–13,000 ^{14}C BP (~13,900–16,000 cal BP) and may have hosted a very low density of human visitors for nearly a millennium before the Younger Dryas began around 10,900 ^{14}C BP (~12,850 cal BP).

Whatever happened during the first, brief exploratory pulse of human dispersal is poorly understood. The sequence of events in the later phases is also not clear because of stratigraphic discontinuities and radiocarbon plateaus that confound our ability to do accurate dating. The first people may have made extensive use of organic materials, although the Paisley Cave dates on possible fiber threads (Jenkins, this volume) are Younger Dryas in age; we do know, however, that one of the earliest known people ate at least one meal of bison. They left too few patterned archaeological remains, especially lithics, in too few sites to shed light on their foraging strategies. The Paisley Cave lithics that have been subjected to obsidian hydration study indicate that a locally available obsidian from lake gravels was used in the earliest assemblage, which might indicate to some scholars a familiarity with resources reflecting human residency for a considerable time already.

In most of the Great Basin, the first people were so few in number and so scattered they often did not return to sites after initial visits. The later (Younger Dryas) settlers often brought exotic toolstone bifaces to the sites to resharpen as they sat around small cooking fires on return visits.

No obvious settlement or establishment period resulted from the earliest explorations. The pre-fluted-point phase might have been a pioneering dispersal by small groups that were nonreproducing, as in many parts of western North America. Every archaeological assemblage they left is therefore a trace of a separate stochastic event rather than a part of a colonization process. There is no uniformity or easily defined patterning in the earliest phase of human presence. This is not to say that the first people did not "enter" the country and make their own places on it. Every time they marked a unit of space, memorized and inspected and learned by walking through the landscape, or made a windbreak, or scraped a hollow in the ground to build a fire, they must have known they were taming unmarked and wild country.

In the Great Basin, the archaeological footprint of the first explorations was small-scale and was followed after a brief time by what looks like a less tentative phase of human dispersal, the stemmed-point phase, which can be said to have truly colonized the region. Stemmed-point makers came on the scene in the early Younger Dryas and gravitated to marshlands, high-quality lithic sources, and the uplands linking them. They successfully exploited the early Holocene landscape for many generations.

Both phases are part of a Paleoindian stage that preceded the Archaic and differed from it in the most important ways. Taken together, the chapters in this volume prominently lead us to that conclusion, as seen in the findings from different sites and the different modes of analysis. The contributors to this volume have answered the question posed in the book's title. The Great Basin's Paleoindian stage—a discontinuous, multiphase chapter in prehistory—is complex but no longer quite so bewildering, and the next round of thoughtful research will provide ever more revealing leads to get us past the remaining interpretive obstacles.

NOTES

1. This coined word is made from two Greek roots that just cannot be assembled comfortably: *paleo* means ancient, and *archaic* means early or old. Thus, *paleoarchaic* literally means "ancient old."
2. It would be useful to apply the same taphonomic caution (about bioturbation) to the pollen study that has been interpreted to mean that mountain mahogany and chokecherry grew next to sagebrush at the end of the Allerød.

Appendix

Selected Attributes and Contents of Fully Recorded Sites in the Old River Bed Delta

Site No. (42To)	Primary Landform[a]	Site Area (m^2)	Proj. Point	Point Types[b]	Knife/ Biface	Scraper	Graver/ Drill	Flake Tool	Core	Debitage
1000	Sa Ch	71,718	17	CDS*	11	6	2	1	—	~300
1153	Sa Ch	78,333	13	ACS	4	5	3	5	—	~200
1157	De Fl	2,480	1	S	2	—	1	4	—	~50
1161	Sa Ch	13,745	4	DMS*	1	1	2	1	—	~75
1162	Sa Ch	37,812	—	—	2	—	—	—	—	> 500
1163	De Fl	76,970	11	S*	1	2	1	2	—	~250
1164	Gr Ch	2,450	—	—	—	—	—	1	—	~250
1172	Sa Ch	57,207	4	PS	5	4	—	3	1	~100
1182	Gr/Sa Ch	159,970	17	ACMS*	2	3	1	5	1	> 500
1352	Sa Ch	37,800	3	S*	6	—	—	1	—	> 500
1353	Gr Ch	1,414	1	S	2	1	—	1	—	19
1354	Gr Ch	2,089	5	CS	6	1	—	—	1	27
1356	Gr Ch	3,162	1	S	—	—	—	—	—	33
1357	Gr Ch	12,155	1	C	4	—	—	—	—	~80
1358	Gr/Sa Ch	201,972	17	CS*	36	10	3	10[c]	7	> 500
1368	Gr Ch	9,257	1	C	7	2	—	5	—	20
1369	Gr Ch	12,172	6	CDS	8	—	—	1	—	~200
1370	Gr Ch	3,981	4	ACS	3	—	—	4	—	~250
1371	Gr Ch	43,933	16	DLPRS	34	4	1	44	4	~100
1383	Sa Ch	83,483	12	CS	12	—	—	1	—	~125
1385	Gr Ch	250	—	—	1	—	—	—	—	~50
1666	Gr Ch	11,914	4	S*	4	—	—	5	2	~60
1667	Gr Ch	1,022	—	—	—	—	—	1	—	19
1668	Gr Ch	4,203	2	CP	4	1	1	3	—	16
1669	Gr Ch	1,816	3	CD	—	—	—	2	—	5
1670	Gr Ch	5,465	—	—	1	1	—	1	—	28
1671	Gr/Sa Ch	8,538	5	S	1	1	—	—	1	~150
1672	Gr Ch	8,048	6	S*	2[d]	—	—	—	—	27
1673	Gr Ch	3,489	2	CS	—	—	—	—	—	20
1674	Sa Ch	11,238	1	D	—	2	—	2	—	7
1675	Sa Ch	5,942	1	S	1	1	—	—	—	16

SITE NO. (42TO)	PRIMARY LANDFORM[a]	SITE AREA (M^2)	PROJ. POINT	POINT TYPES[b]	KNIFE/ BIFACE	SCRAPER	GRAVER/ DRILL	FLAKE TOOL	CORE	DEBITAGE
1676	Sa Ch	2,776	1	S	3	—	—	—	—	16
1677	Sa Ch	16,922	3	CP	5	2	3	1	—	25
1678	Sa Ch	14,301	3	PR	7	1	1	—	—	38
1679	Sa Ch	9,571	5	DPS	4	—	—	1	—	26
1680	Gr Ch	751	—	—	3	1	—	1	—	10
1681	Gr Ch	619	4	PS	2	—	—	—	—	10
1682	Gr Ch	308	2	S	3	1	—	1	—	5
1683	Gr Ch	6,319	6	CLM*	3	1	—	2	—	35
1684	Gr Ch	11,808	8	CDR	3	2	—	2	—	~50
1685	Gr Ch	33,233	15	CDLRS	15[e]	4	—	4[f]	1	> 500
1686	Gr Ch	19,109	30	CDLMPS	16	14	3	7	—	~500
1687	Gr Ch	1,157	8	DLM	1	1	—	—	—	8
1688	Gr Ch	16,121	16	DLRS	12[g]	3	2	29	—	~200
1689	Sa Ch	16,520	4	S*	2	—	1	2	—	~150
1858	Gr Ch	8,898	1	S	1	1	—	—	1	~75
1859	Gr Ch	1,793	3	DPS	3	—	—	—	—	17
1860	Gr Ch	4,242	1	S	—	3	—	3	—	~100
1861	Gr Ch	25,837	8	CDLS	5	2	—	4	—	~50
1862	Gr Ch	13,543	10	DS	2	4	—	3	—	~70
1872	Gr Ch	41,792	23	CDLMRS	14[h]	3	1	21	1	~100
1873	Gr Ch	16,829	16	DLMPR	17	1	1	4	1	~40
1874	Gr Ch	4,744	3	CD	1	1	—	—	—	16
1875	Gr Ch	10,932	10	CDS	5	—	1	—	—	~50
1876	Gr Ch	15,432	2	DS	5	—	1	5	—	3
1877	Gr Ch	1,516	1	S	2	—	—	3[f]	—	9
1878	Gr Ch	2,043	7	CLM	2	—	1	5	—	14
1920	Gr Ch	24,001	4	CS	13	6	1	20[f]	1	~70
1921	Gr Ch	2,587	6	DLS	3	—	—	3	1	26
1922	Gr/Sa Ch	8,289	2	LS	3	1	—	12	—	19
1923	Gr Ch	1,083	1	S	2	2	—	7	1	1
1924	Gr Ch	99,462	15	CDMS	45[g]	8	—	98	1	~150
2551	In Ch	47,757	21	CDS	17	1	1	18	1	~125
2552	In Ch	30,749	5	S	1	1	—	6	—	~125
2553	In Ch	3,146	5	ST	—	—	—	3	—	13
2554	In Ch	52,087	18	DS	19	3	—	11	—	~100
2555	In Ch	23,335	12	DS	5	2	—	10	—	28
2556	In Ch	22,070	22	DS	10	4	3	12	—	~125
2557	In Ch	861	3	DS	2	—	1	2	—	8
2558	In Ch	9,204	7	DS	2	—	—	1	—	~150
2559	In Ch	90,672	37	CDLPS	28	2	4	33	2	~450

Note: Artifact types include field identifications and large collections analyzed by Jones et al. (2002).

[a] Gr Ch = gravel channel; In Ch = intermediate channel; Sa Ch = sand channel; De Fl = delta floodplain; Gr/Sa Ch = gravel and sand channels.

[b] Late Pleistocene/early Holocene types only; asterisks denote sites with Archaic/Fremont period points. A = Haskett; C = Cougar Mountain; D = Dugway Stubby; L = Silver Lake; M = Lake Mohave; P = Pinto; R = Parman; S = stemmed (includes lanceolate, square-based, contracting and expanding stems, and small indeterminate stem fragments); T = possibly fluted.
[c] Includes one spokeshave.
[d] Includes one possible crescent.
[e] Includes two possible crescents.
[f] Includes one chisel.
[g] Includes one crescent.
[h] Includes one Cody knife.

References

Adovasio, J. M., J. D. Gunn, J. Donahue, R. Stuckenrath, J. Guilday, and K. Lord

1978 Meadowcroft Rockshelter. In *Early Man in America from a Circum-Pacific Perspective*. Occasional Paper No. 1. Department of Anthropology, University of Alberta, Edmonton.

Adovasio, J. M., and D. R. Pedler

2004 Pre-Clovis Sites and Their Implications for Human Occupation Before the Last Glacial Maximum. In *Entering America: Northeast Asia and Beringia Before the Last Glacial Maximum*, edited by D. B. Madsen, pp. 139–158. University of Utah Press, Salt Lake City.

Agte, V., K. Tarwadi, S. Mengale, and S. Chiplonkar

2000 Potential of Traditionally Cooked Green Leafy Vegetables as Natural Sources for Supplementation of Eight Micronutrients in Vegetarian Diets. *Journal of Food Composition and Analysis* 13:885–891.

Aikens, C. M.

1970 *Hogup Cave*. University of Utah Anthropological Papers No. 93. Salt Lake City.

Aikens, C. M., and D. B. Madsen

1986 Prehistory of the Eastern Area. In *Great Basin*, edited by W. d'Azevedo, pp. 149–160. Handbook of North American Indians, Vol. 11 (W. C. Sturtevant, general editor). Smithsonian Institution Press, Washington, D.C.

Allen, L.

2005 Multiple Micronutrients in Pregnancy and Lactation: An Overview. *American Journal of Clinical Nutrition* 81:1206S–1212S.

Alley, R., and A. Agústsdóttir

2005 The 8k Event: Cause and Consequences of a Major Holocene Abrupt Climate Change. *Quaternary Science Reviews* 24:1123–1149.

Amick, D. S.

1995 Raw Material Selection Patterns Among Paleoindian Tools from the Black Rock Desert, Nevada. *Current Research in the Pleistocene* 12:55–57.

1997 Geochemical Source Analysis of Obsidian Paleoindian Points from the Black Rock Desert, Nevada. *Current Research in the Pleistocene* 14:97–99.

1999 Using Lithic Artifacts to Explain Past Behavior. In *Models for the Millennium: Great Basin Anthropology Today*, edited by C. Beck, pp. 161–170. University of Utah Press, Salt Lake City.

2004 A Possible Ritual Cache of Great Basin Stemmed Bifaces from the Terminal Pleistocene–Early Holocene Occupation of NW Nevada, USA. *Lithic Technology* 29:119–145.

Anderson, D. G.

1990 The Paleoindian Colonization of Eastern North America: A View from the Southeastern United States. *Research in Economic Anthropology*, Supplement 5:163–216.

Anderson, D. G., and M. K. Faught

2006 Paleoindian Database of the Americas. Online Web site accessed 8 February 2006 (http://pidba.utk.edu/main.htm).

Anderson, D. G., L. D. O'Steen, and K. E. Sassaman
1996 Environmental and Chronological Considerations. In *The Paleoindian and Early Archaic Southeast*, edited by D. G. Anderson and K. E. Sassaman, pp. 3–15. University of Alabama Press, Tuscaloosa.
Anderson, D. G., and K. E. Sassaman (editors)
1996 *The Paleoindian and Early Archaic Southeast*. University of Alabama Press, Tuscaloosa.
Anderson, J., and B. Deskins
1995 *The Nutrition Bible*. William Morrow and Company, Inc., New York.
Andrefsky, W., Jr.
1994 Raw-Material Availability and the Organization of Technology. *American Antiquity* 59:21–34.
Antevs, E.
1948 Climatic Changes and Pre-white Man. In The Great Basin, with Emphasis on Postglacial Times. *Bulletin of the University of Utah* 38(20), *Biological Series* 10(7):167–191.
Arkush, B.S., and B. L. Pitblado
2000 Paleoarchaic Surface Assemblages in the Great Salt Lake Desert, Northwestern Utah. *Journal of California and Great Basin Anthropology* 22:12–42.
Ashworth, C., and C. Antipatis
2001 Micronutrient Programming of Development Throughout Gestation. *Reproduction* 122:527–535.
Bamforth, D. B.
1986 Technological Efficiency and Tool Curation. *American Antiquity* 51:38–50.
1991 Technological Organization and Hunter-Gatherer Land Use: A California Example. *American Antiquity* 56:216–234.
Basgall, M. E.
1988 The Archeology of CA-MNO-679: A Pre-Archaic Site in Long Valley Caldera, Mono County, California. In *Early Human Occupation in Far Western North America: The Clovis–Archaic Interface*, edited by J. A. Willig, C. M. Aikens, and J. L. Fagan, pp. 103–120. Nevada State Museum Anthropological Papers No. 21. Carson City.
Beaton, J. M.
1991 PaleoIndian Occupation Greater than 11,000 Years B.P. at Tule Lake, Northern California. *Current Research in the Pleistocene* 8:5–7.
Beck, C.
1984 Steens Mountain Surface Archaeology: The Sites. Ph.D. dissertation, University of Washington.
1996 Dating the Archaeological Record and Modeling Chronology. Paper presented at the 25th Great Basin Anthropological Conference, Kings Beach, California.
Beck, C., and G. T. Jones
1990a Toolstone Selection and Lithic Technology in Early Great Basin Prehistory. *Journal of Field Archaeology* 17:283–299.
1990b The Late Pleistocene/Early Holocene Archaeology of Butte Valley, Nevada: Three Seasons' Work. *Journal of California and Great Basin Anthropology* 12:231–261.
1992 Paleoindian-Archaic Range Shifts in Eastern Nevada. *Current Research in the Pleistocene* 9:1–2.
1997 The Terminal Pleistocene/Early Holocene Archaeology of the Great Basin. *Journal of World Prehistory* 11:161–236.
n.d. *The Paleoarchaic Occupation of Eastern Nevada: The Sunshine Locality*. University of Utah Press, Salt Lake City (in review).
Beck, C., G. T. Jones, D. L. Jenkins, C. E. Skinner, and J. J. Thatcher
2004a Fluted or Basally-Thinned? Re-Examination of a Lanceolate Point from the Connley Caves in the Fort Rock Basin. In *Early and Middle Holocene Archaeology of the Northern Great Basin*, edited by D. L. Jenkins, T. J. Connolly, and C. M. Aikens, pp. 281–294. University of Oregon Anthropological Papers No. 62. Eugene.

Beck, C., G. T. Jones, and A. K. Taylor
2004b Is there Clovis in the Great Basin? Paper presented at the 69th meeting of the Society for American Archaeology, Montreal.
Beck, C., A. Taylor, G. T. Jones, C. M. Fadem, C. R. Cook, and S. A. Milward
2002 Rocks are Heavy: Transport Costs and Paleoarchaic Quarry Behavior in the Great Basin. *Journal of Anthropological Archaeology* 21:481–507.
Bedwell, S. F.
1973 *Fort Rock Basin: Prehistory and Environment*. University of Oregon Books, Eugene.
Behrensmeyer, A. K.
1978 Taphonomic and Ecologic Information from Bone Weathering. *Paleobiology* 4:150–162.
Behrensmeyer, A. K., and A. P. Hill
1980 *Fossils in the Making: Vertebrate Taphonomy and Paleoecology*. University of Chicago Press, Chicago.
Belinsky, D., and H. Kuhnlein
2000 Macronutrient, Mineral, and Fatty Acid Composition of Canada Goose (*Branta canadensis*): An Important Traditional Food Resource of the Eastern James Bay Cree of Quebec. *Journal of Food Composition and Analysis* 13:101–115.
Bendich, A.
2001 Micronutrients in Women's Health and Immune Function. *Nutrition* 17:858–867.
Benson, L. V.
1999 Records of Millennial-Scale Climate Change from the Great Basin of the Western United States. In *Mechanisms of Global Change at Millennial Time Scales*, edited by P. Clark, R. Webb, and L. Keigwin, pp. 203–225. Geophysical Monograph Series 112. American Geophysical Union, Washington, D.C.
Benson, L. V., J. Burdett, S. Lund, M. Kashgarian, and S. Mensing
1997 Nearly Synchronous Climate Change in the Northern Hemisphere During the Last Glacial Termination. *Nature* 388:263–265.
Benson, L. V., D. R. Currey, R. I. Dorn, K. R. Lajoie, C. G. Oviatt, S. W. Robinson, G. I. Smith, and S. Stine
1990 Chronology of Expansion and Contraction of Four Great Basin Lake Systems During the Past 35,000 Years. *Palaeogeography, Palaeoclimatology, Palaeoecology* 78:241–286.
Benson, L. V., M. Kashgarian, R. Rye, S. Lund, and F. Paillet
2002 Holocene Multidecadal and Multicentennial Droughts Affecting Northern California and Nevada. *Quaternary Science Reviews* 21:659–682.
Bettinger, R. L., and M. A. Baumhoff
1982 The Numic Spread: Great Basin Cultures in Competition. *American Antiquity* 47:485–503.
Bettinger, R. L., and D. A. Young
2004 Hunter-Gatherer Population Expansion in North Asia and the New World. In *Entering America: Northeast Asia and Beringia Before the Last Glacial Maximum*, edited by D. B. Madsen, pp. 239–251. University of Utah Press, Salt Lake City.
Bever, M. R.
2001 Stone Tool Technology and the Mesa Complex: Developing a Framework of Alaskan Paleoindian Prehistory. *Arctic Anthropology* 38:98–118.
Binford, L. R.
1981 *Bones: Ancient Men and Modern Myths*. Academic Press, Orlando.
Bischoff, J., and K. Cummins
2001 Wisconsin Glaciation of the Sierra Nevada (79,000–15,000 yr B.P.) as Recorded by Rock Flour in Sediments of Owens Lake, California. *Quaternary Research* 55:14–24.
Brain, C. K.
1981 *The Hunters or the Hunted? An Introduction to African Cave Taphonomy*. University of Chicago Press, Chicago.

Broadhurst, C. L., Y. Wang, M. Crawford, S. Cunnane, J. Parkington, and W. Schmidt
2002 Brain-Specific Lipids from Marine, Lacustrine, or Terrestrial Food Resources: Potential Impact on Early African *Homo sapiens*. *Comparative Biochemistry and Physiology Part B* 131:653–673.

Brooks, D. R., and D. A. McLennan
1991 *Phylogeny, Ecology, and Behavior: A Research Program in Comparative Biology*. University of Chicago Press, Chicago.

Broughton, J. M., D. B. Madsen, and J. Quade
2000 Fish Remains from Homestead Cave and Lake Levels of the Past 13,000 Years in the Bonneville Basin. *Quaternary Research* 53:392–401.

Bryan, A. L.
1979 Smith Creek Cave. In *The Archaeology of Smith Creek Canyon*, edited by D. R. Tuohy, pp. 164–251. Nevada State Museum Anthropological Papers No. 17. Carson City.
1980 The Stemmed Point Tradition: An Early Technological Tradition in Western North America. In *Anthropological Papers in Memory of Early H. Swanson, Jr.*, edited by L. B. Harten, C. N. Warren, and D. R. Tuohy, pp. 77–107. Special publication of the Idaho State University Museum of Natural History, Pocatello.
1988 The Relationship of the Stemmed Point and Fluted Point Traditions in the Great Basin. In *Early Human Occupation in Far Western North America: The Clovis-Archaic Interface*, edited by J. A. Willig, C. M. Aikens, and J. L. Fagan, pp. 53–74. Nevada State Museum Anthropological Papers No. 21. Carson City.

Butler, B. R.
1963 An Early Man Site at Big Camas Prairie, South-Central Idaho. *Tebiwa* 6:22–33.
1965 A Report on Investigations of an Early Man Site near Lake Channel, Southern Idaho. *Tebiwa* 8:1–21.
1968 An Introduction to Archaeological Investigations in the Pioneer Basin Locality of Eastern Idaho. *Tebiwa* 11:1–14.
1973 Folsom and Plano Points from the Peripheries of the Upper Snake Country. *Tebiwa* 16:69–72.
1978 *A Guide to Understanding Idaho Archaeology: The Upper Snake and Salmon River Country*. 3rd ed. Special publication of the Idaho Museum of Natural History, Pocatello.

Byers, D. A., and A. Ugan
2005 Should We Expect Large Game Specialization in the Late Pleistocene? An Optimal Foraging Perspective on Early Paleoindian Prey Choice. *Journal of Archaeological Science* 32:1624–1640.

CalPal
2006 CalPal2005 ver. 1.4 software. Available online at http://www.calpal.de/ (accessed 25 October 2006).

Campbell, E. W. C., W. H. Campbell, E. Antevs, C. A. Amsden, J. A. Barbieri, and F. D. Bode
1937 *The Archaeology of Pleistocene Lake Mohave: A Symposium*. Southwest Museum Papers No. 11. Los Angeles.

Cannon, M. D., and D. J. Meltzer
2004 Early Paleoindian Foraging: Examining the Faunal Evidence for Large Mammal Specialization and Regional Variability in Prey Choice. *Quaternary Science Reviews* 23: 1955–1987.

Carter, J. A.
1999 *TS-5 Target Complex Cultural Resource Inventory, Wendover Air Force Range, Tooele County, Utah*. Historical Research Associates, Inc., Seattle. Submitted to the U.S. Air Force, Hill Air Force Base, Utah, and National Park Service, Midwest Archeological Center, Lincoln, Nebraska.

Carter, J. A., D. Duke, and D. C. Young
2004 *Limited Site Testing at 22 Archaeological Sites in the Wild Isle Project Area, Utah Test and Training Range, Tooele County, Utah*. Historical Research Associates, Inc., Seattle, Washington. Submitted to the U.S. Air Force, Hill Air Force Base, Ogden, Utah.

Carter, J. A., and D. C. Young Jr.

2001 *TS-5 Central Area and Craners Cultural Resource Inventory, Wendover and Hill Air Force Ranges, Tooele and Box Elder Counties, Utah.* Historical Research Associates, Inc., Seattle, Washington. Submitted to the U.S. Air Force, Hill Air Force Base, Ogden, Utah.

2002 *TS-5 Central Area and Craners Cultural Resources Inventory, Utah Test and Training Range, Tooele and Box Elder Counties, Utah.* Historical Research Associates, Inc., Seattle, Washington. Submitted to the U.S. Air Force, Hill Air Force Base, Utah, and National Park Service, Midwest Archeological Center, Lincoln, Nebraska.

Chamberlin, R.

1911 The Ethno-botany of the Gosiute Indians. *Proceedings of the Academy of Natural Sciences of Philadelphia* 63:24–99.

Christian, L.

1997 *Early Holocene Typology, Chronology, and Mobility: Evidence from the Northern Great Basin.* Department of Anthropology, Sundance Archaeological Research Fund Report No. 3. University of Nevada, Reno.

Clewlow, C. W., Jr.

1968 *Surface Archaeology on the Black Rock Desert.* University of California Archaeological Survey Reports No. 73. Berkeley.

Collins, M. B.

1999a Clovis and Folsom Lithic Technology on and Near the Southern Plains. In *Folsom Lithic Technology: Explorations in Structure and Variation*, edited by D. S. Amick, pp. 12–38. International Monographs in Prehistory, Archaeological Series 12. University of Ann Arbor, Michigan.

1999b *Clovis Blade Technology.* University of Texas Press, Austin.

Collins, M. B., and T. D. Dillehay

1986 The Implications of the Lithic Assemblage from Monte Verde for Early Man Studies. In *New Evidence for the Pleistocene Peopling of the Americas*, edited by A. L. Bryan, pp. 339–335. Center for the Study of Early Man, University of Maine, Oronto.

Collins, M. B., and J. C. Lohse

2004 The Nature of Clovis Blades and Blade Cores. In *Entering America: Northeast Asia and Beringia Before the Last Glacial Maximum*, edited by D. B. Madsen, pp. 159–183. University of Utah Press, Salt Lake City.

Connolly, T. J.

1999 Newberry Crater. A Ten-Thousand-Year Record of Human Occupation and Environmental Change in the Basin-Plateau Borderlands. University of Utah Anthropological Papers No. 121. Salt Lake City.

Connolly, T. J., and P. Barker

2004 Basketry Chronology of the Early Holocene in the Northern Great Basin. In *Early and Middle Holocene Archaeology of the Northern Great Basin*, edited by D. L. Jenkins, T. J. Connolly, and C. M. Aikens, pp. 241–250. University of Oregon Anthropological Papers No. 62. Eugene.

Connolly, T. J., C. S. Fowler, and W. J. Cannon

1998 Radiocarbon Evidence Relating to Northern Great Basin Basketry Chronology. *Journal of California and Great Basin Anthropology* 20:88–100.

Connolly, T. J., and D. A. Trieu

2006 Fiber Artifacts from Paisley Caves, Oregon: Terminal Pleistocene and Early Holocene Assemblage. Paper presented at the 30th Great Basin Anthropological Conference, Las Vegas, Nevada.

Cressman, L. S.

1940 Studies on Early Man in South Central Oregon. In *Carnegie Institution of Washington Year Book* No. 39:300–306. Washington, D.C.

1942 *Archaeological Researches in the Northern Great Basin.* Carnegie Institution of Washington Publication No. 538. Washington, D.C.

1966 *The Sandal and the Cave: The Indians of Oregon*. Oregon State University Press, Corvallis.

1986 Prehistory of the Northern Area. In *Great Basin*, edited by W. L. d'Azevedo, pp. 120–126. Handbook of North American Indians, Vol. 11 (W. C. Sturtevant, general editor). Smithsonian Institution Press, Washington, D.C.

1988 *Golden Journey: Memoirs of an Archaeologist*. University of Utah Press, Salt Lake City.

Cressman, L. S., and H. Williams

1940 Early Man in Southcentral Oregon: Evidence from Stratified Sites. In *Early Man in Oregon: Archaeological Studies in the Northern Great Basin*, edited by L. S. Cressman, H. Williams, and A. Krieger, pp. 53–78. University of Oregon Monographs, Studies in Anthropology No. 3. Eugene.

Cronquist, A., A. H. Holmgren, N. H. Holmgren, and J. L. Reveal.

1972 *Intermountain Flora: Vascular Plants of the Intermountain West, U.S.A.* Hafner Publishing Company, New York.

Cummings, L., and K. Puseman

2005 *Pollen and Macrofloral Analysis of a Soil Column from Kelvin's Cave (Site 10LN93), South-central Idaho, Compared to a Local Archaeoclimatic Model.* Report on file, Bureau of Land Management, Shoshone, Idaho.

Currey, D. R.

1982 *Lake Bonneville: Selected Features of Relevance to Neotectonic Analysis*. U. S. Geological Survey Open-File Report 82-1070. Boulder, Colorado.

1991 *Hemiarid Lake Basins: Hydrographic and Geomorphic Patterns*. Limnotectonics Laboratory Technical Report 91-2. University of Utah, Salt Lake City.

Currey, D. R., and C. G. Oviatt

1985 Durations, Average Rates, and Probable Causes of Lake Bonneville Expansions, Stillstands, and Contractions During the Last Deep-lake Cycle, 32,000 to 10,000 Years Ago. *Geographical Journal of Korea* 10:1085–1099.

Davis, E. L.

1978 *The Ancient Californians: Rancholabrean Hunters of the Mojave Lakes Country*. Natural History Museum of Los Angeles County Science Series 29. Los Angeles.

Davis, E. L., C. W. Brott, and R. Shutler Jr.

1969 *The Western Lithic Co-Tradition*. San Diego Museum Papers No. 6. San Diego.

Davis, L. G.

2006 A Preliminary Geoarchaeological Perspective on Late Pleistocene–Early Holocene Cultural Occupation at the Paisley Five Mile Point Caves. Paper presented at the 30th Great Basin Anthropological Conference, Las Vegas, Nevada.

Davis, L. G., and C. E. Schweger

2004 Geoarchaeological Context of Late Pleistocene and Early Holocene Occupation at the Cooper's Ferry Site, Western Idaho, USA. *Geoarchaeology* 19:685–704.

Dean, E.

1994 Faunal Remains from Structure 3 at the Big M Site, Fort Rock Basin, Oregon. In *Archaeological Researches in the Northern Great Basin: Fort Rock Archaeology Since Cressman*, edited by C. M. Aikens and D. L. Jenkins, pp. 505–529. University of Oregon Anthropological Papers No. 50. Eugene.

DeQuille, D.

1963 *Washoe Rambles.* Westernlore Press, Los Angeles.

Dillehay, T. D.

1997 *Monte Verde: A Late Pleistocene Settlement in Chile,* Vol. 2: *The Archaeological Context and Interpretation*. Smithsonian Institution Press, Washington, D.C.

2002 *Monte Verde: A Late Pleistocene Settlement in Chile,* Vol. 2: *The Archaeological Context and Interpretation. Errata*. Smithsonian Institution Press, Washington, D.C.

Dillehay, T. D., and J. Rossen

2002 Plant Food and Its Implications for the Peopling of the New World: A View from

South America. In *The First Americans: The Pleistocene Colonization of the New World*, edited by N. G. Jablonski, pp. 237–253. Wattis Symposium Series in Anthropology, Memoirs of the California Academy of Sciences No. 27. San Francisco.

Dincauze, D. F.

1993a Fluted Points in the Eastern Forests. In *From Kostenki to Clovis*, edited by O. Soffer and N. D. Praslov, pp. 279–292. Plenum Press, New York.

1993b Pioneering in the Pleistocene: Large Paleoindian Sites in the Northeast. In *Archaeology of Eastern North America: Papers in Honor of Stephen Williams*, edited by J. B. Stoltman, pp. 43–60. Mississippi Department of Archives and History Archaeological Report No. 25. Jackson.

Dort, W., Jr.

1968 Paleoclimatic Implications of Soil Structures at the Wasden Site (Owl Cave). *Tebiwa* 11:31–36.

Douglas, C. L., D. L. Jenkins, and C. N. Warren

1988 Spatial and Temporal Variability in Faunal Remains from Four Lake Mojave-Pinto Period Sites in the Mojave Desert. In *Early Human Occupation in Far Western North America: The Clovis-Archaic Interface*, edited by J. A. Willig, C. M. Aikens, and J. L. Fagan, pp. 273–301. Nevada State Museum Anthropological Papers No. 21. Carson City.

Duke, D.

1998 *Basalt Resource Use and Technological Organization in the North-Central Sierra Nevada*. M.A. thesis, University of Nevada, Reno. University Microfilms, Ann Arbor, Michigan.

2003 *Cultural Resources Inventory of the South Route to Wild Isle and TS-5, Utah Test and Training Range, Tooele County, Utah*. Geo-Marine, Inc., Las Vegas. Submitted to the U.S. Army Corps of Engineers, Fort Worth, Texas, and U.S. Air Force, Hill Air Force Base, Ogden, Utah.

Duke, D., D. C. Young, and J. A. Carter

2004 A Unique Example of Early Technology and Land Use in the Eastern Great Basin. *Current Research in the Pleistocene* 21:32–34.

Dunnell, R. C.

1971 *Systematics in Prehistory*. Free Press, New York.

1980 Evolutionary Theory and Archaeology. In *Advances in Archaeological Method and Theory*, Vol. 3, edited by M. B. Schiffer, pp. 35–99. Academic Press, New York.

1986 Methodological Issues in Americanist Artifact Classification. *Advances in Archaeological Method and Theory*, Vol. 9, edited by M. B. Schiffer, pp. 149–207. Academic Press, New York.

Eiselt, B. S.

1997a *Defining Ethnicity in Warner Valley: An Analysis of House and Home*. Department of Anthropology Technical Report 97-2. University of Nevada, Reno.

1997b Fish Remains from the Spirit Cave Paleofecal Material: 9,400 Year Old Evidence for Great Basin Utilization of Small Fishes. *Nevada Historical Quarterly* 40:117–139.

Elston, R. G.

1982 Good Times, Hard Times: Prehistoric Culture Change in the Western Great Basin. In *Man and Environment in the Great Basin*, edited by D. B. Madsen and J. F. O'Connell, pp. 186–206. SAA Papers No. 2. Society for American Archaeology, Washington, D.C.

1986 Prehistory of the Western Area. In *Great Basin*, edited by W. L. d'Azevedo, pp. 135–148. Handbook of North American Indians, Vol. 11 (W. C. Sturtevant, general editor). Smithsonian Institution Press, Washington, D.C.

1994 Prehistoric Strategies for Living Behind the Argenta Rim. In *Behind the Argenta Rim: Prehistoric Land Use in Whirlwind Valley and the Northern Shoshone Range*, edited by R. G. Elston and M. Bullock, pp. 351–359. Intermountain Research, Silver City, Nevada. Submitted to the Bureau of Land Management, Battle Mountain District, Nevada.

2005 Flaked- and Battered-Stone Artifacts. In *Camels Back Cave*, by D. N. Schmitt and D. B. Madsen, pp. 92–119. University of Utah Anthropological Papers No. 125. Salt Lake City.

Elston, R. G., and D. W. Zeanah

2002 Thinking Outside the Box: A New Perspective on Diet Breadth and Sexual Division of Labor in the Prearchaic Great Basin. *World Archaeology* 34:103–130.

Elston, R. G., D. W. Zeanah, J. A. Carter, and D. P. Dugas

1995 Paleoenvironmental Variation and Human Land Use in Toedokado Territory. In *An Optimal Foraging Model of Hunter-Gatherer Land Use in the Carson Desert*, edited by D. W. Zeanah, J. A. Carter, D. P. Dugas, R. G. Elston, and J. E. Hammett, pp. 293–326. Intermountain Research, Silver City, Nevada. Report submitted to the U.S. Fish and Wildlife Service, Portland, Oregon.

Fagan, J. L.

1988 Clovis and Western Pluvial Lakes Tradition Lithic Technologies at the Dietz Site in South-central Oregon. In *Early Human Occupation in Far Western North America: The Clovis-Archaic Interface*, edited by J. A. Willig, C. M. Aikens, and J. L. Fagan, pp. 389–416. Nevada State Museum Anthropological Papers No. 21. Carson City.

Fall, C., S. Chittaranjan, S. Rao, A. Davies, N. Brown, and J. Farrant

2003 Micronutrients and Fetal Growth. *Journal of Nutrition* 133:1747S–1756S.

Farrand, W. R.

2001 Sediments and Stratigraphy in Rockshelters and Caves: A Personal Perspective on Principles and Pragmatics. *Geoarchaeology* 16:537–557.

Farris, J. S.

1989 The Retention Index and the Rescaled Consistency Index. *Cladistics* 5:417–419.

Fawcett, W. B., and S. R. Simms

1993 *Archaeological Test Excavations in the Great Salt Lake Wetlands and Associated Analyses*. Utah State University Contributions to Anthropology No. 14. Logan.

Felsenstein, J.

2004 PHYLIP (Phylogeny Inference Package), Version 3.6. Distributed by the author. Department of Genome Sciences, University of Washington, Seattle.

Fiedel, S. J.

1999 Older Than We Thought: Implications of Corrected Dates for Paleoindians. *American Antiquity* 64:95–116.

Ford, J. A.

1952 Measurements of Some Prehistoric Design Developments in the Southeastern States. American Museum of Natural History Anthropological Papers Vol. 44, Pt. 1. New York.

1962 *A Quantitative Method for Deriving Cultural Chronology*. Technical Bulletin No. 1. Pan American Union, Washington, D.C.

Fowler, C. S.

1982 Settlement Patterns and Subsistence Systems in the Great Basin: The Ethnographic Record. In *Man and Environment in the Great Basin*, edited by D. B. Madsen and J. F. O'Connell, pp. 121–138. SAA Papers No. 2. Society for American Archaeology, Washington, D.C.

1986 Subsistence. In *Great Basin*, edited by W. L. d'Azevedo, pp. 64–97. Handbook of North American Indians, Vol. 11 (W. C. Sturtevant, general editor). Smithsonian Institution Press, Washington, D.C.

1992 *In the Shadow of Fox Peak: An Ethnography of the Cattail-Eater Northern Paiute People of Stillwater Mars*h. U.S. Fish and Wildlife Service Cultural Resource Series No. 5. Portland, Oregon.

Fowler, D. D.

1986 History of Research. In *Great Basin*, edited by W. L. d'Azevedo, pp. 15–30. Handbook of North American Indians, Vol. 11 (W. C. Sturtevant, general editor). Smithsonian Institution Press, Washington, D.C.

Fowler, D. D., and C. S. Fowler
1990 A History of Wetlands Anthropology in the Great Basin. In *Wetland Adaptations in the Great Basin*, edited by J. C. Janetski and D. B. Madsen pp. 5–16. Occasional Papers of the Museum of Peoples and Culture No. 1. Brigham Young University, Provo, Utah.
Franzen, J.
1980 *A Class II Cultural Resource Inventory of Areas in and Adjacent to the Big Desert Planning Unit, Idaho.* Report on file, Bureau of Land Management, Idaho Falls.
Frémont, J. C.
1845 *Report of the Exploring Expedition to the Rocky Mountains in the Year 1842, and to Oregon and California in the Years 1843–1844.* Goles and Seaton, Washington, D.C.
Friedel, D.
2001 Pleistocene Lake Chewaucan: Two Short Pieces on Hydrological Connections and Lake-level Oscillations. In *Quaternary Studies near Summer Lake, Oregon*, edited by R. Negrini, S. Pezzopane, and T. Badger. Friends of the Pleistocene, Ninth Annual Pacific Northwest Cell Field Trip.
Friedman, I.
1968 Hydration Rind Dates Rhyolite Flows. *Science* 159:878–880.
1977 Hydration Dating of Volcanism at Newberry Crater, Oregon. *U.S. Geological Survey Journal of Research* 5:337–342.
Friedman, I., and W. Long
1976 Hydration Rate of Obsidian. *Science* 191:347–352.
Friedman, I., and R. L. Smith
1960 A New Dating Method Using Obsidian, Part 1: The Development of the Method. *American Antiquity* 25:476–522.
Friedman, I., and F. Trembour
1983 Obsidian Hydration Dating Update. *American Antiquity* 48:544–547.
Fry, G. F.
1976 *Analysis of Prehistoric Coprolites from Utah.* University of Utah Anthropological Papers No. 93. Salt Lake City.
Gambling, L., and H. McArdle
2004 Iron, Copper, and Fetal Development. *Proceedings of the Nutrition Society* 63:553–562.
Gifford-Gonzalez, D. P., D. B. Damrosch, D. R. Damrosch, J. Pryor, and R. L. Thunen
1985 The Third Dimension in Site Structure: An Experiment in Trampling and Vertical Dispersal. *American Antiquity* 50:803–818.
Godsey, H. S., D. R. Currey, and M. A. Chan
2005 New Evidence for an Extended Occupation of the Provo Shoreline and Implications for Regional Climate Change, Pleistocene Lake Bonneville, Utah, USA. *Quaternary Research* 63:212–223.
Goebel, T.
2005 Terminal Pleistocene Technological Activities at Bonneville Estates Rockshelter: A First Look at the Artifact Record. Paper presented at the 70th meeting of the Society for American Archaeology, Salt Lake City.
Goebel, T., K. Adams, K. E. Graf, P. Bruce, and Z. Winter
2004 Jakes Depression: A New Clovis Site in the Great Basin. Paper presented at the 69th meeting of the Society for American Archaeology, Montreal.
Goebel, T., K. E. Graf, B. S. Hockett, and D. Rhode
2003 Late-Pleistocene Humans at Bonneville Estates Rockshelter, Eastern Nevada. *Current Research in the Pleistocene* 20:20–23.
Goloboff, P. A.
1991 Homoplasy and the Choice Among Cladograms. *Cladistics* 7:215–232.
Goodyear, A. C.
1979 *A Hypothesis for the Use of Cryptocrystalline Raw Materials Among Paleo-Indian Groups of North America.* Institute of Archaeology and Anthropology, Research Manuscripts Series 156. University of South Carolina, Columbia.

1989 A Hypothesis for the Use of Crypto-Crystalline Raw Materials Among Paleoindian Groups of North America. In *Eastern Paleoindian Lithic Resource Use*, edited by C. Ellis and J. Lothrop, pp. 1–9. Westview Press, Boulder.

Gosse, J., E. Evenson, J. Klein, B. Lawn, and R. Middleton

1995 Precise Cosmogenic ^{10}Be Measurements in Western North America: Support for a Global Younger Dryas Cooling Event. *Geology* 23:877–880.

Graf, K. E.

2001 *Paleoindian Technological Provisioning in the Western Great Basin*. M.A. thesis, University of Nevada, Las Vegas.

2002 Paleoindian Obsidian Procurement and Mobility in the Western Great Basin. *Current Research in the Pleistocene* 19:87–89.

Graf, K. E., T. Goebel, and B. Hockett

2004 Bonneville Estates Rockshelter: An Update on the Terminal Pleistocene. Paper presented at the 29th Great Basin Anthropological Conference, Sparks, Nevada.

Graf, K. E., T. Goebel, B. Hockett, and D. Rhode

2002a Late Pleistocene to Early Holocene Archaeology at Bonneville Estates Rockshelter, Eastern Great Basin. Paper presented at the 28th Great Basin Anthropological Conference, Elko, Nevada.

Graf, K. E., T. Goebel, B. Hockett, and D. Rhode

2002b *Recent Archaeological Investigations at Bonneville Estates Rockshelter, Eastern Great Basin*. Paper presented at the 67th meeting of the Society for American Archaeology, Denver, Colorado.

Gramly, R. M.

1993 *The Richey Clovis Cache*. Persimmon Press Monographs in Archaeology. Belmont, California.

Grayson, D. K.

1977 Paleoclimatic Implications of the Dirty Shame Rockshelter Mammalian Fauna. *Tebiwa* 9:1–26.

1979 Mount Mazama, Climatic Change, and Fort Rock Basin Archaeofaunas. In *Volcanic Activity and Human Ecology*, edited by P. D. Sheets and D. K. Grayson, pp. 427–458. Academic Press, New York.

1984 *Quantitative Zooarchaeology: Topics in the Analysis of Archaeological Faunas*. Academic Press, Orlando.

1988 *Danger Cave, Last Supper Cave, and Hanging Rock Shelter: The Faunas*. Anthropological Papers of the American Museum of Natural History, Vol. 66, Pt. 1. New York.

1993 *The Desert's Past: A Natural Prehistory of the Great Basin*. Smithsonian Institution Press, Washington, D.C.

1998 Moisture History and Small Mammal Community Richness During the Latest Pleistocene and Holocene, Northern Bonneville Basin, Utah. *Quaternary Research* 49:330–334.

2000a The Homestead Cave Mammals. In *Late Quaternary Paleoecology in the Great Basin*, edited by D. Madsen, pp. 67–89. Utah Geological Survey Bulletin Vol. 130. Salt Lake City.

2000b Mammalian Responses to Middle Holocene Climatic Change in the Great Basin of the Western United States. *Journal of Biogeography* 27:181–192.

Grayson, D. K., and D. B. Madsen

2000 Biogeographic Implications of Recent Low-elevation Recolonization by *Neotoma cinerea* in the Great Basin. *Journal of Mammalogy* 81:1100–1105.

Grayson, D. K., and D. J. Meltzer

2002 Clovis Hunting and Large Mammal Extinction: A Critical Review of the Evidence. *Journal of World Prehistory* 16:313–359.

Greeley, R., and J. King

1977 *Volcanism on the Snake River Plain, Idaho: A Comparative Planetary Geology Guidebook*. Prepared for the Office of Planetary Geology. National Aeronautics and Space Administration, Washington, D.C.

Greenspan, R. L.

1993 Analysis of the Buffalo Flat Vertebrate Faunal Remains. In *The Archaeology of Buffalo Flat: Cultural Resources Investigations for the Conus OTH-B Buffalo Flat Radar Transmitter Site, Christmas Lake Valley, Oregon,* Vol. 2: *Analyses, Synthesis, Appendices,* edited by A. C. Oetting, pp. 613–674. Historical Research Associates Report No. 151. Eugene, Oregon. Submitted to the U.S. Army Corps of Engineers, Seattle, Washington.

1995a Appendix A, Ancillary Studies and Tables from the Dunn Site (35HA1261): The Vertebrate Faunal Assemblage. In *Adaptive Transitions and Environmental Change in the Northern Great Basin: A View from Diamond Swamp,* edited by R. R. Musil, R. L. Greenspan, B. E. Hemphill, P. F. McDowell, and N. A. Stenholm, pp. 245–252. University of Oregon Anthropological Papers No. 51. Eugene.

1995b Appendix B, Ancillary Studies and Tables from the McCoy Creek Site (35HA1263): The Vertebrate Faunal Assemblage. In *Adaptive Transitions and Environmental Change in the Northern Great Basin: A View from Diamond Swamp,* edited by R. R. Musil, R. L. Greenspan, B. E. Hemphill, P. F. McDowell, and N. A. Stenholm, pp. 279–299. University of Oregon Anthropological Papers No. 51. Eugene.

Gruhn, R.

1961 *The Archaeology of Wilson Butte Cave, South-Central Idaho.* Occasional Papers of the Idaho State College Museum 6. Pocatello.

Guerrero, J., J. Martinez, and M. Isasa

1998 Mineral Nutrient Composition of Edible Wild Plants. *Journal of Food Composition and Analysis* 11:322–328.

Harper, A.

1999 Defining the Essentiality of Nutrients. In *Modern Nutrition in Health and Disease,* edited by M. Shils, J. Olson, M. Shike, and A. Ross, pp. 3–10. Lippincott, Williams and Wilkins, Philadelphia.

Harper, K. T., and G. M. Alder

1972 Paleoclimatic Inferences Concerning the Last 10,000 Years from a Resampling of Danger Cave, Utah. In *Great Basin Cultural Ecology: A Symposium,* edited by D. Fowler, pp. 13–23. Desert Research Institute Publications in the Social Sciences No. 8. Reno.

Harrington, M. R., and R. D. Simpson

1961 *Tule Springs, Nevada, with Other Evidence of Pleistocene Man in North America.* Southwest Museum Papers No. 18. Los Angeles.

Haws, J., and B. Hockett

2004 Theoretical Perspectives on the Dietary Role of Small Animals in Human Evolution. In *Petits Animaux et Societes Humaines du Complement Alimentaire aux Ressources Utilitaires,* edited by J.-P. Brugal and J. Desse, pp. 173–184. APDCA, Antibes.

Haynes, C. V., Jr.

1982 Were Clovis Progenitors in Beringia? In *Paleoecology of Beringia,* edited by D. M. Hopkins, J. V. Matthews, C. E. Schweger, and S. B. Young, pp. 383–398. Academic Press, New York.

1992 Contributions of Radiocarbon Dating to the Geochronology of the Peopling of the New World. In *Radiocarbon After Four Decades,* edited by R. E. Taylor, A. Long, and R. S. Kra, pp. 355–374. Springer-Verlag, New York.

Haynes, G.

1980 Evidence of Carnivore Gnawing on Pleistocene and Recent Mammalian Bones. *Paleobiology* 6:341–351.

1988 Spiral Fractures, Cutmarks, and Other Myths about Early Bone Assemblages. In *Early Human Occupation in Far Western North America: The Clovis-Archaic Interface,* edited by J. A. Willig, C. M. Aikens, and J. L. Fagan. Nevada State Museum Anthropological Papers No. 21. Carson City.

2002 *The Early Settlement of North America: The Clovis Era.* Cambridge University Press, Cambridge.

Heizer, R. F.
1967 Analysis of Human Coprolites from a Dry Nevada Cave. *University of California Archaeological Survey Reports* 70:1–20. Berkeley.
Heizer, R. F., and M. A. Baumhoff
1970 Big Game Hunters in the Great Basin: A Critical Review of the Evidence. In *Papers on the Anthropology of the Western Great Basin*, pp. 1–12. Contributions of the University of California Archaeological Research Facility No. 7. Berkeley.
Heizer, R. F., and L. K. Napton
1970 Archaeology and the Prehistoric Great Basin Lacustrine Subsistence Regime as Seen from Lovelock Cave, Nevada. *Contributions of the University of California Archaeological Research Facility* 10:1–202. Berkeley.
Hennig, W.
1950 *Grundzuge einer Theorie der Phylogenetischen Systematik*. Deutscher Zentralverlag, Berlin.
1966 *Phylogenetic Systematics*. University of Illinois Press, Urbana.
Hildebrant, W. R., and J. H. King
2002 Projectile Point Variability along the Northern California–Great Basin Interface: Results from the Tuscarora-Alturas Projects. In *Boundary Lands: Archaeological Investigations along the California-Great Basin Interface*, edited by K. R. McGuire, pp. 5–28. Nevada State Museum Anthropological Papers No. 24. Carson City.
Hockett, B.
1991 Toward Distinguishing Human and Raptor Patterning on Leporid Bones. *American Antiquity* 56:667–679.
1994 A Descriptive Reanalysis of the Leporid Bones from Hogup Cave, Utah. *Journal of California and Great Basin Anthropology* 16(1):106–117.
2000 Paleobiogeographic Changes at the Pleistocene-Holocene Boundary near Pintwater Cave, Southern Nevada. *Quaternary Research* 53:263–269.
2005 Middle and Late Holocene Hunting in the Great Basin: A Critical Review of the Debate and Future Prospects. *American Antiquity* 70:713–731.
Hockett, B., and J. Haws
2002 Taphonomic and Methodological Perspectives of Leporid Hunting During the Upper Paleolithic of the Western Mediterranean Basin. *Journal of Archaeological Method and Theory* 9:269–302.
2003 Nutritional Ecology and Diachronic Trends in Paleolithic Diet and Health. *Evolutionary Anthropology* 12:211–216.
2005 Nutritional Ecology and the Human Demography of Neandertal Extinction. *Quaternary International* 137:21–34.
Hoffecker, J. F., W. R. Powers, and T. Goebel
1993 The Colonization of Beringia and the Peopling of the New World. *Science* 259:46–53.
Holmer, R. N.
1986 Common Projectile Points of the Intermountain West. In *Anthropology of the Desert West: Essays in Honor of Jesse D. Jennings*, edited by C. J. Condie and D. D. Fowler, pp. 89–115. University of Utah Anthropological Papers No. 110. Salt Lake City.
Huckleberry, G., C. Beck, G. T. Jones, A. Holmes, M. Cannon, S. Livingston, and J. M. Broughton
2001 Terminal Pleistocene/Early Holocene Environmental Change at the Sunshine Locality, North-Central Nevada, U.S.A. *Quaternary Research* 55:303–312.
Hughes, R. E.
1984 Obsidian Sourcing Studies in the Great Basin: Problems and Prospects. In *Obsidian Studies in the Great Basin*, edited by R. E. Hughes, pp. 1–20. Contributions of the University of California Archaeological Research Facility No. 45. Berkeley.
1986 *Diachronic Variability in Obsidian Procurement Patterns in Northeastern California and South Central Oregon*. University of California Publications in Anthropology No. 17. Berkeley.

2002 Geochemical Research Laboratory Report on Source Analysis. In *Analysis of Artifacts from the Old River Bed Delta, Dugway Proving Ground, Utah*, by G. T. Jones, C. Beck, and R. A. Kessler, Appendix A, pp. 1–3. Hamilton College, Clinton, New York. On file, Directorate of Environmental Programs, U.S. Army Dugway Proving Ground, Utah.

Hunt, C. B.

1967 *Physiography of the United States*. W. H. Freeman and Co., San Francisco.

Janetski, J. C.

1986 The Great Basin Lacustrine Subsistence Pattern: Insights from Utah Valley. In *Anthropology of the Desert West: Essays in Honor of Jesse D. Jennings*, edited by C. J. Condie and D. D. Fowler, pp. 145–168. University of Utah Anthropological Papers No. 116. Salt Lake City.

Janetski, J. C., and D. B. Madsen (editors)

1990 *Wetland Adaptations in the Great Basin*. Museum of Peoples and Cultures Occasional Papers No. 1. Brigham Young University, Provo.

Jefferson, G., H. McDonald, W. Akersten, and S. J. Miller

2002 Catalogue of Late Pleistocene and Holocene Fossil Vertebrates from Idaho. In *And Whereas... Papers on the Vertebrate Paleontology of Idaho Honoring John A. White*, Vol. 2, edited by W. Akersten, M. Thompson, D. Meldrum, R. Rapp, and H. McDonald, pp. 157–192. Idaho Museum of Natural History Occasional Paper 37. Pocatello.

Jenkins, D. L.

1994 Settlement-Subsistence Patterns in the Fort Rock Basin: A Cultural-Ecological Perspective on Human Responses to Fluctuating Wetlands Resources of the Last 5,000 Years. In *Archaeological Researches in the Northern Great Basin: Fort Rock Archaeology Since Cressman*, edited by C. M. Aikens and D. L. Jenkins, pp. 599–628. University of Oregon Anthropological Papers No. 50. Eugene.

2000 Early to Middle Holocene Cultural Transitions in the Northern Great Basin of Oregon: The View from Fort Rock. In *Archaeological Passages: A Volume in Honor of Claude Nelson Warren*, edited by J. S. Schneider, R. M. Yohe II, and J. Gardner, pp. 69–109. Western Center for Archaeology and Paleontology, Publications in Archaeology Vol. 1. Hemet, California.

2004 The Grasshopper and the Ant: Middle Holocene Occupations and Storage Behavior at the Bowling Dune Site in the Fort Rock Basin, Oregon. In *Early and Middle Holocene Archaeology of the Northern Great Basin*, edited by D. L. Jenkins, T. J. Connolly, and C. M. Aikens, pp. 123–156. University of Oregon Anthropological Papers No. 62. Eugene.

2005 Distribution and Dating of Pleistocene Fauna and Cultural Remains at Paisley 5 Mile Point Caves, South Central Oregon. Paper presented at the 70th meeting of the Society for American Archaeology, Salt Lake City.

Jenkins, D. L., T. J. Connolly, and C. M. Aikens

2004 Early and Middle Holocene Archaeology in the Northern Great Basin: Dynamic Natural and Cultural Ecologies. In *Early and Middle Holocene of the Northern Great Basin*, edited by D. L. Jenkins, T. J. Connolly, and C. M. Aikens, pp. 1–20. University of Oregon Anthropological Papers No. 62. Eugene.

Jenkins, D. L., and C. E. Skinner

2004 Obsidian Source Characterization and Hydration Analyses at the Paisley 5 Mile Point Caves (35LK3400): A Pleistocene-Holocene Site in the Northern Great Basin. Paper presented at the 29th Great Basin Anthropological Conference, Sparks, Nevada.

Jennings, J. D.

1957 *Danger Cave*. University of Utah Anthropological Papers No. 27. Salt Lake City.

1978 *Prehistory of Utah and the Eastern Great Basin*. University of Utah Anthropological Papers No. 98. Salt Lake City.

1986 Prehistory: Introduction. In *Great Basin*, edited by W. L. d'Azevedo, pp. 113–119. Handbook of North American Indians, Vol. 11 (W. C. Sturtevant, general editor). Smithsonian Institution, Washington, D.C.

1989 *Prehistory of North America.* 3rd ed. Mayfield Publishing Company, Mountain View, California.

Jochim, M., C. Herhahn, and H. Starr

1999 The Magdalenian Colonization of Southern Germany. *American Anthropologist* 101: 129–142.

Jones, G. T., and C. Beck

1999 Paleoarchaic Archaeology in the Great Basin. In *Models for the Millennium: Great Basin Anthropology Today*, edited by C. Beck, pp. 83–95. University of Utah Press, Salt Lake City.

Jones, G. T., C. Beck, E. E. Jones, and R. E. Hughes

2003 Lithic Source Use and Paleoarchaic Foraging Territories in the Great Basin. *American Antiquity* 68:5–38.

Jones, G. T., C. Beck, and R. A. Kessler

2002 *Analysis of Artifacts from the Old River Bed Delta, Dugway Proving Ground, Utah.* Hamilton College, Clinton, New York. Submitted to the Directorate of Environmental Programs, U.S. Army Dugway Proving Ground, Utah.

Jones, G. T., C. Beck, and P. D. LeTourneau

1996 A Possible Association Between *Camelops* cf. *hesternus* and Lithic Artifacts from the Sunshine Locality in Eastern Nevada. *Current Research in the Pleistocene* 13: 27–29.

Jones, K. T., and D. B. Madsen

1989 Calculating the Cost of Resource Transportation: A Great Basin Example. *Current Anthropology* 30:529–534.

Jull, A. J. T., D. J. Donahue, and T. H. Zabel

1983 Target Preparation for Radiocarbon Dating by Tandem Accelerator Mass Spectrometry. *Nuclear Instruments and Methods in Physics Research* 218:509–514.

Keen, C., M. Clegg, L. Hanna, L. Lanoue, J. Rogers, G. Daston, P. Oteiza, and J. Uriu-Adams

2003 The Plausibility of Micronutrient Deficiencies Being a Significant Contributing Factor to the Occurrence of Pregnancy Complications. *Journal of Nutrition* 133:1597S–1605S.

Kelly, I. T.

1964 *Southern Paiute Ethnography.* University of Utah Anthropological Papers No. 69. Salt Lake City.

Kelly, R. L.

1988 Three Sides of a Biface. *American Antiquity* 53:717–734.

1992 Mobility/Sedentism: Concepts, Archaeological Measures, and Effects. *Annual Review of Anthropology* 21:43–66.

1995 *The Foraging Spectrum: Diversity in Hunter-Gatherer Lifeways.* Smithsonian Institution Press, Washington, D.C.

1996 Ethnographic Analogy and Migration to the Western Hemisphere. In *Prehistoric Mongoloid Dispersals*, edited by T. Akazawa and E. J. E. Szathmáry, pp. 228–240. Oxford University Press, Oxford.

2001 *Prehistory of the Carson Desert and Stillwater Mountains: Environment, Mobility, and Subsistence in a Great Basin Wetland.* University of Utah Anthropological Papers No. 123. Salt Lake City.

Kelly, R. L., and L. C. Todd

1988 Coming into the Country: Early Paleoindian Hunting and Mobility. *American Antiquity* 53:231–244.

Kilby, J. D., and B. B. Huckell

2003 A Comparison of Caches: An Initial Look at Regional Variation in Clovis Caching. Paper presented at the 68th meeting of the Society for American Archaeology, Milwaukee.

n.d. A Comparison of Caches: An Initial Look at Regional Variation in Clovis Caching. In *The Clovis Age Continent: Early Paleoindian Foragers in North America*, edited by D. Kilby, B. Buchannan, and M. J. Hamilton. University of New Mexico Press, Albuquerque (in review).

King, G. Q.
1986 Morphological Regions of the Great Basin. *Forum of the Association for Arid Land Studies* 2:65–72.
Kitching, I. J., P. L. Forey, C. J. Humphries, and D. M. Williams
1998 *Cladistics: The Theory and Practice of Parsimony Analysis*. Oxford University Press, Oxford.
Krieger, A. D.
1944 Review of *Archaeological Researches in the Northern Great Basin*, by L. S. Cressman, F. C. Baker, P. S. Conger, H. P. Hanson, and R. F. Heizer. *American Antiquity* 9:351–359.
1964 Early Man in the New World. In *Prehistoric Man in the New World*, edited by J. D. Jennings and E. Norbeck, pp. 23–81. University of Chicago Press, Chicago.
Kuhn, S. L.
1990 A Geometric Index of Reduction for Unifacial Stone Tools. *Journal of Archaeological Science* 17:583–593.
1994 A Formal Approach to the Design and Assembly of Mobile Toolkits. *American Antiquity* 59:426–442.
1995 *Mousterian Lithic Technology: An Ecological Perspective*. Princeton University Press, Princeton.
Kuhnlein, H., H. Chan, D. Leggee, and V. Barthet
2002 Macronutrient, Mineral, and Fatty Acid Composition of Canadian Arctic Traditional Food. *Journal of Food Composition and Analysis* 15:545–566.
Kuntz, M., H. Covington, and L. Schorr
1992 An Overview of Basaltic Volcanism of the Eastern Snake River Plain, Idaho. In *Regional Geology of Eastern Idaho and Western Wyoming*, edited by P. Link, M. Kuntz, and L. Platt, pp. 227–267. Geological Society of America Memoir 179. Washington, D.C.
Lagiou, P., L. Mucci, R. Tamimi, H. Kuper, A. Lagiou, C. Hsieh, and D. Trichopoulos
2005 Micronutrient Intake During Pregnancy in Relation to Birth Size. *European Journal of Nutrition* 44:52–59.
Larson, C. S., C. B. Rugff, and R. L. Kelly
1995 Structural Analysis of the Stillwater Postcranial Human Remains: Behavioral Implications of Articular Joint Pathology in Long Bone Diaphyseal Morphology. In *Bioarchaeology of the Stillwater Marsh: Prehistory Human Adaptation in the Western Great Basin*, by C. S. Larson and R. L. Kelly, pp. 107–133. Anthropological Papers of the American Museum of Natural History Vol. 77, Pt. 1. New York.
Layton, T. N.
1970 High Rock Archaeology: An Interpretation of the Prehistory of the Northwestern Great Basin. Ph.D. dissertation, Harvard University, Cambridge.
1979 Archaeology and Paleo-Ecology of Pluvial Lake Parman, Northwestern Great Basin. *Journal of New World Archaeology* 3(3):41–56.
Layton, T. N., and J. O. Davis
n.d. Last Supper Cave: Early Post-Pleistocene Cultural History and Paleoecology in the High Rock Country of the Northwestern Great Basin. Unpublished manuscript.
Lee, R. B., and I. DeVore
1968 *Man the Hunter*. Aldine, Chicago.
Leonardy, F. C., and D. G. Rice
1970 A Proposed Culture Typology for the Lower Snake River Region, Southeastern Washington. *Northwest Anthropological Research Notes* 4:1–29.
Lockett, C., C. Calvert, and L. Grivetti
2000 Energy and Micronutrient Composition of Dietary and Medicinal Wild Plants Consumed During Drought. Study of rural Fulani, northeastern Nigeria. *International Journal of Food Sciences and Nutrition* 51:195–208.
Long, A., and B. Rippeteau
1974 Testing Contemporaneity and Averaging Radiocarbon Dates. *American Antiquity* 39: 205–215.
Long, A., A. T. Wilson, R. D. Ernst, and B. H. Gore
1989 AMS Radiocarbon Dating of Bones in Arizona. *Radiocarbon* 31:231–238.

Lupo, K. D., and D. N. Schmitt
1997 Experiments in Bone Boiling: Nutritional Returns and Archaeological Reflections. *Anthropozoologica* 25:137–144.

Lyman, R. L., and M. J. O'Brien
2000 Chronometers and Units in Early Archaeology and Paleontology. *American Antiquity* 65:691–707.

Lyman, R. L., M. J. O'Brien, and R. C. Dunnell
1997 *The Rise and Fall of Culture History*. Plenum, New York.

MacArthur, R. H., and E. R. Pianka
1966 On Optimal Use of a Patchy Environment. *American Naturalist* 100:603–609.

Madsen, D. B.
1982 Get It Where the Gettin's Good: A Variable Model of Great Basin Subsistence and Settlement Based on Data from the Eastern Great Basin. In *Man and Environment in the Great Basin*, edited by D. B. Madsen and J. F. O'Connell, pp. 207–226. SAA Papers No. 2. Society for American Archaeology, Washington, D.C.
1999 The Nature of Great Basin Environmental Change During the Pleistocene/Holocene Transition and Its Possible Impact on Human Populations. In *Models for the Millennium: The Current Status of Great Basin Anthropological Research*, edited by C. Beck, pp. 75–82. University of Utah Press, Salt Lake City.
2000 *Late Quaternary Paleoecology in the Bonneville Basin*. Utah Geological Survey Bulletin 130. Salt Lake City.
2002 Great Basin Peoples and Late Quaternary Aquatic History. In *Great Basin Aquatic Systems History*, edited by R. Hershler, D. B. Madsen, and D. R. Currey, pp. 387–405. Smithsonian Contributions to Earth Sciences No. 33. Smithsonian Institution Press, Washington, D.C.
2004 Colonization of the Americas Before the Last Glacial Maximum: Issues and Problems. In *Entering America: Northeast Asia and Beringia Before the Last Glacial Maximum*, edited by D. B. Madsen, pp. 1–26. University of Utah Press, Salt Lake City.

Madsen, D. B., and D. R. Currey
1979 Late Quaternary Glacial and Vegetation Changes, Little Cottonwood Canyon Area, Wasatch Mountains, Utah. *Quaternary Research* 12:254–270.

Madsen, D. B., R. Hershler, and D. R. Currey
2002 Introduction. In *Great Basin Aquatic Systems History,* edited by R. Hershler, D. B. Madsen, and D. R. Currey, pp. 1–10. Smithsonian Contributions to the Earth Sciences No. 33. Smithsonian Institution Press, Washington, D.C.

Madsen, D. B., and J. Kirkman
1988 Hunting Hoppers. *American Antiquity* 53:593–604.

Madsen, D. B., D. Page, and D. N. Schmitt
2005 *Archaeological Survey of Additional Channel Remnants in the Old River Bed Delta, Dugway Proving Ground, Western Utah*. Desert Research Institute, Reno, Nevada. Submitted to the Directorate of Environmental Programs, U.S. Army Dugway Proving Ground, Utah.

Madsen, D. B., and D. R. Rhode
1990 Early Holocene Pinon (*Pinus monophylla*) in the Northeastern Great Basin. *Quaternary Research* 33:94–101.

Madsen, D. B., D. Rhode, D. K. Grayson, J. M. Broughton, S. D. Livingston, J. M. Hunt, J. Quade, D. N. Schmitt, and M. W. Shaver III
2001 Late Quaternary Environmental Change in the Bonneville Basin, Western U.S.A. *Palaeogeography, Palaeoclimatology, Palaeoecology* 167:243–271.

Madsen, D. B., and D. N. Schmitt
2005 *Buzz-Cut Dune and Fremont Foraging at the Margin of Horticulture*. University of Utah Anthropological Papers No. 124. Salt Lake City.

Madsen, D. B., D. N. Schmitt, and J. M. Hunt
2000 *Archaeological Evaluation of Areas Associated with the Gilbert Shoreline and Old River Bed Delta, Dugway Proving Ground, Utah*. Utah Geological Survey, Salt Lake City.

Submitted to the Directorate of Environmental Programs, U.S. Army Dugway Proving Ground, Utah.

Marks, A.

2000 Upper Paleolithic Occupation of Portugal: Residents vs. Visitors. In *Paleolitico da Peninsula Iberica*, Vol. 2, edited by V. Jorge, pp. 341–349. ADECAP, Porto.

Mayewski, P. A., L. D. Meeker, S. Whitlow, M. S. Twickler, M. C. Morrison, R. B. Alley, P. Bloomfield, and K. Taylor

1993 The Atmosphere During the Younger Dryas. *Science* 261:195–197.

McGuire, K. R.

2002 Obsidian Production in Northeastern California and the Northwestern Great Basin: Implications for Land Use. In *Boundary Lands: Archaeological Investigations Along the California–Great Basin Interface*, edited by K. R. McGuire, pp. 85–103. Nevada State Museum Anthropological Papers No. 24. Carson City.

Meatte, D., J. Woods, and G. Titmus

1989 *Kelvin's Cave Preliminary Report, May 1989*. On file, Bureau of Land Management, Shoshone, Idaho.

Mehringer, P. J.

1977 Great Basin Late Quaternary Environments and Chronology. In *Models and Great Basin Prehistory: A Symposium*, edited by D. D. Fowler, pp. 113–167. Desert Research Institute Publications in the Social Sciences No. 12. University of Nevada, Reno.

Mehringer, P. J., Jr., and F. F. Foit Jr.

1990 Volcanic Ash Dating of the Clovis Cache at East Wenatchee, Washington. *National Geographic Research* 6:495–503.

Meltzer, D. J.

1984 Late Pleistocene Human Adaptations in Eastern North America. Ph.D. dissertation, University of Washington, Seattle.

1988 Late Pleistocene Human Adaptations in Eastern North America. *Journal of World Prehistory* 2:1–52.

1993 Is There a Clovis Adaptation? In *From Kostenki to Clovis*, edited by O. Soffer and N. D. Praslov, pp. 293–310. Plenum, New York.

Metcalfe, D., and K. R. Barlow

1992 A Model for Exploring the Optimal Trade-off Between Field Processing and Transport. *American Anthropologist* 94:340–356.

Mifflin, M. D., and M. M. Wheat

1979 *Pluvial Lakes and Estimated Pluvial Climates of Nevada*. Nevada Bureau of Mines and Geology Bulletin 94. University of Nevada, Reno.

Milewski, A., and R. Diamond

2000 Why Are Very Large Herbivores Absent from Australia? A New Theory of Micronutrients. *Journal of Biogeography* 27:957–978.

Miller, S. J.

1979 The Archaeological Fauna of Four Sites in Smith Creek Canyon. In *The Archaeology of Smith Creek Canyon, Eastern Nevada*, edited by D. R. Tuohy and D. L. Rendall, pp. 272–329. Nevada State Anthropological Papers No. 17. Carson City.

1982 The Archaeology and Geology of an Extinct Megafauna/Fluted-point Association at Owl Cave, the Wasden Site, Idaho: A Preliminary Report. In *Peopling of the New World*, edited by J. Ericson, R. Taylor, and R. Berger, pp. 81–95. Ballena Press Anthropological Papers 23. Malki Museum, Banning, California.

Miller, S. J., and W. Dort Jr.

1978 Early Man at Owl Cave: Current Investigations at the Wasden Site, Eastern Snake River Plain, Idaho. In *Early Man in America from a Circum-Pacific Perspective*, edited by A. Bryan, pp. 129–139. Department of Anthropology, University of Calgary Occasional Papers 1. Calgary.

Minor, R., and L. Spencer

1977 *Site of a Probable Camelid Kill at Fossil Lake, Oregon: An Archaeological Evaluation*. Department of Anthropology, University of Oregon, Eugene. Submitted to the Bureau of Land Management, Lakeview, Oregon.

Moore, M.

1993 Faunal Analysis. In *Archaeological Investigations in Warner Valley, Oregon, 1989–1992 An Interim Report*, edited by D. D. Fowler, pp. 67–80. Department of Anthropology Technical Report 93-1. University of Nevada, Reno.

Moore, M. J.

1999 *Prehistoric Land-use Pattern Changes in the Vicinity of Beatty's Butte, Southeastern Oregon*. Department of Anthropology, Sundance Archaeological Research Fund Technical Paper No. 6. University of Nevada, Reno.

Morrow, J. E., and T. A. Morrow

1999 Geographic Variation in Fluted Projectile Points: A Hemispheric Perspective. *American Antiquity* 64:215–231.

Mosimann, J. E., and P. S. Martin

1975 Simulating Overkill by Paleoindians. *American Scientist* 63:304–313.

Napton, L. K.

1997 The Spirit Cave Mummy: Coprolite Investigations. *Nevada Historical Quarterly* 40: 97–107.

Negrini, R. M.

2002 Pluvial Lake Sizes in the Northwestern Great Basin Throughout the Quaternary Period. In *Great Basin Aquatic Systems History,* edited by R. Hershler, D. B. Madsen, and D. Currey, pp. 11–52. Smithsonian Contributions to the Earth Sciences No. 33. Smithsonian Institution Press, Washington, D.C.

O'Brien, M. J., and R. L. Lyman

2003 *Cladistics and Archaeology*. University of Utah Press, Salt Lake City.

O'Brien, M. J., J. Darwent, and R. L. Lyman

2001 Cladistics Is Useful for Reconstructing Archaeological Phylogenies: Palaeoindian Points from the Southeastern United States. *Journal of Archaeological Science* 28:1115–1136.

O'Connell, J. F., and C. M. Inoway

1994 Surprise Valley Projectile Points and Their Chronological Implications. *Journal of California and Great Basin Anthropology* 16:162–198.

O'Connell, J. F., K. T. Jones, and S. R. Simms

1982 Some Thoughts on Prehistoric Archaeology in the Great Basin. In *Man and Environment in the Great Basin*, edited by D. B. Madsen and J. F. O'Connell, pp. 227–240. SAA Papers No. 2. Society of American Archaeology, Washington, D.C.

O'Connor, T.

2000 *The Archaeology of Animal Bones*. Texas A&M University Anthropology Series No. 4. University Press, College Station.

Odell, G. H.

1996 Economizing Behavior and the Concept of "Curation." In *Theoretical Insights into Human Prehistory*, edited by G. H. Odell, pp. 51–80. Plenum Press, New York.

Oetting, A. C.

1989 Villages and Wetlands Adaptations in the Northern Great Basin: Chronology and Land Use in the Lake Abert–Chewaucan Marsh Basin, Lake County, Oregon. Ph.D. dissertation, University of Oregon.

1993 *The Archaeology of Buffalo Flat: Cultural Resources Investigations for the CONUS OTH-B Buffalo Flat Radar Transmitter Site, Christmas Lake Valley, Oregon*. Historical Research Associates Report No. 151. Eugene, Oregon. Submitted to the U.S. Army Corps of Engineers, Seattle, Washington.

O'Grady, P.

2004 Zooarchaeological Analysis of Cultural Features from Four Early to Middle Holocene Sites in the Fort Rock Basin. In *Early and Middle Holocene Archaeology of the Northern Great Basin*, edited by D. L. Jenkins, T. J. Connolly, and C. M. Aikens, pp. 187–208. University of Oregon Anthropological Papers No. 62. Eugene.

Orr, P. C.

1956 *Pleistocene Man in Fishbone Cave, Pershing County, Nevada*. Nevada State Museum, Department of Anthropology Bulletin No. 2. Carson City.

Overpeck, J., D. Anderson, S. Trumbore, and W. Prell
1996 The Southwest Indian Monsoon Over the Last 18,000 Years. *Climate Dynamics* 12:213–225.
Oviatt, C. G.
1988 Late Pleistocene and Holocene Lake Fluctuations in the Sevier Lake Basin, Utah, USA. *Journal of Paleolimnology* 1:9–21.
1997 Lake Bonneville Fluctuations and Global Climate Change. *Geology* 25:155–158.
1999 *Surficial Deposits and Late Quaternary History of Dugway Proving Ground, Utah: Preliminary Notes.* Utah Geological Survey, Salt Lake City. Submitted to the Directorate of Environmental Programs, Dugway Proving Ground, Utah.
Oviatt, C. G., D. R. Currey, and D. Sack
1992 Radiocarbon Chronology of Lake Bonneville, Eastern Great Basin, USA. *Palaeogeography, Palaeoclimatology, Palaeoecology* 99:225–241.
Oviatt, C. G., D. B. Madsen, and D. N. Schmitt
2003 Late Pleistocene and Early Holocene Rivers and Wetlands in the Bonneville Basin of Western North America. *Quaternary Research* 60:200–210.
Oviatt, C. G., D. M. Miller, J. P. McGeehin, C. Zachary, and S. Mahan
2005 The Younger Dryas Phase of The Great Salt Lake, Utah, USA. *Palaeogeography, Palaeoclimatology, Palaeoecology* 219:263–284.
Page, D., S. Ferguson, K. Varley, S. Rumsey, D. Rhode, D. N. Schmitt, and C. Martin
2003 *A Class III Cultural Resource Inventory in Support of Stryker OT-Phase II, Dugway Proving Ground, Utah.* Desert Research Institute, Las Vegas, Nevada. Submitted to the Directorate of Environmental Programs, U.S. Army Dugway Proving Ground, Utah.
Parr, R. E.
2006 *Protein Residue Analysis of Coprolites from the Paisley 5 Mile Point Cave Site (35LK3400) Lake County, Oregon.* Laboratory of Archaeological Sciences Report LAS-104. California State University, Bakersfield.
Pendleton, L. S.
1979 Lithic Technology in Early Nevada Assemblages. M.A. thesis, California State University, Long Beach.
Pettigrew, R. M.
1984 Prehistoric Human Land-use Patterns in the Alvord Basin, Southeastern Oregon. *Journal of California and Great Basin Anthropology* 6:61–90.
Pettigrew, R. M., and C. M. Hodges
1995 Prehistoric Hunter-Gatherer Land-Use Systems: Pacific Northwest. In *Archaeological Investigations PGT-PG&E Pipeline Expansion Project Idaho, Washington, Oregon, and California,* Volume IV: *Synthesis and Findings*, edited by R. F. Schalk, pp. 2-1–2-70. M. J. Moratto, general editor. Pacific Gas Transmission Company, Portland, Oregon.
Pinson, A. O.
1999 *Foraging in Uncertain Times: The Effects of Risk on Subsistence Behavior During the Pleistocene-Holocene Transition in the Oregon Great Basin.* Ph.D. dissertation, University of New Mexico, Albuquerque. University Microfilms, Ann Arbor, Michigan.
2004 Of Lakeshores and Dry Basin Floors: A Regional Perspective on the Early Holocene Record of Environmental Change and Human Adaptation at the Tucker Site. In *Early and Middle Holocene Archaeology of the Northern Great Basin*, edited by D. L. Jenkins, T. J. Connolly, and C. M. Aikens, pp. 53–76. University of Oregon Anthropological Papers No. 62. Eugene.
Pitblado, B. L.
2003 *Late Paleoindian Occupation of the Southern Rocky Mountains: Early Holocene Projectile Points and Land Use in the High Country.* University Press of Colorado, Boulder.
Platnick, N. I., and D. Cameron
1977 Cladistic Methods in Textual, Linguistic, and Phylogenetic Analysis. *Systematic Zoology* 26:380–385.
Preston, R. S., E. Person, and E. S. Deevey
1955 Yale Natural Radiocarbon Measurements, Pt. 2. *Science* 122:954–960.

Ramakrishnan, U., R. Manjrekar, J. Rivera, T. Gonzales-Cossia, and R. Martorell

1999 Micronutrients and Pregnancy Outcome: A Review of the Literature. *Nutrition Research* 19(1):103–159.

Ramos-Elorduy, J., J. Moreno, E. Prado, M. Perez, J. Otero, and O. Guevara

1997 Nutritional Value of Edible Insects from the State of Oaxaca, Mexico. *Journal of Food Composition and Analysis* 10:142–157.

Raven, C.

1990 *Prehistoric Human Geography in the Carson Desert, Part 2: Archaeological Field Tests of Model Predictions*. U.S. Fish and Wildlife Service Cultural Resource Series No. 4. Portland, Oregon.

1993 Prehistoric Wetland Adaptations in the Great Basin and the Special Case of Malheur Lake. In *Land and Life at Malheur Lake: Preliminary Geomorphological and Archaeological Investigations*, by C. Raven and R. G. Elston, pp. 5–12. U.S. Fish and Wildlife Service Cultural Resource Series No. 6. Portland, Oregon.

Raven, C., and R. G. Elston (editors)

1988 *Preliminary Investigations in Stillwater Marsh: Human Prehistory and Geoarchaeology*. U.S. Fish and Wildlife Service Cultural Resource Series No. 1. Portland, Oregon.

Reimer, P.

2004 INTCAL04 Terrestrial Radiocarbon Age Calibration, 0-26 CAL kyr BP. *Radiocarbon* 46:1029–1058.

Rhode, D.

1990 Transportation Costs of Great Basin Resources: An Assessment of the Jones-Madsen Model. *Current Anthropology* 31:413–419.

2000 Holocene Vegetation History in the Bonneville Basin. In *Late Quaternary Paleoecology in the Bonneville Basin*, by D. B. Madsen, pp. 149–164. Bulletin 130. Utah Geological Survey, Salt Lake City.

2002 *Native Plants of Southern Nevada: An Ethnobotany*. University of Utah Press, Salt Lake City.

2006 Prehistoric Dietary Plant Use in the Lead Mine Hills: The Record from Hearths and Other Features at Bonneville Estates Rockshelter. Paper presented at the 30th Great Basin Anthropological Conference, Las Vegas, Nevada.

Rhode, D., T. Goebel, K. Graf, B. Hockett, K. T. Jones, D. B. Madsen, C. G. Oviatt, and D. N. Schmitt

2005 Latest Pleistocene–Early Holocene Human Occupation and Paleoenvironmental Change in the Bonneville Basin, Utah-Nevada. In *Interior Western United States*, edited by J. Pederson and C. Dehler, pp. 211–230. Geological Society of America Field Guide 6. Boulder, Colorado.

Rhode, D., and D. B. Madsen

1998 Pine Nut Use in the Early Holocene and Beyond: The Danger Cave Archaeobotanical Record. *Journal of Archaeological Science* 25: 1199–1210.

Rhode, D., D. B. Madsen, and K. T. Jones

2006 Antiquity of Early Holocene Small-Seed Consumption and Processing at Danger Cave. *Antiquity* 80:328–339.

Ricks, M. F., and W. J. Cannon

1998 Rock Art as an Indicator of Early Upland Aggregation Sites in the Northern Great Basin. Paper presented at the 63rd meeting of the Society for American Archaeology, Seattle, Washington.

Ridings, R.

1991 Obsidian Hydration Dating: The Effects of Mean Exponential Ground Temperature and Depth of Artifact Recovery. *Journal of Field Archaeology* 18:77–85.

1996 Where in the World Does Obsidian Hydration Dating Work? *American Antiquity* 61: 136–148.

Rogers, M. J.
1958 San Dieguito Implements from the Terraces of the Rincon-Pantano and Rillito Drainage System. *The Kiva* 24:1–23.
Ross, M.
1989 Early Oceanic Linguistic Prehistory. *Journal of Pacific History* 24:135–149.
Sanchez, M. G.
2001 A Synopsis of Paleo-indian Archaeology in Mexico. *The Kiva* 67:119–136.
Schmidt, G., and A. LeGrande
2005 The Goldilocks Abrupt Climate Change Event. *Quaternary Science Reviews* 24:1109–1110.
Schmitt, D. N.
1986 Faunal Analysis. In *The Archaeology of the Vista Site, 26Wa3017*, edited by C. D. Zeier and R. G. Elston, pp. 209–239. Intermountain Research, Silver City, Nevada. Submitted to the Nevada Department of Transportation, Carson City.
1988 Some Observations on Vertebrate Taphonomy and Site Formational Processes in Stillwater Marsh. In *Preliminary Investigations in Stillwater Marsh: Human Prehistory and Geoarchaeology*, edited by C. Raven and R. G. Elston, pp. 359–372. U.S. Fish and Wildlife Service Cultural Resource Series No. 1. Portland, Oregon.
Schmitt, D. N., and D. B. Madsen
2005 *Camels Back Cave*. University of Utah Anthropological Papers No. 125. Salt Lake City.
Schmitt, D. N., D. B. Madsen, K. Callister, and R. Quist
2003 *Archaeological Survey of Areas Associated with Lake Öferneet and Additional Old River Bed Deltaic Channel Features, Dugway Proving Ground, Utah*. Desert Research Institute, Reno, Nevada. Submitted to the Directorate of Environmental Programs, U.S. Army Dugway Proving Ground, Utah.
Schmitt, D. N., D. B. Madsen, J. M. Hunt, K. Callister, K. Jensen, and R. Quist
2002a *Archaeological Inventories of Areas Associated with West Baker Dunes in the Old River Bed Delta at U.S. Army Dugway Proving Ground, Utah*. Utah Geological Survey. Submitted to the Directorate of Environmental Programs, U.S. Army Dugway Proving Ground, Utah.
Schmitt, D. N., D. B. Madsen, and K. D. Lupo
2002b Small-Mammal Data on Early and Middle Holocene Climates and Biotic Communities in the Bonneville Basin, U.S.A. *Quaternary Research* 58:255–260.
2004 The Worst of Times, the Best of Times: Jackrabbit Hunting by Middle Holocene Human Foragers in the Bonneville Basin of Western North America. In *Colonisation, Migration, and Marginal Areas: A Zooarchaeological Approach*, edited by M. Mondini, S. Munoz, and S. Wickler, pp. 86–95. Oxbow Books, Oxford.
Schott, M. J.
1993 *The Leavitt Site: A Parkhill Phase Paleo-Indian Occupation in Central Michigan*. Memoirs of the Museum of Anthropology No. 2. University of Michigan, Ann Arbor.
Schroedl, A. R., and N. J. Coulam
1989 *Bonneville Estates Rockshelter: Cultural Resources Report 435-01-8906*. P-III Associates, Salt Lake City, Utah. Submitted to Bureau of Land Management, Elko, Nevada.
Schwarcz, H., and M. Schoeninger
1991 Stable Isotope Analyses in Human Nutritional Ecology. *Yearbook of Physical Anthropology* 34:283–321.
Shutler, R., Jr.
1967 Archaeology of Tule Springs. In *Pleistocene Studies in Southern Nevada*, edited by H. M. Wormington and D. Ellis, pp. 298–303. Nevada State Museum Anthropological Papers No. 13. Carson City.
Simms, S. R.
1985 Acquisition Cost and Nutritional Data on Great Basin Resources. *Journal of California and Great Basin Anthropology* 7:117–126.

1987 *Behavioral Ecology and Hunter-Gatherer Foraging*. BAR International Series 318. British Archaeological Reports, Oxford.

1988 Conceptualizing the Paleo-Indian and Archaic in the Great Basin. In *Early Human Occupation in Far Western North America: The Clovis-Archaic Interface*, edited by J. A. Willig, C. M. Aikens, and J. L. Fagan, pp. 41–52. Nevada State Museum Anthropological Papers No. 21. Carson City.

Simpson, R. D.

1956 An Introduction to Early Western American Prehistory. *Bulletin of Southern California Academy of Sciences* 55:61–71.

Singer, V. J.

2004 Faunal Assemblages of Four Early to Middle Holocene Marsh-side Sites in the Fort Rock Valley, South Central Oregon. In *Early and Middle Holocene Archaeology of the Northern Great Basin*, edited by D. L. Jenkins, T. J. Connolly, and C. M. Aikens, pp. 167–185. University of Oregon Anthropological Papers No. 62. Eugene.

Skinner, C. E., J. J. Thatcher, D. L. Jenkins, and A. C. Oetting

2004 X-Rays, Artifacts, and Procurement Ranges: A Mid-Project Snapshot of Prehistoric Obsidian Procurement Patterns in the Fort Rock Basin. In *Early and Middle Holocene Archaeology of the Northern Great Basin*, edited by D. L. Jenkins, T. J. Connolly, and C. M. Aikens, pp. 221–232. University of Oregon Anthropological Papers No. 62. Eugene.

Smith, G. I., and F. A. Street-Perrott

1983 Pluvial Lakes of the Western United States. In *Late Quaternary Environments of United States*, Vol. 1: *The Late Pleistocene*, edited by S. C. Porter, pp. 190–212. University of Arizona Press, Tucson.

Smith, G. M.

2005 The Paleoarchaic Occupation of Moonshine Spring South and Moonshadow Spring, Pershing County, Nevada: Implications for Early-Period Mobility in the Great Basin. *Nevada Archaeologist* 20–21:57–70.

2006 Pre-Archaic Mobility, Settlement Systems, and Technological Organization: A View from the Parman Localities, Humboldt County, Nevada. M. A. thesis, University of Nevada, Reno.

Smith, G. M., M. Melson, and T. Goebel

2004 *Parman Lake Revisited: The Paleoarchaic Occupation of Pluvial Lake Parman, Humboldt County, Nevada*. Sundance Archaeological Research Fund, Department of Anthropology, University of Nevada, Reno. Submitted to the Bureau of Land Management, Winnemucca.

Sokal, R. R., and P. H. A. Sneath

1963 *Principles of Numeric Taxonomy*. Freeman, San Francisco.

Solomons, N.

2000 Micronutrients and Infection. *Nutrition* 16:1093–1095.

Stein, J. K., and W. R. Farrand

2001 *Sediments in Archaeological Context*. University of Utah Press, Salt Lake City.

Stevenson, C. M., M. Gottesman, and M. Macko

2000 Redefining the Working Assumptions of Obsidian Hydration Dating. *Journal of California and Great Basin Anthropology* 2:223–236.

Steward, J. H.

1938 *Basin-Plateau Aboriginal Sociopolitical Groups*. Bureau of American Ethnology Bulletin No. 120. Washington, D.C.

Stuiver, M., and P. M. Grootes

2000 GISP2 Oxygen Isotope Ratios. *Quaternary Research* 53:277–284.

Stuiver, M. P., J. Reimer, and R. Reimer

2006 calib.qub.ec.uk\calib\execution version 5.0.2 html.

Swanson, E. H.
1972 *Birch Creek: Human Ecology in the Cool Desert of the Northern Rocky Mountains 9,000 B.C.–A.D. 1850*. Idaho State University Press, Pocatello.

Taylor, A.
2002 Results of a Great Basin Fluted Point Survey: Chronological and Functional Relationships Between Fluted and Stemmed Points. Archaeology senior thesis, Hamilton College, Clinton, New York.
2003 Results of a Great Basin Fluted-Point Survey. *Current Research in the Pleistocene* 20: 77–79.

Taylor, R. E., C. V. Haynes Jr., and M. Stuiver
1996 Clovis and Folsom Age Estimates: Stratigraphic Context and Radiocarbon Calibration. *Antiquity* 70:515–525.

Thomas, D. H.
1969 Great Basin Hunting Patterns: A Quantitative Method for Treating Faunal Remains. *American Antiquity* 34:392–401.
1985 *The Archaeology of Hidden Cave, Nevada*. Anthropological Papers of the American Museum of Natural History Vol. 61, Pt. 1. New York.

Tuohy, D. R., and T. N. Layton
1977 Towards the Establishment of a New Series of Great Basin Projectile Points. *Nevada Archaeological Survey Reporter* 10(6):1–5.

Underwood, B.
2002 Health and Nutrition in Women, Infants, and Children: Overview of the Global Situation and the Asian Enigma. *Nutrition Reviews* 60:S7–S13.

U.S. Department of Agriculture
2005 USDA National Nutrient Database for Standard Reference (http://www.nal.usda.gov/fnic/foodcomp/search/).

Villa, P.
1982 Conjoinable Pieces and Site Formation Processes. *American Antiquity* 47:276–290.

Viteri, F., and H. Gonzalez
2002 Adverse Outcomes of Poor Micronutrient Status in Childhood and Adolescence. *Nutrition Reviews* 60:S77–S83.

Ward, R., and K. Weiss
1976 The Demographic Evolution of Human Populations. In *The Demographic Evolution of Human Populations*, edited by R. Ward and K. Weiss, pp. 2–23. Academic Press, London.

Warren, C. N.
1967 The San Dieguito Complex: A Review and Hypothesis. *American Antiquity* 32:168–185.
1990 *Archaeological Investigations at Nelson Wash, Fort Irwin, California*. Dames & Moore, San Diego, California. Submitted to the National Park Service, Western Region, San Diego.

Warren, C. N., and C. Phagan
1988 Fluted Points in the Mojave Desert: Their Technology and Cultural Context. In *Early Human Occupation in Far Western North America: The Clovis-Archaic Interface*, edited by J. A. Willig, C. M. Aikens, and J. L. Fagan, pp. 121–130. Nevada State Museum Anthropological Papers No. 21. Carson City.

Watters, D. R.
1979 On the Hunting of "Big-Game" by Aboriginal Populations. *Journal of New World Archaeology* 3(3):57–64.

Wigand, P. E., and D. Rhode
2002 Great Basin Vegetation History and Aquatic Systems: The Last 150,000 Years. In *Great Basin Aquatic Systems History*, edited by R. Hershler, D. B. Madsen and D. R. Currey,

pp. 309–367. Smithsonian Contributions to Earth Sciences No. 33. Smithsonian Institution Press, Washington, D.C.

Wiley, E. O., D. J. Siegel-Causey, D. R. Brooks, and V. A. Funk

1991 *The Complete Cladist: A Primer of Phylogenetic Procedures*. University of Kansas, Museum of Natural History, Lawrence.

Willerslev, E., N. Naveran, T. Gilbert, J. Sanchez, D. L. Jenkins, J. Binladen, A. Götherström, and M. Hofreiter

2006 Ancient DNA Analyses of Pre-Clovis Human Coprolites. Paper presented at the 30th Great Basin Anthropological Conference, Las Vegas, Nevada.

Williams, T. R., and M. S. Bedinger

1984 Selected Geologic and Hydrologic Characteristics of the Basin and Range Province, Western United States: Pleistocene Lakes and Marshes. *Miscellaneous Investigation of the United States Geological Survey*: Map I-1522-4. Washington, D.C.

Willig, J. A.

1988 Paleo-Archaic Adaptations and Lakeside Settlement Patterns in the Northern Alkali Basin. In *Early Human Occupation in Far Western North America: The Clovis-Archaic Interface*, edited by J. A. Willig, C. M. Aikens, and J. L. Fagan, pp. 417–482. Nevada State Museum Anthropological Papers No. 21. Carson City.

1989 Paleo-Archaic Broad Spectrum Adaptations at the Pleistocene-Holocene Boundary in Far Western North America. Ph.D. dissertation, University of Oregon, Eugene.

1991 Clovis Technology and Adaptation in Far Western North America: Regional Pattern and Environmental Context. In *Clovis: Origins and Adaptations*, edited by R. Bonnichsen and K. L. Turnmire, pp. 91–118. Peopling of the Americas Publications. Center for the Study of the First Americans, Corvallis, Oregon.

Willig, J. A., and C. M. Aikens

1988 The Clovis-Archaic Interface in Far Western North America. In *Early Human Occupation in Far Western North America: The Clovis-Archaic Interface*, edited by J. A. Willig, C. M. Aikens, and J. L. Fagan, pp. 1–40. Nevada State Museum Anthropological Papers No. 21. Carson City.

Willig, J. A., C. M. Aikens, and J. L. Fagan (editors)

1988 *Early Human Occupation in Far Western North America: The Clovis-Archaic Interface*. Nevada State Museum Anthropological Papers 21. Carson City.

Wingard, G. F.

2001 *Carlon Village: Land, Water, Subsistence, and Sedentism in the Northern Great Basin*. University of Oregon Anthropological Papers No. 57. Eugene.

Woods, J.

1988 Traces of Early Settlers 12,000 Years Ago: Lincoln County Cave Yields Bones, Stone Chips. *The Times News* No. 236:A1.

Woods, J., and G. Titmus

1985 A Review of the Simon Clovis Collection. *Idaho Archaeologist* 8:3–8.

Wormington, H. M.

1957 *Ancient Man in North America*. Denver Museum of Natural History, Popular Series No. 4. Denver.

Wriston, T.

2003 The Weed Lake Ditch Site: An Early Holocene Occupation on the Shore of Pluvial Lake Malheur, Harney Basin, Oregon. M.A. thesis, University of Nevada, Reno.

Yoshida, S., C. Keen, A. Ansari, and M. Gerschwin

1999 Nutrition and the Immune System. In *Modern Nutrition in Health and Disease*, edited by M. Shils, J. Olson, M. Shike, and A. Ross, pp. 725–750. Lippincott, Williams and Wilkins, Philadelphia.

Young, D. C.

2000 Distributary Channel System and the Late Pleistocene/Early Holocene Transition at TS-5, Great Salt Lake Desert, Utah. Paper presented at the 27th meeting of the Great Basin Anthropological Conference, Ogden.

2002 Secondary Obsidian Sources of the Madeline Plains: Paleolandscapes and Archaeological Implications. In *Boundary Lands: Archaeological Investigations along the California-Great Basin Interface*, edited by K. R. McGuire, pp. 75–84. Nevada State Museum Anthropological Papers No. 24. Carson City.

Zeanah, D. W.

2004 Sexual Division of Labor and Central Place Foraging: A Model for the Carson Desert of Western Nevada. *Journal of Anthropological Archaeology* 23:1–32.

Zeanah, D. W., J. A. Carter, D. P. Carter, R. G. Elston, and J. E. Hammett

1995 *An Optimal Foraging Model of Hunter-Gatherer Land Use in the Carson Desert.* Intermountain Research, Silver City, Nevada. Submitted to the U.S. Fish and Wildlife Service, Portland, Oregon.

Zic, M., R. M. Negrini, and P. E. Wigand

2002 Evidence of Synchronous Climate Change Across the Northern Hemisphere Between the North Atlantic and the Northwestern Great Basin, United States. *Geology* 30:635–638.

Zoback, M. L., R. E. Anderson, and G. A. Thompson

1981 Cenozoic Evolution of the State of Stress and Style of Tectonism of the Basin and Range Province of the Western United States. *Philosophical Transactions of the Royal Society of London,* Series A 300:407–434.

Contributors

Charlotte Beck, Department of Anthropology, Hamilton College, 198 College Hill Road, Clinton, NY 13323

Daron G. Duke, Department of Anthropology, University of Nevada, Reno, NV 89557; Far Western Anthropological Research Group Inc., P.O. Box 758, Virginia City, NV 89440

Ted Goebel, Center for the Study of the First Americans, Department of Anthropology, Texas A&M University, College Station, TX 77843

Kelly E. Graf, Department of Anthropology, University of Nevada, Reno, NV 89557; Center for the Study of the First Americans, Department of Anthropology, Texas A&M University, College Station, TX 77843

Gary Haynes, Department of Anthropology, University of Nevada, Reno, NV 89557

L. Suzann Henrikson, Center for Archaeological Research, California State University, Bakersfield, CA 93311

Bryan Hockett, Bureau of Land Management, Elko Field Office, 3900 East Idaho Street, Elko, NV 89801

Dennis L. Jenkins, Museum of Natural and Cultural History, 1224-University of Oregon, Eugene, OR 97403-1224

George T. Jones, Department of Anthropology, Hamilton College, 198 College Hill Road, Clinton, NY 13323

Montana M. Long, Museum of Natural and Cultural History, 1224-University of Oregon, Eugene, OR 97403-1224

Lisbeth Louderback, Department of Anthropology, University of Nevada, Reno, NV 89557

David B. Madsen, Mercyhurst Archaeological Institute, Mercyhurst College, Erie, PA 16546; Texas Archeological Research Laboratory, University of Texas at Austin, 1 University Station, Austin, TX 78712

Charles G. Oviatt, Department of Geology, Kansas State University, Manhattan, KS 66506

Ariane Pinson, University of New Mexico, Albuquerque; 1817 32nd Street SE, Rio Rancho, NM 87124

Rachel Quist, Directorate of Environmental Programs, U.S. Army Dugway Proving Ground, CSTE-DTC-DP-EP-CP, Building 5330, Dugway, UT 84022

David Rhode, Earth and Ecosystem Sciences, Desert Research Institute, 2215 Raggio Parkway, Reno, NV 99512

Dave N. Schmitt, Department of Anthropology, Washington State University, P.O. Box 644910, Pullman, WA 99164; Desert Research Institute, Division of Earth and Ecosystem Sciences, 2215 Raggio Parkway, Reno, NV 89512

Geoffrey Smith, Desert Research Institute, Division of Earth and Ecosystem Sciences, 2215 Raggio Parkway, Reno, NV 89512

D. Craig Young, Far Western Anthropological Research Group Inc., P.O. Box 758, Virginia City, NV 89440

Index

Numbers in *italics* refer to figures and tables.